Urban Homesteading

Heirloom Skills for Sustainable Living

RACHEL KAPLAN

with

K. RUBY BLUME

Skyhorse Publishing

Skyhorse Publishing books may be purchased in bulk at special discounts for sales promotion, corporate gifts, fund-raising, or educational purposes. Special editions can also be created to specifications. For details, contact the Special Sales Department, Skyhorse Publishing, 307 West 36th Street, 11th Floor, New York, NY 10018 or info@skyhorsepublishing.com.

Skyhorse® and Skyhorse Publishing® are registered trademarks of Skyhorse Publishing, Inc.®, a Delaware corporation.

www.skyhorsepublishing.com

10 9 8 7 6 5 4

Library of Congress Cataloging-in-Publication Data is available on file.
ISBN: 978-1-61608-054-9

Printed in China

This book is dedicated to the children who follow after us.

May they inherit a fertile and abundant world filled with people who honor the diversity of life teeming around us: from the tiniest microbe, to the wondrous chicken, to the beauty of human community.

Contents

EATING CLOSE TO THE GROUND

THE HOUSE

SELF-CARE, CITY-CARE

Acknowledgments

Abook like this is never written by one person, or even by two. Collaborating on this project has been a good experiment for us, even when we wished we were in the garden rather than in front of the computer. We are grateful for the support and encouragement we received while working on it, and want to acknowledge it here.

For their time and commitment to the path: Aaron Ableman, Patricia Algara, Jane Allard Allen, Laura Allen, Kevin Bayuk, Ellen Beeler, George Beeler, Rachel Brinkerhoff, Maryann Brooks, Massey Burke, Alli Chagi-Starr, Lindsey Dailey, Trilby Dupont, Lauren Elder, Ashel Eldridge, Michael Erlich, Tom Ferguson, Barbara Finnen, Trathen Heckman, Jenifer Kent, Ben Macri, Jeannie McKenzie, Daniel Miller, Evelina Molina, Frank Morrow, Jim Montgomery, Giancarlo Muscadini, Hank Obermayer, Erik Ohlsen, Cris Oseguera, Sasha Rabin, Kitty Sharky, Christopher Shein, Troy Silviera, Rick Taylor, Andree Thompson and Tina Wilder. Many of these fine folks are linked, directly or indirectly, through two organizations we love: The Institute of Urban Homesteading in Oakland, California and Daily Acts, a person-powered organization for sustainability education and inspiration in Sonoma County, California.

Others who chimed in with thoughts, photos and support: Karen Lamphear, Michael and Dana Yares, Jennifer McWilliams, Diane Dew, Toby Hemenway, Daily Acts, Ellen Bicheler, Annie Deichmann, Dan Kaplan, Karen Romanowski, Carl Shuller, Karen Lyons, Molly Goulet Bolt, Karen Erlichman, Jennifer Lindsey, Seth Zuckerman, Victoria Temple, Stacey Evans, Keith Hennessy, Michael-Medo Whitson, Michelle Lujan, Jonathon Gavzer, Glenn Caley Bachmann, Stacey Meinzen, Kelli Loux, Sabrina Kahn, Chris at the Farmer's Market, Ruth Persselin, Patty Sherwood, Bart Anderson, Willi Paul and Planetshifter.com.

Readers extraordinaire who made it so much better: Tracy Theriot, Miguel Micah Elliott, Laura Allen, Tamar Bland, and Marion Kaplan. Elinor Burnside's contribution as reader, proofreader, sounding board and general all around inspiration for living was extraordinary and much appreciated.

For their long-term, salt-of-the-earth support, their continued faith in her work on the planet and for simply being amazing and inspiring each in their own way, Ruby wants to thank her personal angels: her mom Jacki Fox Ruby, her partner Erik Bjorkquist, and her friend Allyson Steinberg.

Rachel: My gratitude extends to my dear neighbors Peter, Lisa, Gabriel, and Elias Stein for blessing our family with more family (and to Peter especially for being there for the illustrious eleventh-hour save); to Esmé Kaplan-Kinsey who inspires me in the work of repairing the earth and tending to the future; and to Adam Kinsey—first, last, and best reader, in life and in art—I'm so glad you are walking by my side.

Autumn, 2010

Why We're Here

We've been friends for nearly twenty-five years, sharing a life as community artists and activists in San Francisco's Mission District, and finding ourselves evolving toward the same urban homesteading lifestyle grounded in the urgency of the moment and the need to create real cultural change. We're neither partners nor roommates; we don't even live in the same city. But we share a love of the earth and a creative spirit, as well as our practices as body-centered healers, teachers, and activists.

Ruby created the Institute for Urban Homesteading in 2008 in Oakland, California, as a venue for sharing the homegrown wisdom she's gathered over the years. Rachel lives in Sonoma County with her partner and daughter, works as a somatic psychotherapist, and teaches homesteading skills. She also helps coordinate a group of homesteaders and backyard gardeners into the Homegrown Guild, an action-oriented project of Daily Acts, a nonprofit with a mission to transform communities through inspired action and education.

We wanted this book to represent voices other than our own because we find ourselves part of an outpouring of energy toward a diversified, healthy ecosystem in the midst of crowded urban intensity. We are part of an urban homesteading *movement*. All the people we interviewed live in the urban or suburban Bay Area. Our choice to restrict our interviews to homesteaders in our area reflects our lack of a travel stipend and not the reach of urban homesteading in this country, which is growing rapidly and expressing itself in diverse ways in different places, meeting the requirements of bioregion, economic necessity, and local sensibility. Each person or family we interviewed inspired us, and represented a foray into some part of the homesteading lifestyle we think is important. We chose homesteads that were small enough in scale to apply to a diversity of cities across the country, and captivated us with their creativity, beauty, or verve. We are grateful to everyone we spoke with for the generosity of their time, and for their ongoing and embodied commitment to birthing a regenerative culture.

As we interviewed different homesteaders, we found that no one has a handle on every aspect of homegrown sustainability. Each place is marked by the limits of space and time and skill and affinity. Some people focus on growing food and learning how to preserve it. Others have a leaning toward water, or compost, or recycling the waste stream. Some people have fully devoted themselves to permaculture as a way of making a living. We are landscapers, nonprofit workers, students, teachers, greywater experts, architects, stonemasons, mothers, and fathers. These homesteaders are all homegrown urban farmers, busy experimenting with the space they have, and building their toolbox of sustainable living skills.

City people grow and butcher animals for food, milk the goats, and gather the honey, just like homesteaders everywhere. Everyone is trying to grow as much food, save as many resources, and connect as much with their neighbors as possible. We are all motivated by concern for our cultural moment and a desire to live the change we want to see, to be part of crafting a solution rather than perpetuating the problem. As you will see, there are some limits to our success, and some spectacular unfolding social experiments.

Throughout the book, unless otherwise noted, all photographs are by K. Ruby Blume, who also created the original art for the section headings and maps, and was the visual wizard for the entire book. Drawings throughout are by Marco Aidala.

A note on the inclusion of wise words, and on pronouns: Sometimes, the stories of these homesteaders are represented directly in the interviews we did with them. Sometimes we represent their voices by simply saying, "Trathen said," or "Jane said" because their words were the wisest ones we could find. Sometimes when we say "we," it means Rachel and Ruby. Sometimes, it means Rachel and her family of three. Sometimes "we" refers to the movement of urban homesteaders. We have done our best to clarify the use of this ubiquitous pronoun throughout the text. Whenever the "I" pronoun is used and no one else is credited, this is Rachel's voice. We've avoided the gendered pronouns "he" and "she" out of long years of practice in the gender-blendy west coast, and chose instead the more inclusive "you" whenever possible. Please take this as our personal invitation for your own participation in the practices offered in this book.

Living in the garden
of earthly delights,
demonstration
garden of the
Institute of Urban
Homesteading,
Oakland, California.

Start Where You Are

Knit It Up

My heart is moved by all I cannot save
So much has been destroyed
I have cast my lot with those
Who, age after age, perversely
With no extraordinary power
Reconstitute the world

—Adrienne Rich[1]

The weed growing up through the cracks in a city sidewalk—that sharp green shard of life persisting against all odds—reflects nature's resilience. It's also a metaphor for the uprising earth consciousness growing in our cities—small, surprising, commonplace. Spreading. Across the country, citizens are looking for solutions to the seemingly intractable problems of our time, and evolving new ways to live. Picking up the shovel and the hoe, turning their closets and roofs and backyard decks into places to grow food and their yards into chicken coops, urban farmers are reclaiming heirloom agrarian practices as strategies for artful living. This book tells the story of this grassroots do-it-yourself cultural explosion rooted in the urban earth, a homegrown guild of people generating resilient, local culture in response to the urgency of the moment and a collective awareness of our need to be the change we want to see.

The more I know, the less I sleep. There is something decidedly brinkish about our era. We are bombarded by desperate stories—collapse of the Arctic ice, clear-cutting of the forests, massive oil spills, catastrophic droughts and floods, volatile nation states, dangerous levels of CO_2 in the air, the depletion of oil, and the overwhelming power of corporations to devour the world at will—all conspiring to create fear and dread. We are told we are powerless until we begin to believe it. The convergence of the seemingly unstoppable forces of climate change, the savagery of global corporate capitalism, and the downward spiral of our predatory economy all lead to an inevitable conclusion: We are coming undone. We are unraveling.

Knitters know all about unraveling. You knit along for a while, until you drop a stitch or add a stitch or do something else peculiar that just doesn't work. If you want it right, you have to unravel, and knit it up again. Or sometimes you unravel by choice because it's just not coming out quite the way you planned. Re-knitting always takes less time than you think,

A family of strangers travels along the city streets, seeking the heart of people and place. A circle is drawn on the pavement in flowers. The world is an altar. Our actions are sacred. (Blume Family ritual procession, collaboration between Keith Hennessy and K. Ruby Blume, 1992.)

and there you are again at the place you left off, with a piece of fabric that looks and feels right.

This homegrown metaphor takes us only so far; we can't unravel back to the beginning of our disastrous misalignments with people and place that, in our country, permitted the genocide of the first peoples; the dispossession, oppression, and slavery of others; and the short-sighted desecration of natural resources leading up to our current environmental and economic predicaments. Unlike strands of wool on our needles, we are people who have to work from where we are. But there are lessons about process and outcome in the knitting metaphor and operating instructions for how to proceed—when we make mistakes, it's best to go back, sort out what's worth saving from what needs to be let go, and get back to the work of stitching together again. The urban earth has been shattered by hundreds of years of neglect and abuse; our relationships are fractured and deformed by long stories of hate and race and class. We've made something lopsided and misshapen, and it is time to weave another tapestry, tell another story about how we can live together with this planet.

Urban homesteading is happening in small and large cities across the country, a homegrown response to this potent need for a new, life-giving story. Urban homesteaders are relearning heirloom skills that have been abandoned in the relentless march toward convenience; valuing thrift and community self-reliance; and tending to our home places in an intentional repudiation of the cultural forces of speed, need, and greed. Urban home-steading is also part of a global movement for change rooted in respect for indigenous peoples and values, a cadre of environmental first responders and a network of progressive social change organizations seeking peace and reconciliation at every level. All of these together forge an opportunity to rewrite the story of our relationship to the earth and the possibility of remaking culture around an ethic of care and stewardship for this place that is our shared home.

This is a David and Goliath story—backyard gardens competing with Monsanto's patents on the gene pool, rain barrels and greywater versus the worldwide privatization of natural resources, bicycles against the power of Big Oil. Garden by garden, block by block, neighborhood by neighborhood, each partial effort is a step in the right direction, our participation in the human immunological response to our diseased world.[2] Will it work? The outcome is uncertain. Is it worth trying? Without a doubt. The tragedy of living in a "Christian" country that repeatedly rapes God's creation can be combated only by learning to cherish and tend the kingdom of God, which is among us, now.

Battles like this have been won before—the tobacco industry once ruled the roost, and now it's the chicken no one wants in their backyard coop. Despite what politicians and commercials teach us at every turn, our daily actions can remake the world. Urban homesteading is a story of the

Bringing up our children to respect life is central to the homesteading way. *Photo by Lauren Elder*

Photo by Dafna Kory.

power of community and the joy in following an artful course of living in a time of contraction and fear. This is a book about action, about things we can do to reweave the web by living in place, but it's also a book about how we think about our actions during this time of unraveling. It's a hopeful vision in a hopeless time. Urban homesteading is a proactive response, a series of earth-based actions that make an immediate difference in the places we call home.

When we speak of a system being unsustainable, what we're really saying is: This cannot last. Sustainability is the ability to exist over time. We have to get real about our current lack of sustainability as we face the end of many systems in the twenty-first century, and take up the good work of re-imagining ones that can endure. The urban earth practices in this book show how we can radically reduce consumption and maximally increase community self-reliance and joy in living, both of which are necessary for positive social change and progressive human evolution. They will help ground you in the reality and freedom of limits. Some say this is an "old" way, and it is. But it's also a new way of relating based on reciprocity and cooperation, of living in partnership with the earth rather than treating it like our personal garbage can.

Now is the time to answer the call. You've probably picked up this book because you hear it, and because you want to answer in your own way and bring your best offering forward at this time of unraveling. Ultimately, this is a book about reverence and our love for our beautiful world, about our grief in seeing her die and our complicity in her dying.

This book weaves in the voices of many people who live an urban homesteading lifestyle. While there are differences between us all, everyone feels creatively motivated and spiritually connected on our urban farms, and in the midst of breakdowns of all kinds we write to remind you that the force of necessity motivating these practices is beautiful, raw, and vivid, and that god is in the broccoli. Our work reflects a commitment toward a regenerative, living culture, rather than the consumptive consumerism our country has refined to a sick art. We opt out by digging in.

And so we find ourselves in our backyards fighting gophers, pulling carrots, harvesting rabbits and eggs, tending bees, and gathering raspberries, grapes, broccoli, and kale. We save our seeds. We pee in a bucket and dump it on the compost bin. We harvest our rainwater and drain our bathtubs into the garden. On hot summer afternoons you'll find us preserving jars of peaches, plums, and nectarines that have fallen from the trees. We bring people together to learn how to can, make yogurt, hold a meeting, or turn a lawn into a garden. We experience our practices in the urban earth with the bees and the animals and the things we make with our hands as spiritual, a prayer to the life force and a vehicle for our own connectivity and sense of purpose.

We are here to say: there is a life in the earth for you. There is birth and abundance and death and regeneration and joy. We are on the side of the seed, growing through the cracks of our profound and tragic mistakes, the particulate scaffolding of the natural world still calling to us, teaching us the true order of things. We are sitting. We are listening. Knitting, unraveling, and knitting again.

A family tends their garden plot at the La Tercera Community Garden, Petaluma, California.
Photo courtesy of Terry Hankins/Petaluma Argus Courier.

The Empire Has No Clothes

The only recognizable feature of hope is action.

—Grace Paley[3]

All the systems that sustain us—food, water, shelter, medicine, family, and community—are at risk from the ongoing disintegration of life brought about by global capitalism's profound disrespect for natural limits. In this past decade alone, the decay of basic support systems has been staggering. With corporate control of our government becoming more entrenched, it is hard to imagine a future that will hold the processes of life and the needs of humans and other living creatures as its guide. It's past time for us to redesign our cities and our lives with an ethic of care at its core, remaking local systems based on the model of the earth itself—adaptive, lush with diversity, and fertile with possibility. Rather than continuing to direct our life energies toward a system that is degenerate on every level—personal, social, and environmental—we advocate the relearning of skills and strategies to maximize interdependence, community resilience, and a sense of sufficiency in living locally.

The dangers we face are large and undeniable. Our situation is urgent—and growing more so—but it is not too late to change direction. Escalating climate instability, along with peak oil and its relationship to the current economic redistribution of wealth, underscore the urgency of our situation. The processes of recovery and change, localism, and the empowerment of do-it-yourself culture support the practices of urban homesteading, which are a direct response to the cultural challenges we face.

Climate Change and Peak Oil

Climate change and global warming are finally front and center in mainstream consciousness as emergencies warranting immediate attention. It is no longer debated by any reputable source whether human activity is the culprit, specifically the use of nonrenewable fossil fuels, industrial agriculture, and the ongoing devastation of the earth's forests and waters. Yet despite

The patterns of nature reflect resilience, cooperation, integration, and beauty.

the clear and present danger, global carbon emissions per capita continue to rise, increasing the likelihood of large-scale climate catastrophe and putting our lives and the lives of other beings at risk.

There is an immediate need for citizens, especially in countries that consume the vast majority of resources, to sharply limit the carbon in the environment by employing conservation strategies and curtailing energy use wherever possible.[4] Global industry, the prime culprit in generating carbon emissions, has a huge role to play, but enforcing systemic change is remarkably difficult as long as industry controls the mechanisms of political power and a majority of resources. So although change is needed at all levels and should be fought for, actions taken by individuals and local communities are more immediate and will show more tangible consequences in our lives. It seems clear that change in this arena will have to come from the grassroots, or not at all.

"Peak oil" refers to a point in time when the maximum rate of global extraction of the nonrenewable resource of fossil fuel is reached. H. King Hubbard predicted in the 1970s that we would shortly reach the point of peak oil in the United States. He was laughed out of the room, but had the last laugh himself as his predictions proved correct. Evidence suggests that global oil production has already peaked and begun its inevitable decline. Many former oil-producing nations, including our own, reached peak oil production decades ago. Others are approaching peak oil while our global needs for this nonrenewable resource continue to increase unabated. While there is still plenty of oil to be pulled from the earth, extraction is becoming more difficult and costly, making inevitable further disasters like the Gulf of Mexico oil spill of 2010. Unless we wean ourselves from oil, the decline in extraction will lead to a precipitous rise in oil prices and the fuel that runs our entire economy will become more and more scarce. It is easy to see how this combination of factors is a disaster in the making for individuals and communities, particularly marginalized and impoverished ones.

The significance of peak oil is not that fossil fuel is going to "run out" tomorrow or next year or even in ten years. The significance lies in the fact that as fossil fuel gets more difficult to extract (already happening), prices rise (already happening), political instability arises (already happening), the gap between rich and poor widens (already happening), and our ability to continue with business as usual is compromised (already happening). Compounding the environmental impacts of global warming, we face a potential cascade of economic and political catastrophes. This is the story of apocalypse that our fear and habituation feeds.

But we have a choice about whether or not to contribute our life force energy to this story or to direct our will to the world we want to bring into being. This is nothing short of an initiatory moment for humanity: Will we grow up out of our need to consume whatever we want when we want it (like a child or an adolescent) or learn to care for the earth, the source of all nourishment (like an adult)? It's time to stop pretending that each of us doesn't have a role to play, and to tend the piece of earth we've been given.

The contraction of energy availability and its ramifications throughout our world are referred to as "energy descent." Understanding the impact of energy descent leads to four important conclusions.

1. The future is local. Reduced fossil fuel and a reversion to a renewable energy lifestyle will radically change our systems of food production and distribution, transport, communication, energy, medicine, and government, bringing them closer to home. The sooner we get a handle on how to generate equitable resource production and distribution at the local level the better.

2. If we wait for government action before jumping on board, it will be too late. Change like this has to begin. In Congress. In the boardroom. In your home. You only have control over one of those things. Exert it.

3. Our actions are more powerful close to home. Thinking locally and acting locally works.

4. Working with others toward shared goals is more effective than working alone.

Climate change and peak oil together should be a profound wake-up call compelling the need to redesign our human systems toward resilience (our ability to recover) rather than toward our current mode of addiction (our tendency to do the same thing over and over again, even if it kills us). Each of us needs to embody true change at the level of our beliefs, our attitudes, and our actions. This is a process that can be learned, and is available to all of us.

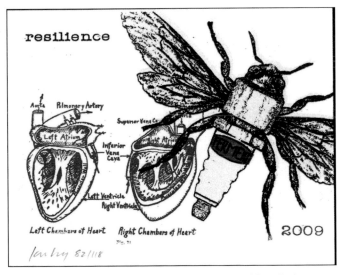

Original silkscreen by K. Ruby Blume. Bee courtesy of Evan Barbour.

Resilience and Recovery

Resilience is the ability of a system to recover from shock, trauma, or change. The more resilient a system, the more shocks and impacts it can withstand and still recover. As systems—cultural or ecological—lose the strength of diversity, they become vulnerable to disruption or collapse. Nature is the ultimate example of resilience, with its systems of multiple planned redundancies and complex relationships between organisms responding in different ways to threat. Fungi have the ability to begin the regenerative processes within a landscape after fire, paving the way for other microorganisms and animals to return to the devastated area and continue the repair work. Animals contain population through the checks and balances of the food chain. Nature grows through an understanding of limits and through the conservation and recycling of resources. We must learn to do the same. Inevitably, nature will be our strongest teacher in the process of change, or the agent of our harshest consequences. To quote Paul Hawken, "There are no economies of scale; there is only nature's economy."[5]

While individuals and sometimes communities possess resiliency in the face of difficulties, the more common human reaction to threat is a frozen or traumatized state of fight, flight, or freeze. People (and cultures) in this state can't make good choices or think clearly through a problem or creatively get out of a box. This frozen reactivity keeps us repeating the nightmares of the past, unable to see what is really happening in front of us, doing the same things and imagining a different future. Yet the imperative is clear. We need to find a way beyond our terrifying possibility—the collapse of our environment and our civilization—and we need our thinking to be crystal clear, creative, and responsive to the challenge facing us.

Can We Change?

Even as global consciousness about our situation rises, it remains difficult to harness our energies toward cultural regeneration. We see this especially when we look at our social institutions, but also when we look at ourselves. What is it that makes it so hard to change, especially when the problems we face are so serious, and have been so well articulated? Part of our limitation is our understanding of change as something that just "happens," as opposed to a process that requires our participation, awareness, and agreement. Denial, addiction, and a lifestyle of affluence also insulate people from the need (or desire) to change. And finally, a pervasive sense of pessimism about the powerlessness of our actions immobilizes many of us. If we are to make sense of the situation we are in, each of us has to go through our own individual process, confronting our habitual mechanisms of avoidance and denial to overcome our fear and conditioned cynicism. This process can only happen in stages, and will require patience, cooperation, and a little bit of humor.[6]

The Stages of Change, highly successful with addicts in recovery, seems particularly apt for our relationship to fossil fuels and our inflated sense of planetary entitlement. [7] The Stages include *recognition of a problem*, a *willingness to contemplate change*, *planning for possible new behaviors*, and a time for both *activating a plan* and *integrating the changes*. Within the process lies the inevitability of relapses and cycling back again. This model requires a shift in awareness and a personal desire to participate in making change

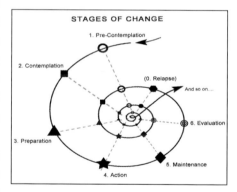

The Stages of Change, a spiral journey. People progress through different stages at their own pace, which might include relapsing, on their way to making successful change. (From Prochaska and DiClemente.)

happen. It works best within a context of community support, over time. An awareness of the cyclical nature of the process helps us keep renewing our commitment toward new behaviors, which cannot happen overnight. Change really is two steps forward, one step back.

In terms of the ecological and cultural problems we face, *pre-contemplation* on a social level began about forty years ago with the publication of Rachel Carson's *Silent Spring* and the emergence of a broader ecological movement. *Contemplation* of the problem followed, and beginning steps toward change were enacted: early attempts to reduce our dependence on fossil fuel, the back-to-the-land movement, and the inevitable pushback from industry. The cyclical and recursive nature of the process is evident in the progression of these cultural movements.

Gleaning for apples in a neighbor's backyard.
Photo by Petaluma Bounty

Our culture is growing now into the next level of *change*, evidenced by social movements addressing the environmental crisis with direct actions toward revising how we live today. Urban homesteading is just one of many creative approaches to this problem. The Transition Town movement is another, as are the growing numbers of young people from all strata of society training to become organic farmers, solar installers, and water conservation experts. There will undoubtedly be setbacks along the path, but as the diagram illustrates, this kind of change is cyclical and continues to ripple outward, especially as new habits are created and maintained. Engaging in the cycle of change with compassion for oneself and acknowledging the magnitude of the problem will be necessary to successfully take on an urban homesteading lifestyle. No one can do all of this, but everyone can do something. Don't worry about how or where to start. Just pick something you love, and do it.

In the wake of recent natural and man-made disasters like Hurricane Katrina and 9/11, research has been done on human resilience that reveals interesting trends for recovery. Dr. Alicia Lieberman's studies on the brain development of young people who witnessed trauma or violence show that their experiences of spirituality, animals, nature, and creativity were instrumental in sparking their recovery.[8] Judith Lewis Herman's research into trauma reveals the following resilience factors: the ability to help someone else during the trauma (taking action, rather than fleeing or freezing); the ability to make meaning and purpose out of the experience, to understand its history and context; and the ability to stay connected to at least one other person.[9] Recent studies of resilient people reveal some additional working strategies for recovery, including optimism, a sense of playfulness and generosity, the ability to "pick your battle," and the ability to focus on things over which you have some influence. Staying healthy is important, as is the skill of finding a silver lining. In a recent article on resilience, Beth Howard writes, "Resilient people convert misfortune into good luck and gain strength from adversity. They see negative events as opportunities for change and growth."[10]

These strategies mirror basic homesteading practices as steps toward healing and change: our renewed relationship to animals and the earth; our sense of meaning and purpose in the work we do; our connections to one another in community; and a spiritual understanding of our actions. A sense of creativity, play, generosity, and optimism are all activated as well. Urban homesteading is a battle that can be picked—actions bearing on our local economies and our homes have real influence, and are a wonderful example of converting adversity into possibility.

Grow It Local

Restructuring local economies to protect the earth and evolve our culture is central to the homesteading path. We are currently enmeshed in what has been called the extractive economy, where corporate wealth is regarded as the foundation for economic health; where mining our earth's resources and exploiting our citizens and international neighbors is accepted as the cost of doing business. The urban homesteading way seeks a local life-serving economy that creates, as David Korten artfully said, "a living for all, rather than a killing for a few." These practices protect our common inheritance of clean water, breathable air, and a life of joy and meaning for our families.[11]

BIOREGIONAL QUIZ: WHERE ARE YOU?

Urban homesteading is grounded in place. How familiar are you with the place you call home? Get curious. If you don't have all the answers, take some time to find them. Knowing these things is fundamental to everything that follows. It will help you become a better steward of the place where you live.

1. Can you trace the water you drink from precipitation to tap?
2. How many days from today until the moon is full and/or new?
3. Describe the soil around your house.
4. What were the primary subsistence techniques of the cultures(s) that lived in your area before you?
5. Name five edible plants in your bioregion and their seasons of availability.
6. From what direction do winter storms generally come?
7. Where does your garbage go?
8. Where does your sewage go?
9. How long is your growing season?
10. Name five resident and migratory birds in your area.
11. Name five resident and migratory human beings in your area.
12. What is the land use history by humans in your bioregion in the past century?
13. What primary geological events and processes influenced the landforms of your bioregion?
14. What animal or plant species have become extinct in your region?
15. From where you are, point to the north.
16. Name one of the first spring wildflowers to bloom in your area.
17. What kinds of rocks and minerals are found in your area?
18. Were the stars out last night?
19. Name some non-human being with whom you share your space.
20. Do you celebrate the turning of the winter and summer solstice?
21. How many people live next door to you? What are their names?
22. How much gasoline do you use, on average, in a week?
23. What form of energy costs you the most money?
24. What is the largest wilderness area in your bioregion?
25. What are the greatest threats to the integrity of the ecosystem in your bioregion?
26. What is the name of the creek or river that defines your watershed?
27. What geographic and/or biotic features define your bioregion?
28. What particular place or places have special meaning for you?

The best way to participate in changing the world is to change our own personal practices, including how we live, how we eat, how we travel, and how we relate to others. Re-inventing our relationship to the places we call home can significantly impact change. The home has been moved from the center of culture by the force of the marketplace. This devolutionary move has robbed us of the means of production, and the ability to care in simple, basic ways for ourselves, our families, our communities, and the earth. Bringing the home back to the center of culture where it belongs will create a meaningful path toward a regenerative future.

One of the central ethics of homesteading is a sense of bioregionalism, our awareness of, and commitment to, the place where we live. Bioregionalism teaches us about the specific ecological and cultural relationships happening around us, engaging a process of asking simple questions about moonrise and moonset, about soil, about air and wind, about where our water comes from and where our waste goes. This way of becoming native to place, of living within nature's limits and gifts, is a way of creating a life that can be shared by all and passed on to future generations. As Paul Hawken said, "We must know our place in a biological and cultural sense, and reclaim our role as engaged agents of our continued existence… Concern for the well-being of others is bred in the bone. We became human by working together and helping one another, and what it takes to arrest our descent into chaos is one person after another remembering who and where they really are."[12]

Bioregionalism values home above all else because home is where values and behaviors are learned before they move out into the world. In the home, alternatives can root and flourish and become deeply embedded in our way of being. The word *ecology* points us in this direction: *oikos*, the Greek root of "eco," means home. Hearth and home provide the theater of our human ecology, the place where we can relearn how to think with our hearts, to embody what we know to be true: that tending to our environment is the same as tending to ourselves, and we ignore this true work at our peril.

The Homegrown Guild

One of the great losses to culture in the last sixty years has been the ability of people to be even modestly self-sufficient at home. Homesteading in the city is a land-based, action-oriented *YES!* to the possibility of remaking culture with people and planet in mind, bringing back some of this lost power of doing it ourselves. We make no claims toward self-sufficiency: we can bake our own bread, but we cannot grow the wheat. But self-sufficiency, like independence, isn't a true goal. Our greatest need at this time is to learn to work together, to form guilds of differently abled farmers, blacksmiths, renegade plumbers, solar installers, beekeepers, mycologists, fermenting fetishists, somatic healers, technology wizards, performance artists, alternative educators, and herbal potion–brewers to remake our cities.

A guild is an alliance of craftspeople or artisans from a more traditional time. An early form of the union, its primary benefit was camaraderie and support for best practices, as well as a source for learning more skills and expanding support for the profession. Guilds also had the conservative function of slowing down the processes of innovation generated by industrialization that often resulted in a loss of quality and right livelihood. We need homegrown guilds today, as we relearn skills we have forgotten and redesign our cities toward sustainability.

Here's an example of what that can look like. In 2009, six households in Daily Acts' Homegrown Guild produced more than 3,000 pounds of food; foraged another ton of local fruit; harvested more than 4,000 pounds of urban waste to be composted and mulched; planted more than 185 fruit trees and hundreds of varieties of edible and habitat plants; installed five greywater and rainwater catchment systems that saved and recycled tens of thousands of gallons of water; tended to bees, chickens, quail, ducks, and rabbits; and worked toward reducing energy use and

Grow it local: bounty from the birds. Chicken, duck, and quail eggs.

Fused glass art by K. Ruby Blume.

enhancing commuting and transportation goals. All this from *six* households! Imagine a city where a majority of people tended to many of their daily needs in this way—the amount of food and water and energy and waste that could be managed sustainably is incredible.

Our small daily actions toward the things that nourish us have an enormous impact. We have to shake off the trance that tells us this is not so. Now is the time to experiment, maybe fail, but always learn some more. We cannot remake the world in whole, only in part. We have at hand old and new technologies we can harness in remaking the world. Resourceful participation in the big work of repositioning ourselves in a swiftly changing world, learning skills we can use at home, is the way of the future. We offer these technologies as spiritual practices in an incredibly challenging time and are here to report that in many ways that are good for planet and people, they work.

Urban farming is nothing new; in many parts of the world, it's a way of life. Cuba has an active urban farming movement, initiated when the USSR collapsed and precipitously stopped oil exports to the country. In Shanghai, residents produce 85 percent of their vegetables within city limits. The government of Tanzania encourages the cultivation of every piece of land in Dar es Salaam. Homesteaders around this country are engaged with the differing realities that their watersheds, climates, and history demand. Austin, Philadelphia, Newark, Brooklyn, Oakland, Portland, Los Angeles, and Detroit are all centers of rapid agricultural growth and production, each with their own place-based expression and local, evolving economies.

Some of the central urban homesteading practices are the same as homesteading practices everywhere—growing and preserving food, caring for and harvesting animals, foraging, making medicine, tending to the resources of water and waste and energy. But a city's unique and abundant resource is human energy—the intelligence, creativity, needs, hurts, history, and future of a city's people converging in exciting and sometimes destructive ways. Learning to harvest this energy and direct it toward community projects is a central survival strategy of the twenty-first century. The land frontiers have been conquered. The final frontier is learning how to live in harmony with one another and the world around us. Rebuilding a network of relationships between the earth and its inhabitants will be key to human evolution and survival.

Do-It-Yourself (DIY) Culture

DIY is an alternative culture strategy that helps us thrive outside the confines of the capitalist machine. It is an ethic of curiosity, exploration, and empowerment that can be applied to many aspects of our lives—growing food, sewing clothes, creating homegrown entertainment, writing books, fermenting vegetables, educating children. It feels good to do it yourself. This is a sane way to reorient our living toward a more just and equitable distribution of limited natural resources, and it supports the goal of sustainability through a maximum reduction in consumption and an expansion of creativity, and personal and community empowerment.

It's important for each of us to have a physical skill that is satisfying as well as sustaining—knitting or sewing or blacksmithing or canning or gardening. A "can do" attitude about all the activities people mastered as a matter of course in the past is required. It's important to remember how to be resourceful and figure out how to do something yourself. Collapsing at the mere thought of failure is no longer an option. Standing up and doing it yourself is a core homesteading way, something to relearn in our buy-it-yourself culture.

Many of the solutions in this book are simple, affordable, transportable, and good to do with others. Homesteading practices are not about austerity or apocalypse; they're about living a simpler, more joyful, more effective life. Homesteading is not a replay of a Depression-era mentality. It is a series of skills and practices that lift us out of a culture of inaction and cynicism and into a culture of abundance, care, and possibility. So this isn't a book about canning or making a nice pie out of foraged apples, at least not directly. It's about shifting consciousness toward a conservation and care-based ethic, which will undoubtedly manifest in many creative ways in your own life. In the name of limiting consumption and finding ways to break our addiction to needing and buying, many of the how-tos are a bit more intangible (like finding a Sit Spot in nature, or creating a community tool shed, or planning a potluck). When we do share a how-to of a more material nature, it will almost always include instructions on how to do it yourself on the cheap.

The Territory Ahead

This book is a map to the territory of urban homesteading. There are many awesome, time-honored practices in the art of living, which we have mostly forgotten and collectively need to remember: organic gardening, tending an orchard, beekeeping, fermenting, jamming, herbalism, self-care, community relations, and land, energy, and water stewardship. These all deserve (and have) many specific volumes dedicated to the intricacies of their art and this one book cannot do each of them justice. We can be definitive perhaps in only one sense—the necessity of reclaiming these heirloom skills for living in the twenty-first century. The resource list in the back of the book will point you toward other excellent books specific to the different practices highlighted here, so as you read and track your own interests, you can find your path through the woods. We also recommend, whenever possible, the practice of finding an experienced teacher to take you on the journey.

As you read, remember that one practice leads to another, and that having one skill will always lead you to someone with another. Perhaps you choose to become a beekeeper. Soon you have more honey and beeswax than you need, and something to trade with your neighbor, that fantastic tomato grower. Your fantastic tomato-growing friend trades with her greywater plumber, who trades her time for fresh goat milk. Perhaps your small beekeeping experiment grows into a cottage industry, further evolving your community network and economic center. These are all strands in a growing web of local culture happening all around the country, an alternative, restorative economy existing beyond and separate from the economic mudslide of dominant culture. It is this cultural growth—from the one to the many, from our homes to our communities—that this book and this movement is really all about.

When visiting and speaking with people about their choices for living, we noticed a few themes that may be some evolving principles of urban homesteading. Embodying these principles will take time and commitment and, for some of us, represents a big change in lifestyle. For others, they are already second nature. If they are new to you, remember that lifestyle changes can be challenging but are reinforced over time through practice and support from others.

Simplify. Our lives are complex, over-consumptive, harried. Choosing a path of voluntary simplicity *is* possible, and feels good.

Use Less. We consume more than we need. Curb the habit. Break the addiction.

Share More. Many of us have more than we need, and some not enough. Give it away. Share it with friends, neighbors, and strangers.

Localize. Commit your time and energy to businesses, gardens, organizations, and people in your community to strengthen the financial, biological, and social economy of your place.

Diversify. Ecologically diverse systems that include multiple plants, solutions, and people create more security for all. Apply the metaphor to the ecology of the city where you live.

Pump up the local economy: Biofuels Oasis in Berkeley, California sells biodiesel and urban farm supplies.

Do It Yourself. If you want it, make it happen. If you can do it yourself, do it.

Indigenate. Belong to your place.

Embody. Let your body's wisdom motivate and inform your actions.

Relate. Making connections between people and things in our environments makes us stronger and more effective.

Forgive. Clear your body of old anger and hurt so you can do your best work today. Forgiveness is an individual and communal act.

Listen and Observe. We are in constant conversation with life. Slow down and pay attention.

Create and Renew. Our planetary culture is calling for renewal. Use your creativity to find a way to participate in answering the call.

Begin. Start where you are. Make mistakes. Begin again.

Permaculture— Peace with Creation

All aspects of our current crisis reflect the same mistake, setting ourselves apart, and using others for our gain. So to heal one aspect helps the others to heal as well. Just find what you love to work on and take joy in that.

—Joanna Macy[13]

Learning to live within the limits of nature and reconnecting to its rhythms is key to resolving some of our dilemmas. Permaculture is a cultural design method that offers a set of principles and practices for creating stable agricultural and human settlements within nature's limits. One aspect of permaculture design is the creation of regenerative and productive designs that attempt to mimic the wisdom and diversity of the natural world as the ultimate model for sustainable living. Permaculture applies nature's principles to address human needs in a way that renews the earth rather than depleting it. The word combines two roots, "permanent" and "culture," suggesting a practice that sustains itself over time. Permaculture in an urban setting can be applied to creating strong communities, as well as strategies for small-scale and ongoing production and management of food, energy, water, and waste.

The term "permanent agriculture" was coined in 1911 by Franklin Hiram King in his classic book, *Farmers of Forty Centuries: Or Permanent Agriculture in China, Korea and Japan*, and was understood as agriculture that can be sustained indefinitely.[14] In 1929, Joseph Russell Smith took up the term as the subtitle for *Tree Crops: A Permanent Agriculture*. He saw the world as an inter-related whole and suggested mixed systems of trees and under-story crops.[15] The work of Howard T. Odum was another early influence with its focus on systems ecology.[16]

Playing with plaster during an earthen plastering workshop. Natural building made from cob, earthen plaster, and recycled glass.   *Photo by Sasha Rabin/Vertical Clay*

Other influences include the work of Esther Deans, who pioneered the No-Dig Gardening method, and Masanobu Fukuoka, who began advocating no-till orchards and gardens in Japan in the late 1930s. His classic work, *The One Straw Revolution*, describing this method, is still timely today. His understanding of the intrinsic power of agriculture to create and be created by culture is summed up in this beloved quote: "The ultimate goal of farming is not the growing of crops but the cultivation and perfection of human beings."[17]

In the mid-1970s, Australians Bill Mollison and David Holmgren began developing strategies for stable agricultural systems in response to the rapid growth of destructive industrial-agricultural methods in their country. They saw these methods poisoning the land and water, reducing biodiversity, and removing billions of tons of topsoil from previously fertile landscapes. They announced their approach with the publication of *Permaculture One* in 1978. The term permaculture initially meant "permanent agriculture" but was quickly expanded to also mean "permanent culture" when it became clear that social aspects were integral to a truly sustainable system. By the early 1980s, the concept had broadened from agricultural systems design towards complete, sustainable human habitats, and by the mid-1980s, many of Mollison and Holmgren's students had become successful practitioners and began teaching the techniques they had learned. In a short period of time permaculture groups, projects, associations, and institutes were established in more than one hundred countries.

While permaculture began developing among Westerners in the beginning of the twentieth century, many of its core ideas root back to indigenous land-based practices around the world. Indigenous people tended the "wild" for millennia so that it remained healthy and yielded plentiful food and habitats for a diversity of living creatures. California's first peoples were experts at managing the water-challenged landscape, selecting some plants and removing others through fire and hand seeding, taking no more than they needed, and thanking the spirits for what they received. They experienced the land and its creatures as living relations, and themselves within the circle of that family.[18] Explorers who "discovered" this land noted its wild beauty, but were unable to perceive the land-nurturing patterns of care the first peoples had developed over thousands of years.

Some of the foundational ethics of permaculture are rooted in this indigenous awareness of the earth as an animate, sacred world upon whom we depend. The premise of Ecology 101—the intrinsic connections between all parts of a system, implying that "we are all connected"—is indigenous wisdom translated into modern scientific language. From this native way, we learn to think like a tree and work for the long term, the seventh generation, rather than always satisfying our personal needs in the moment. We cultivate right relationship to all the creatures around us to honor the life force. We bow to the complex order of nature and seek a deeper sense of connection with it.

The aboriginal ways are long gone on the North American continent, but permaculture practices bow toward this sense of relatedness and care that, in its duration and flexibility, served the land and its people so well for so long. Urban homesteading, in its evolutionary, fragmentary way, gestures toward this kind of sustainability and seeks to relearn the skills of listening, observing, and responding to nature's ways as the means for creating healthy, enduring human settlements.

Permaculture Ethics

Permaculture is based on three ethical principles:

Earth care—recognizing that the earth is the source of all life (and is itself a living entity). We are a part of the earth, not apart from it.

People care—supporting and helping each other change to ways of living that do not harm us or the planet, including developing healthy societies that prioritize the first principle, earth care.

Fair share—placing limits on consumption of the earth's limited resources, ensuring they are used in ways that are equitable and wise. Acknowledging the history of oppression and genocide that has distorted all the systems of our culture, fair share seeks a just outcome.

Reading this book you will encounter these themes again and again—caring for the earth, caring for people,

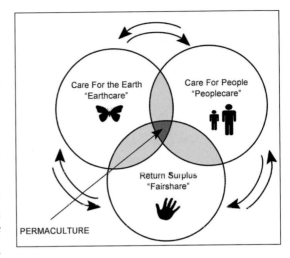

There is a synergistic relationship among the three permaculture ethics. Each one feeds the other and can't exist without the third.

and creating systems that limit consumption and ensure the fair distribution of the earth's resources. Our practices at home are based on the unique needs and resources of each location, but the basic ethics are the same. Some people focus on reworking policy at the local level; others are installing greywater systems wherever they can; still others are fantastic beekeepers or teach gardening to elementary school children. The goal in every case is a renewed home for all beings, a life lived in peace with creation, and there is more than enough work for us all.

on the ground: principles applied

Kevin Bayuk is a permaculture teacher and activist living in San Francisco's Haight Ashbury district in a small rental unit with his wife and daughter. He shares his backyard food forest with the residents of a twelve-unit apartment building. Kevin seems to simply live the values of permaculture in his teaching and thinking and his daily actions. Our conversation circled around permaculture ethics, and how he understands their application to contemporary living.

"I am really inspired by the ethics shared by permaculture—earth care, fair share, and people care," Kevin explained. "There's a lot of waste in the industrial agriculture system, both in production and distribution. I know what it does to the watershed and what it does to the topsoil, so I know that buying industrial food is not providing any earth care. When I am a primary producer of food in a way that builds soil, increases its carbon content, and increases the life of the soil as I create an abundant delicious yield, that's an expression of earth care. Similarly, when I think about water both from a food perspective and the water we use at home, any little bit I can conserve or source from a sustainable supply is a way to take better care of the earth than being dependent on the industrial system.

"How do we know we're doing people care? If I have enough health, ease, leisure, joy, play, I imagine I am engaging in this ethic. And this needs to be focused not just on some people, but everybody, which is what really ties it into the third ethic. When we're doing people care, it's a natural thing to do to conserve energy and materials wherever possible. When we become food producers, we have a direct, intimate relationship with our nutrition. But my family's security and joy in life is interdependent with the security and joy of all beings, so when we have a surplus, we share it. When we recognize how taking care of people has to extend to making sure everyone is cared for, we can see how the second and the third ethic are connected to one another.

Photo by Daniel Miller/Spiral Gardens

"What I love about permaculture is that it's one thing to know about how to get a low flow showerhead and how to install a small greywater system or how to put in compact fluorescent light bulbs or plug strips for all your appliances, but the thought that goes behind it, the question of why you do those things and what you look for and how to look for it, is what's important. The ethics and principles give us the lens or the perspective on how to see: Produce no waste. Observe and interact. Use and value renewable energy. Accept feedback. How many people look at their energy bill and know how many kilowatt-hours they used last month? If you don't know what you're doing, how can you change the habit?

"I tend to hear feedback from people that when they think about a lifestyle that is in harmony with the living systems around them, they think of it as less, that they will have to give something up. For me, it goes the other way—when I have less stuff there is still the possibility of more. There may be fewer iPods, but more live music or more music making with friends. Instead of eating out less often, we get more quality time cooking with friends. I advocate for finding the 'more' of whatever you have a 'less' thought about. I like to think of the cultural shift as not moving back to the Stone Age or moving back at all, but actually moving toward more quality connection, more tolerance, more peace of mind, more acceptance, more leisure, more peace, more wonder, more good food, more community. . .

"I've seen people approach this type of lifestyle or message as something they must do. Climate change, species extinction! Do something now! We must! I've had those feelings of urgency, but when people approach this kind of lifestyle with a sense of mustness, it's just a few years before burnout. That type of energy leads directly to failure; it doesn't fit with the ecology of a healthy system. I advocate for a different metaphor for why you'd live like this. I remember a story that comes from science that says the G-type star we're flying around on is five or six billion years old, and it might live another twelve billion years. If humanity makes it, twelve billion years down the road all the hydrogen will have fused into helium in that star and it's going to erupt and expand and envelop the Earth and all the life on it will be gone. In this story, you can't save the Earth or humanity, so there's no must about it. The story's written; it's just a matter of time. Is it twelve billion years from now, fifteen years from now, 100 years from now? It doesn't matter to me; I just know the story of trying to 'save' the Earth is foolish.

"If you have the perspective that life is going to end, it seems there would be a natural sense of hedonism. But if you go deeply into what you want for yourself, you might find that your ultimate expression of happiness is totally interdependent with the happiness of all life. If you go all the way to the edge of seeing that there's no point in trying to save anything, it leads you into service to everyone else's satisfaction. It's natural to the point of being spontaneous to want to conserve energy, grow your own food, or have good relations with others. It's spontaneous in the way a hand heals itself, and scabs over. It's just what you do.

"There's a tendency to prognosticate in the permaculture movement. As soon as anyone learns about peak oil, it's all about the future. When is it going to happen? The food system is going to fail! Tomorrow? Ten years? Five days? Everything is about the future, and the tragedy of that is we forget about right now. As much as we can bring ourselves to right now, to get focused and not worry about what's going to happen when, we're on a pathway to another way of being."

Practical Designs

Permaculture is a way of looking at a whole system or problem; observing how the parts relate; mending sick structures by applying ideas learned from long-term sustainable working ones; and maximizing connections between key parts. Practitioners observe and imitate the working systems of nature to mend the damaged landscapes of human agriculture and cities. This same thinking can be applied to the design of your backyard, the organization of your kitchen, getting around town, relating to people at home or work, or managing the water that falls on your house.

Here's an example: On a small urban lot, it's simple enough to start growing food by digging up the soil, tossing out some seeds, and seeing what grows. But without observing and tending the soil first, you run the risk of further depleting it. Applying permaculture principles, you would investigate the soil first, and replenish it with any missing nutrients. You might recycle food scraps as compost, urine as a nitrogen rich fertilizer for the garden, and if you have barnyard animals, enrich the soil with their manure as well. In this way you begin to create a closed loop, regenerative system where food production can be enhanced by the "waste" you put into the soil, thereby creating more food. This is how the earth works—no waste, every by-product a source of nourishment somewhere else along the line. Animal and human manure contribute to the healthy soil we need to grow the healthy food we want to eat to restore the world we want to leave to our children, who will teach their children about regenerating the soil and the processes of food production all over again.

Permaculture design attempts to replicate nature by developing edible ecosystems closely resembling their wild counterparts. The prime example of this is the food forest, a seven-level perennial planting design closely mimicking the intelligence of the forest ecology. Starting at the root level and rising to the tree canopy level, this design works well for the rural home site in temperate and tropical regions and has been widely documented. We can plant a modified food forest in our smaller urban places, but the application more unique to urban centers is a redesign of city space using these ecological principles.

Efficient design systems based on the resilient model of the forest ecology can be developed for city infrastructure, food distribution, and human interactions. This means looking at the different elements that make up the city's landscape and beginning to forge new relationships among them. A city is made up of streets, and buildings, and a sewage system, and various laws and regulations and governmental entities. It is also made up of communities of people who have different alliances, histories, and cultural traditions. Cities house people's dreams, their economy, their children's education, their entertainment, and their health. What are the seven levels of the urban ecology?

Local license plate of the urban dirt farmer.

Brooklyn Grange—one rooftop acre of productive farmland—in Queens, part of the biggest metropolis in the United States. If it can happen here, it can happen anywhere. *Photo by Cyrus Dowlatshahi*

Urban permaculture design asks: How can we bring each of these elements into better working relationships with one another? How can we bring each element of what makes up a city—the visible and invisible structures that make up our lives—into a synergistic working alignment? Each element in the redesign of our cities can be looked at in terms of the input of energy and resource, and possible return, just as we do when we look at the forest ecology and notice the closed loop system of nature's design.

Community gardens exemplify a system with balanced input and output (with maybe a bit more on the output side.) The resources of soil, fertilizer, seed, time, and labor go into building these gardens. The output is nourishing food, a network of relationships, educational opportunities for children and adults, and the creation of a commons where people return again and again to enrich their lives.

We can assess our personal relationships in the same way. In order to flourish, a relationship requires attention, time, and a certain level of communication, as well as opportunities for play and relaxation. The output is a sense of connectivity, camaraderie, support, security, pleasure, and joy. If the input in a relationship is greater than the output—too much processing, no pleasure—the relationship is probably ready for a redesign, or for the compost bin.

Although urban density is sometimes hard to endure, it does have some ecological advantages. Human beings living in clustered settlements do less damage to our remaining wild lands; cities, with their already developed infrastructure, are prime targets for intelligent redesign; and because of people's proximity in cities, energy output, especially in the realms of transportation and home heating, can be significantly reduced.

Cities have a vital resource that is less abundant in rural environments: the engines of human creativity, ingenuity, and diversity, resulting in an increased possibility of evolving different solutions to problems that affect us all. The city is where most of us are and will continue to be in the twenty-first century. We obviously have a lot of work to do if we are to re-create our cities on nature's model. Permaculture design is one tool for regenerating urban systems toward greater integration and productivity, reduced waste, and a finding a place for everyone at the table.

The Principles

One of the most important things about permaculture is that it is founded on a series of principles that can be applied to any circumstance—agriculture, urban design, or the art of living. The core of the principles is the working relationships and connections between all things. The focus is on small-scale, energy- and labor-efficient intensive systems that use biological resources instead of fossil fuels. Designs stress ecological connections and closed energy and material loops. A key to efficient design is observation and the replication of natural ecosystems.

Observe and interact.   *Photo by Rachel Kaplan*

In the city, permaculture also makes much use of the synergy of human energies, and works to create generative connections between people. Both Mollison and Holmgren articulated underlying principles for permaculture in their own ways; Holmgren's are outlined here. The title and connected aphorisms are Holmgren's; the descriptions are our own understanding of the principles. Although humorously stated, they provide simple yet profound guidance for people interested in changing their actions and their lives with an eye toward regeneration.[19]

1. **Observe and Interact (Beauty Is in the Eye of the Beholder).** Taking time to observe nature makes it possible to design made-to-fit solutions. An example is the gardener who allows a season to pass before planting her garden, taking the time to observe the arc of the sun and the moon, the direction of the wind, the flow of water, and the impact of neighbors on her garden. While our timing is urgent, it is more urgent to make intelligent choices within the limits of the ecological systems that sustain life through difficult and abundant times. Within the problem lies the solution; observing and interacting slows down the action enough to allow us to meet a problem on its own terms, and gives rise to out-of-the-box thinking.

2. **Catch and Store Energy (Make Hay While the Sun Shines).** Develop systems that collect resources at times of peak abundance for use in times of need. For example, a solar array catches and stores the sun's energy for later use. Or a composting toilet, which, rather than flushing our excrement into the sea, composts it and provides fertilizer for our garden. Contemplative practices (yoga and meditation) catch personal energetic resources and store them up for another time.

3. **Obtain a Yield (You Can't Work On an Empty Stomach).** Good work yields rewards. A garden is an excellent example: we till the soil, sow the seeds, pull the weeds, manage the pests, and harvest the food. The yield is good food, good work, beauty, and a sense of knowledge and relaxation in the environment. Children gain skillful means in the garden for later production of their own food. Another example of practical yield is in our relationships. When we build communities where we rely upon one another to help raise the children, the food, and the management of resources, the yield is in dynamic, interdependent relationships that can sustain us throughout life and up to our deaths.

4. **Apply Self-Regulation and Accept Feedback (The Sins of the Father Are Visited on the Children unto the Seventh Generation).** Modify behaviors that do not work. Enhance behaviors that support the functioning of the system. This reflects the maturation of consciousness from adolescence to adulthood: our actions have consequences. The homeostasis of the earth as a single, self-regulating system is the best example of self-regulation and feedback.

Nature's patterns are the best models for intelligent design. The gills of this mushroom flowing outward from the center tell a unified story of energy stored, recycled, and used again.

5. Use and Value Renewable Resources and Services (Let Nature Take Its Course). Access the resources of nature to reduce excessive consumption and dependence on nonrenewable resources. Solar power, wind power, and intellectual power are all examples of renewable resources.

6. Produce No Waste (Waste Not Want Not). Nature produces no waste—everything is food for someone. Value and make use of all the resources available. Nature's best example of this principle is the earthworm that lives by consuming plant "wastes" and converting them into valuable soil. A compost pile is the garden's best example.

7. Design from Patterns to Details (Can't See the Woods for the Trees). The patterns in society and nature can form the backbone of our designs. Details arise from patterns. Consider the spider web—each one designed to serve a specific function, yet each unique to its location. The human body is another example of this principle—we're all built on the same model, but have individual differences related to experience, family life, and culture. With design solutions, as with people, one size does not fit all.

8.   **Integrate Rather than Segregate (Many Hands Make Light Work).** Don't leave anyone behind. Put elements in the right place, and relationships of support will arise between them. In human communities, we can see how segregation hurts everyone, but most often places the burden on the least powerful member of the equation. Co-housing communities show how much can be accomplished when more people participate, and how much richer our human gardens are when populated by different kinds of people. The segregation of cities, on the other hand, reflects a poverty of design, and the waste of human resource and potential.

Another crucial concept embedded in this principle is that of *stacking functions*, every element in a design doing more than one thing at a time. Take the ubiquitous chicken, which provides eggs, meat, and feathers, while she turns our kitchen scraps into nitrogen-rich fertilizer for our gardens. The chicken is popular because she does so many things at once. Some plants have stacking functions, offering pollination opportunities, and providing food, medicine, and beauty. Built structures can have multiple functions—greenhouses along the south side of a home can access and store passive solar energy, thereby serving as food-growing zones as well as places to share meals. This is an important concept when there is no time to waste.

9.   **Use Small and Slow Solutions (Slow and Steady Wins the Race).** Small and slow systems are easier to maintain than big ones, make better use of local resources, and produce more sustainable outcomes. The snail is nature's exemplar of this principle. Slow and steady, carrying its home wherever it goes, the snail is amazingly versatile and adaptable in a wide range of environments and, as any gardener can tell you, a remarkably powerful creature in the garden.

10.   **Use and Value Diversity (Don't Put All Your Eggs in One Basket).** Diversity provides insurance from the vagaries of nature and everyday life. Diversity reduces vulnerability and utilizes the unique nature of its environment. Diversity maintains and evolves culture and horticulture. Different foods and crops arise in different regions as an expression of cultural, aesthetic, spiritual, and sentimental needs. This diversity brings quality and texture to living and shapes the way people come to understand themselves.

An aligned concept is that *each function is supported by many elements*, which is a fancy way of talking about planned redundancy. Different plants that have the same function provide security for one another. If any one of them fails we still have access to the benefits of the other plants serving that same function. Nature employs planned redundancy as a strategy all the time; the designs of our gardens and our cities should do the same.

11. **Use Edges and Value the Marginal (Don't Think You Are On the Right Track Just Because It's a Well-Beaten Path).** The interface between things is where the most interesting events take place. These are often the most valuable, diverse, and productive elements in the system. "Marginal" subcultures are often the places where the most inventive, ingenious, and creative innovations take place. The long fight for food and environmental justice in impoverished communities of color that predates the current urban homesteading movement by decades is an example of this principle.

12. **Creatively Use and Respond to Change (Vision Is Not Seeing Things as They Are, But as They Will Be).** Change is inevitable; we can have a positive impact on it by carefully observing and intervening at the right time. Evolutionary change impacts stability. While we need to create durable natural living situations, durability paradoxically depends on our capacity to be flexible and to change. This idea is reflected in science and spirituality. Within the center of stillness is a vast, unending motion. We must learn to ride the rapids of change in our little paper canoes, together.

Zones and Sectors in an Urban Setting

Zonation is an important concept in permaculture. It is a way of understanding and organizing the objects, plants, and animals you have most interaction with closest to the center of your life. "Zones are a way to manage energies available on site: people, machines, wastes, and fuels of the family or society," wrote Bill Mollison in *Permaculture*.[20] His model delineated different areas in the homestead itself as Zones 1–5, with Zone 1 being the area right outside your door. Zone 2 moves farther from the home to an area you might visit once a day. Perhaps you place a vegetable garden and your chicken coop in this area. Zone 3 might be where you place the part of the garden that doesn't need daily maintenance, Zone 4 holds the orchard or other parts of your homestead that need less of your daily time, and Zone 5 is for the woodlot, or the windbreak—the outlying areas of your farm.

For Mollison, the golden rule is to develop the nearest area first. We can work with this rule in our small city homesteads, but it is likely that everything will be pretty close in, as space is at a premium, and we will be left to source many of the things we need from beyond our property line (for example, food from the farmers' market, or materials from local businesses). We can focus on zones when we plan our gardens and figure out the best place for

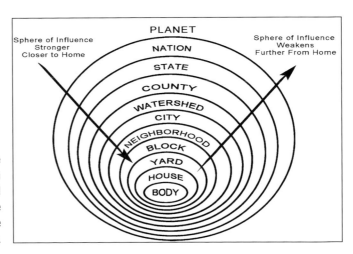

Zonation in the city starts with the body and maps the source of all inputs and influences in our designs. In the city, we gather resources from near and far, and cannot expect to produce everything we need on one site. City zones include all spheres of influence, those closest to home, and those furthest away.

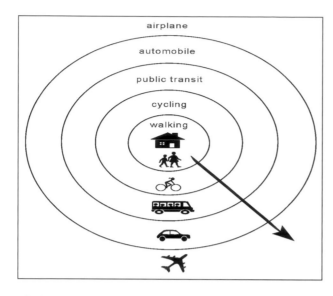

our chickens and bees, but another way to think about zones in the city is to start with the territory closest in—the body. The body becomes Zone 00, the place where you are always at home. Beyond the body, zones spread in concentric circles to include your house or apartment, your garden, the street where you live, your neighborhood, your communities of necessity and affinity, your city, local government, watershed, bioregion, and so on. Delineating these zones can give us a sense of where resources come from, and how we might make new choices to localize our living.

All of these areas offer us different resources—the comfort of home, the diversity of the marketplace, the proximity of parks and recreational areas, the availability of wild land. Notice that as the forces *increase* in scale and distance, our power to influence them *decreases*. With its emphasis on living a local life, permaculture focuses on the zones we can most easily touch and influence, but it is also important to understand how the larger circles of influence affect the choices we make in our intimate spaces.

Because a conversation about zones is always about conserving energy, another way to think about zones in a city has to do with the amount of fossil fuel energy it takes to get to a certain place.[21] Walking then becomes the center of the concentric circles, with cycling, public transit, automobile travel, and air travel as the outer zone.

When we see how much space we occupy in each of these zones, we learn something about our own participation in the fossil fuel debacle. Make your own map by placing the different things you do over the course of a day, week, or month in relationship to the energy it takes to access these resources. For example, a home garden will be in a walking zone; a visit to a national park is likely to be a car ride, unless you live adjacent to one. Can you walk to your grocery store? Work? School? In this way, you will begin to understand how much fossil fuel energy you are using, and how you might begin to curtail your use.

Zones help us understand how to manage on-site energy; sectors are a way to look at natural or wild energy as it flows across the land. In a classic permaculture analysis, we observe wind, water, fire, sun, and weather as they affect our site. In an urban sector analysis, these natural elements are observed, but we also assess the cultural and economic forces and flows affecting our lives. Urban zone and sector analysis reflects the interplay between small personal spaces and communities and the larger social networks that impact us all.

Private property is one of the biggest socioeconomic forces defining sectors in a city. We can delineate seven sectors defined by different types of ownership:

Personal: the household—rent or own.

Family and friends: informal but strong relationships.

Associations: clubs, churches, volunteer groups, etc.

Community: neighborhood, city, county, state, federal.

Local businesses: retailers, professionals, farmers, and craftspeople.

Mega-corporations: conglomerates, chain stores, the Fortune 500.

Undefined: land without clear ownership, such as vacant lots, underpasses, abandoned houses.

Using this new model, map the fossil fuel–based zones as they are influenced by these cultural sectors. Fill in this map for yourself. Where do your resources come from? What kind of energy do you need to access them? What do you notice about your energy and resource use that could change to meet the values of earth care, people care, and fair share? Hopefully, your map demonstrates that there are many ways to meet your needs other than through personal ownership. For example, if you don't have the space to grow your own food at home, you can get nutritious food from a community garden plot or the farmers' market. Use these maps to assess how you might begin to streamline and localize your lifestyle.

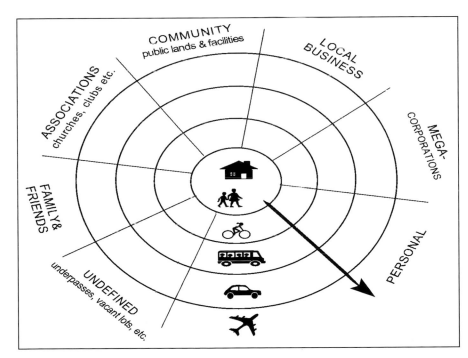

This diagram shows the overlap of city zones and sectors, divided by distance and lines of property. Fill it in for yourself to map your own urban experience.

Getting familiar with permaculture principles will help you embody a set of practices that can powerfully direct you toward sustainability. In so doing, you'll be able to evolve strategies that place you in deeper relationship to your own piece of urban earth. Permaculture offers a template for action and a decision-making protocol, whether we are designing gardens, relationships, community actions, spiritual practice, work, or the space and time for freedom and relaxation in our lives. Practicing these principles within your sphere of influence will help you articulate your own art-in-living, and your personal contribution to the change our culture so clearly needs.

Creating a Personal Sustainability Plan

Living within the biological constraints of the earth may be the most civilized activity a person can pursue, because it enables our successors to do the same. You cannot live within the carrying capacity of a region if you don't know where you are. Most of the developed world lacks this knowledge. We have little understanding of where our water and food comes from, the impacts of our cars and homes, the activities undertaken by others around the globe to support our lifestyle, and the effects we have on the environment and its people. . . . We will never know ourselves until we know where we are on this land.

—*Paul Hawken*[22]

One of the central strategies for ecological living is to reduce consumption (or our "resource footprint") as much as possible in every sector of our lives. To do this you must first understand how much you're using. Just as we assess different elements in the design of our homestead and how they relate to one another, a personal sustainability plan with guidelines for conserving resources can grow out of a true assessment of our needs, our inputs, and our impact.

At present, most of us use the whole world as our resource base and feel entitled to access resources from anywhere, whenever we want. This is an untenable position that needs to be examined. Whenever possible, we need to reduce consumption and localize our resource base. The way to change the world is to change our own practices—in our homes, at work, with how we eat, how we travel, and how we relate to our communities. We should continue to work for institutional change, but such efforts cannot succeed unless we examine how we act in our own lives.

Taking matters into our own hands. The first step in building a new home is to gather and shape the soil that makes it.
Photo by Miguel Micah Elliott

With a personal sustainability plan we can track inefficiencies of use and begin to remove them, step by step, from our lives. There is energy embedded in everything we do—from the toilets we flush to the beer bottles we recycle to the showers we take and the cars we drive. The clothing and products we buy have embedded energy, or an "energy signature." When making choices about how to reduce consumption, considering the embedded energy is an important, often missing, part of the calculation. This is something we can learn to do as we construct a personal sustainability plan.

When we really start to inquire into the impact of our actions, we are following the first permaculture principle: observe and interact. As we gather information from observation, we can begin to make new choices about how we want to live that are in alignment with our values and needs. This application of the principles is a way of looking honestly at how we live, an opportunity to really consider every aspect of your life and whether or not that process aligns with our ethics.[23] Each of us must decide how specific and detailed and rigorous we want our personal sustainability plan to be.

It is best to set reasonable goals rather than extreme and hard-to-achieve ones. Bite-size changes to maximize your success are simply more efficient than committing to a total change makeover, since success inspires more of the same and failure is discouraging and inhibiting. New habits and styles of living take time to assimilate, and change is most likely to establish itself as a new habit when done in moderation. This sounds contradictory because the house *is* burning, but walk, don't run, to get yourself out of there for good. Remember: use small and slow solutions. Once you've assimilated a new habit, you can raise the bar and challenge yourself to the next level of ecological awareness and living.

Sometimes a sustainability plan is best made in cooperation with others. You can organize a neighborhood group committed to performing block-by-block audits of energy and water use, recycling and garbage patterns, land availability, and food production possibilities. Working together on benchmarks for your neighborhood to use less and grow more will enhance community spirit while lessening the group's total impact on the environment. Doing it together is often the key to making things happen.

Sustainability assessments include an inquiry into both the visible/built structures of our lives—energy use, water use, waste production, food needs, healthy soil—and the invisible structures, including our needs and inputs to culture, community, family, economy, health, and spirit. Our impact on the built environment—how much energy and water we use, how much waste we produce, how much food we eat—can be measured in pounds of waste recycled or number of kilowatt hours of electricity used. Objective measurements like these help us track our progress. Measuring the invisible structures is more subjective, and varies according to individual needs.

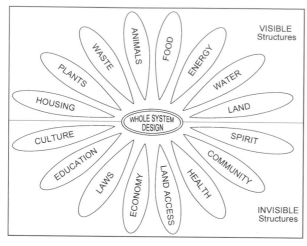

The Carbon Calculator

Tracking our own carbon output is useful for understanding our participation in climate change as well as the complexity of extricating ourselves from it. City dwellers are firmly enmeshed within the systems that exploit carbon-spewing emissions: most of us drive cars, use a refrigerator, turn on the heat, and take a hot shower, all simple everyday activities with carbon impact. While rural homesteads can create an off-the-grid setup (sometimes still relying on nonrenewable resources), going off grid in the city is challenging.

To track your carbon output in relation to averages for the country, as well as averages around the world, check out the carbon calculator online at www.empowermentinstitute.net. The calculator assesses how much energy you use in home and transportation. There are also carbon calculators that track the embedded energy in the food you eat, the clothing you wear, the furniture you buy, the water you use, and the buildings you build. These can be found at www.cleanmetrics.com. More about this in Chapter 15, Powering Down.

How Much Garbage Do You Make?

The noble calling of waste management can begin by tracking your own waste cycle. Analyze your output with simple questions: How many pounds of garbage or recycling do I take to the curb each week? How much organic matter do I compost? What will I do with that old toy or bookshelf or book or piece of clothing I don't need? How many times a day do I flush the toilet? Tracking specific waste outputs is a necessary first step in a waste reduction plan. Easily adopted practices can be structured like a game that even children can understand and appreciate. More about this in Chapter 17, Zero Waste, and Chapter 14, From the Ground Up.

Home Energy Audit

The United States far surpasses every other country on the planet in its per capita consumption of fossil fuel. We use an annual 11.4 kW of power per person, compared to 6 kW/person in Japan and Germany, 1.6 kW/person in China, and 0.2 kW/person in Bangladesh.[24] Even though fossil fuel use is rising alarmingly in parts of the developing world (and people in the United States sometimes use this as an excuse for not changing their own consumptive behavior), we can see that reducing our personal use can make a quantifiable difference. No matter what people in other countries are doing, United States citizens continue to use 25 percent of the world's energy to support only 5 percent of the human population. No amount of math makes that equation work.

If everyone consumed at this level, we would need no fewer than three Earths to support our insatiable appetites.[25] But even the world average isn't compatible with a stable climate. Earth's natural systems can remove only about a third of the excess carbon that humans are emitting daily. The earth simply cannot absorb the amount of pollution created by people. Despite the enormity of the problem, our individual actions taken together can shift the balance to recovery. Reduction of use is the best option we've got.

A useful tracker for understanding home energy efficiency can be found at www.energysavers.gov/your_home/energy_audits. More on these kinds of ecological fixes in Chapter 15, Powering Down.

The sacred beauty of water.

Travel

Transportation has a huge carbon impact. Assess yours by exploring your daily travel; air travel; and the ways you can increase your biking, walking, and use of public transport to get around. This aspect of our lives is often the biggest carbon culprit, and can be challenging to impact significantly, especially when there is no public transportation in the place where you live. More on this in Chapter 15, Powering Down.

Water Footprint

The water footprint is defined as the total volume of fresh water used to produce the goods and services consumed by an individual, community, or business. Just as fossil fuel energy is embedded in the products we use every day, so is water embedded in many of the things we take for granted in our daily lives. For example, 11 gallons of water are needed to irrigate and wash the fruit in one half-gallon container of orange juice. Behind that morning cup of coffee are 37 gallons of water consumed in growing, producing, packaging, and shipping the beans. Two hundred and sixty-four gallons of water are required to produce two pounds of beef. According to the EPA, the average American uses about 100 gallons of water daily, but this number does not factor in these hidden gallons embedded in the products we use.[26]

One flush of a toilet in the developed world uses as much water as the average person in the developing world allocates for an entire day's cooking, washing, cleaning, and drinking. The annual American per capita water footprint is about 8,000 cubic feet—twice the global average. With water use increasing six-fold in the past century, our demands for fresh water are rapidly outstripping the planet's ability to replenish its watersheds. One comprehensive glance at our own physical makeup—70 percent water—gives a frightening reality check about how crucial water is for life. To check out your own water footprint, go to www.h2oconserve.org, or visit www.waterfootprint.org.[27] For more on water, see Chapter 16, Sourcing the River.

Fossil-Fuel Food

The food we eat also has a high-energy impact, especially when it comes from far away from home. Some foods are literally "drenched in oil"—anything that has traveled thousands of miles to reach our tables has embedded in it all the energy it took to get there. Additionally, certain kinds of foods (meat, dairy, and industrially

Backyard chickens are one link in the chain of a localized food system.   *Photo by Dafna Kory*

produced vegetables) have higher carbon impacts stemming from intensive methods of production and their environmental impacts. Meat, for example, is the highest impact food when it comes to carbon, and accounts for about 18 percent of greenhouse gases overall.[28] Food grown in your backyard is nearly carbon neutral, saves water, renews the soil, and mitigates negative land use.

When assessing your carbon impact through the food you eat, look for how much of it is processed, out of season, or grown far from home. Assess your true capacity to grow a portion of your food yourself, or source from environmentally positive producers close to home. More on this in Chapter 10, Food Is a Verb.

Tracking the Invisible Structures

When it comes to measuring the invisible structures of your life, the questions you ask should reflect an understanding of your own needs. For example, what are your energy inputs into community? Do you volunteer your time or only work for money? How much time to you spend in community? What are your needs for community? Consider your family time in the same way. How much family time do you want to factor into your life? How much time do you now give to members of your family? How does that impact your resource use? (For example, driving your children to soccer practice has an energy impact, as well as a time impact for you and your child. Does this line up with your values and needs around sustainability? If not, what could be changed?)

Other invisible structures include health, spiritual life, and money. Here questions are again subjective. What do I consider to be good health? What are the elements of living that lead to good health? What are the inputs and outputs of creating a healthy lifestyle, or maintaining optimum health? For some people this will include an exercise regimen, or an understanding of how much money they need to visit an acupuncturist and herbalist once a month. Factoring in these personal needs is crucial in assessing the overall inputs and yields of your life, as well as what you might do to live more in alignment with your values.

Economic issues are also subjective. Some people choose to use their time working for an income to support an experience of material abundance. Others live more frugally and gain the advantage of time. Still others find they have neither enough time nor money. If you find yourself moving toward an urban homesteading lifestyle, now may be the time to sit down and really think about your priorities. Time? Money? Travel? What calls you? This question could yield highly inclusive results, like deciding to move somewhere with cheaper home prices in order to limit the amount of money you need to earn, or doing with less on the material plane to gain more time every day.

A personal assessment that helps us understand what we actually need and how to mitigate our impact will lead to an assessment of bioregion. What is my impact? What does it mean to live well? What can I provide for myself? What is produced close to home? What can I do without? What can I make or share with others? How much do I really need to feel satisfied? What can I source locally, at thrift, or through trade and barter? Can I provide for my food needs from local producers? What kind of time or interest do I have for food growing? How can I use less energy for transportation? What parts of my life can I streamline or localize?

Remember that everyone's going to come up with different answers to all these questions depending on many different factors, including stage of life, health, economic security or insecurity, etc. Homesteading and living a sustainable lifestyle doesn't offer a cookie-cutter approach to living—it's about finding your own way in the game of planetary restoration. For example, some people don't want to garden. That's okay. Buying from local farmers is a great way to support them while sourcing your food from your own bioregion rather than from thousands of miles away. Some people can replace three car rides a week with bicycle rides or public transport.

Others can commit to buying fewer new products in the stores, instead relying more on secondhand or barter for the things they need. There are many different solutions to the problems that face us. Understand what's true for you, and do it.

On Not Being a Martyr

An ecological lifestyle is often looked on as deprivation, a life without things or comfort or pleasure. Not so. Clearly, people who want to live ecologically out of a sense of fear or guilt often burn out on the feeling of *must* change, *must* act, *must* do more (or less) of whatever it is they are doing to hurt the planet. If this freaked out mantra were the story of our lives, we wouldn't be writing this book. We love our lives as homesteaders. It gives us a sense of personal connection and power and agency. Even though we can be cash poor, we are personally rich. We eat great food, have great relationships, and enjoy the opportunity to embody our values with our actions. We like the fact that we give less to an economy that is destroying the planet, and more to the earth, our friends, and our neighbors. Don't confuse this lifestyle with a fear-driven mentality of scarcity and lack. This kind of living is about the richness of the present moment and the joy in living a simpler, uncluttered life.

If you're feeling pressured to be part of the solution in your own life, find a way to engage comfortably within your personal experience of time, interest, and resources. Start small. Find something you love doing that brings you a sense of satisfaction and joy. We promise that this kind of success will lead to a greater desire and capacity to do more good things that will actualize your ecological values. Don't be a martyr. It won't help you, it won't help your family, and it won't help the world. This is no time to put on your hair shirt for the Earth. This

Is it beautiful?

is the time to imagine a better way, a time to throw your personal energy behind the world you want to create. Don't deprive yourself of a sense of wonder and joy in living.

Start with some simple questions about what you do. Bryan Welsh suggests these four: *Is it beautiful? Could I do it easily again, or teach it to someone else? Does it give back more than it takes? Does it create abundance?*[29] We include this one as well: *Does it honor my values of earth care and people care and fair share?* If you're getting a solid *yes*, you're on the right track. If not, what changes can you make to get to *yes*? The point of living sustainably is not to use any particular technique or become an inflexible ideologue. Rather, a creative assessment and intelligent application of conservation principles can change our use of resources based on our ethics. Renewing the world is a series of small creative and political and spiritual acts that take place day by day.

Many of the DIY projects offered in this book will help you come up with your own solutions to reducing resource use. It's highly unlikely that you will do every one of these projects. They are presented as options, as creative ways to bring your consumption levels in line with the closed-system reality of the earth. Follow your interests and your focus; this will help you find your own way.

Who's Got the Time?

Learning about our impact and all the different things we can do to change our lives can be enlivening, but it can also be intimidating. No one has the time to do everything in this book, and obviously, no one lives impact-free. Luckily, doing it all ourselves isn't the goal. Each of us doing some of the work is what it's going to take to change things. And the more inspired we are to really step up our change, and inspire others, the more we can become leaders in changing our culture and our world.

Does it create abundance?

Joy in the spring garden—earth care, people care, fair share. *Photo courtesy of Argus Courier*

Some of the projects are one-time-only events—you set up a system to manage your greywater and it works for a long time, needing only some minor maintenance or tinkering. Some tasks are more ongoing and repetitive, like work in the garden and the kitchen. Some people will be inclined toward the one-time installation upgrades; others feel most comfortable engaging with the daily or seasonal rhythm brought into our homes through the garden. The best way to manage the demands of homesteading is to share them. In the agrarian past, a homestead wasn't self-sufficient, the community was.

Working toward that sense of camaraderie in the tasks of caring for the earth is a central survival strategy for the twenty-first century. There are so many opportunities to learn to work and create together, to gain skills not only in sufficiency but also in interdependence. If you focus on what you like and go about the task of gathering information and experience, we guarantee you will find a number of fine friends along the way who want to do the same thing.

One More Note on Time: Jumping Ship

The more you work outside the home, the less time you will have for living an ecological lifestyle. This is a trap for too many of us—we need money to live, but we want to live in a way that is better for ourselves and our families and the planet. It might be ridiculous to say, "Quit your job" as an answer to our ecological problems, but it also might not be so ridiculous. It is true that if we find ourselves wanting to live more ecologically, with a lower carbon footprint, or having better food for our family, or more leisure time, we need to start getting curious about different ways of living. At least ask yourself if the money you are paid for your job is "worth" the cost of your time and energy. And if it isn't, what can you do to bring these things into alignment?

Making friends while working in the garden.

What simple things feed me?

We often say, "It has to be this way," but does it, really? What would happen if you quit your job and found a lower-intensity one, closer to home? How can you lower your expenses and therefore work less out of the home at tasks that do not feed you? What can you do to add to the substance and texture of your life? In short, how can you redesign your life to maximize your participation in the values of earth care, people care, and fair share, and minimize the amount of time you give to a system that isn't working for you or your family or the planet?

If you find yourself currently un- or under-employed in the extractive economy, frame it as an opportunity, rather than a problem. Now you have more time to learn skills that will bring value to your life—skills in food growing and preserving, skills in water harvesting and waste reduction, skills in community building. Growing the "green economy" and finding viable productive ways to participate in it is an opportunity for each one of us.

On Learning New Skills

No one is born knowing how to can fruit or build a chicken coop or design a rainwater system. But all these skills are learnable. The only prerequisite is a can-do attitude. *I can get the information I need. I can find the teachers who know. I can study until I understand. I can practice until I learn. I can become a teacher to new homesteaders.* This positive, can-do attitude is perhaps the most important resource for a homesteading lifestyle.

We live in an information-rich world—books, websites, blogs, and other people embody a wealth of information. Many people who master a skill like to teach it to others who really want to learn. Learning from a live human is always best practice; always seek out people where you live who have the knowledge you want.

Everyone starts as a beginner. The important thing is to start where you are and keep going. You will make mistakes. Don't be discouraged. You're bound to do better next time. And before you know it, you'll be passing on your skills to the next wave of urban homesteaders while advising them to start where they are, and just keep going.

Gardening and Growing

The True Growth Economy

The real work of planet saving will be small, humble, and humbling and (insofar as it involves love) pleasing and rewarding. Its jobs will be too many to count, too many to report, too many to be publicly noticed or rewarded, too small to make anyone rich or famous.

—Wendell Berry[30]

The garden is the holy center of the homestead. It's an altar to life and death, an improvisational theater of failure and success, moonrise and sunset, and the cycle of the seasons. In a deeply uncertain time, the garden is a sure bet: a timeless, calming place where your lineage as a human being in relation to the earth is affirmed again and again. Getting in touch with the sow-till-harvest cadence of life is the most satisfying act we know. And growing food is central to the path of stepping toward sustainability—it lowers our carbon footprint, grows us great food, and locates us in place, connected to other people and the complex reality of our human needs. Growing soil, sequestering carbon, providing habitat, sharing with our neighbors, and increasing the health and nutrition of our food supply is a win-win situation on every level: personal, cultural, and political.

Organic gardening is a huge topic, from designing your garden, to growing healthy soil, to learning which plants grow best in your locale, to managing pests and recreating diversity in the fractured urban earth. This long chapter is a short description of a vast subject with an eye on turning small urban spaces into lavish food growing zones.

Start Small

One day you finally knew what you had to do, and began,
though the voices around you kept shouting their bad advice.
—*Mary Oliver*[31]

Here's our first simple idea: Start small. Start with one bed, or one large pot, and see what you can do as you begin. Don't expect to be living on an urban farm in one season. Look at your resources and limitations, and start by growing a few things you want to eat. If you love lettuce, get a big plastic bucket, put some holes in the bottom, fill it with soil, and plant some seeds. You'll love to watch them grow, and you'll love eating them, too. If you've had success with something small, then plan something bigger for your next gardening experiment. If you've made a mistake or lost a crop, learn from it. Maybe you didn't have enough sun; maybe you over-watered, or under-watered. Maybe you have predators you never knew existed before. Maybe you planted too late, or too early, or picked a variety that isn't really suited to your location. Mistakes are inevitable, and are great teachers too. Use them as opportunities to grow on.

SEASON ONE

Easy to grow: Zucchini, squash, beans, potatoes, lettuce, turnip, radish, chard, kale, collards, parsley, strawberries, mint, tomatoes, herbs, broccoli, cucumber, leeks, peas, raspberries.

SEASON TWO

Harder to grow: Carrots, parsnips, corn, celery, yams, onions, garlic, bell peppers, cabbage, spinach, basil, eggplant, beets.

Tools of the Trade: Don't Skimp, Don't Binge, Share

You need good tools to grow a good garden. Even though we're total advocates of the reuse ethic, when it comes to tools, we suggest you buy the best tool you can afford. If you can find it used, grab it. If not, buy it new. Good tools will last you for life, and with care, you'll never need to buy another.

When tooling up, buy what you need for the task at hand. Don't buy more than you need. For a garden, your most essential tools will be a fork, a trowel, and a bypass pruner. This will get you through many gardening seasons. As time goes on, you may need other tools—a spade, a flathead shovel, a hard rake, a soft rake, and a pruning saw—but if you don't need it right away, don't buy it until you do. In an urban space where you are gardening on a deck or in a small yard, you could use just the shovel, pruner, and trowel for years and be more than satisfied.

Another option is to share the tools you have with your neighbors. If you're in excellent communication with them, you can each buy certain tools you'll need and commit to sharing them around. That cuts down on costs for all, and brings the level of neighborly collaboration up a notch. Does everyone really need to own a ladder? A group of gardening friends can also invest in tools that are too expensive to buy alone, and for which you only have sporadic need: chain saw, wood chipper, back hoe, and so on. If you can find a common shed in which to store tools, a community tool co-op is a great urban homesteading asset. Another great option is a tool lending library. Some cities sponsor such libraries that function just like regular libraries. People borrow tools for a short time for free and return them when done. Volunteers or nonprofit organizations often run these lending libraries. If your city doesn't have one, organizing such a valuable resource is great work.

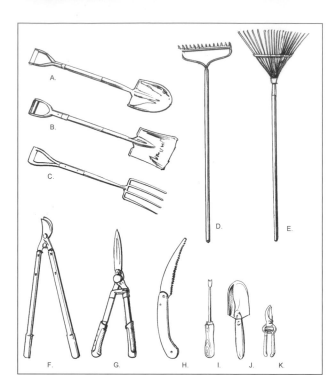

A. Spade
B. Flathead Shovel
C. Garden Fork
D. Hard Rake
E. Soft Rake

F. Loppers
G. Double-cut pruners
H. Pruning saw
I. Hand Weeding Tool
J. Trowel
K. Bypass Pruner

Tools of the Trade

Got Land?

Access to land is not a simple issue, and the widening gap between rich and poor will continue to rein-force the problem of access to private property in all of our cities. Owning or renting the place where you live impacts the kind of choices you'll make there. Ownership usually makes it easier for people to invest time and energy in a long-term project like a garden, and people tend to commit more fully to places where they feel they can stay for a long time, or where they imagine their energy, money, and time will not be wasted. Ideally, people make investments in the places they inhabit because it's how they live and not because they own it, but that's not how the renter's dilemma plays itself out for many people. And some landlords just don't want gardens in the backyard, or greywater systems alongside the house, or composting toilets anywhere at all, which can limit our ability to live ecologically.

It takes a lot of time and love to make a commitment to a garden or to some of the other sustainable systems discussed in this book. But homesteading is about giving more than taking—nothing is lost in your commitment to loving the place where you live and rooting your life there for the duration of your stay. Cultivating this atti-tude of reverence toward the place where you live may give rise to a desire to grow food, or harvest rainwater, or conserve energy whether or not you own your home. It can be as simple as growing something in a small pot outside your window, or learning to recycle your kitchen scraps, catching water in a bucket, or scattering seeds in the backyard. And for many people, it will mean finding ground to grow that isn't necessarily where you live. Much urban homesteading happens off-site, in community with others.

Create a garden in a bale of straw in the median between sidewalk and street. The straw bale will eventually compost, amending the soil, making this marginal space good for growing.
Photo by Trathen Heckman/Daily Acts

Especially in marginalized and impoverished communities, many people do not have access to any kind of soil, or places for growing, and it can be challenging to remediate the soil in some inner city landscapes. In these places, available land is often used for community-based projects, including community gardens, food security projects, gleaning opportunities, youth leadership and job training programs, as well as school gardens.

Work within your limits and your means. Many urban dwellers are renters, and many renters want to be part of the solution, so many of the projects we discuss in this book are small enough to take with you when you go and affordable enough to invest in regardless of your ownership status. Gardens can be constructed in raised beds, or in moveable barrels and pots. You can even grow a garden in a bale of straw on the median between sidewalk and street. Small structures you build on your homestead can come with you when you go, especially if you build them on wheels. And sometimes it's okay to build something or plant something knowing you'll leave it behind for the next renter. Don't wait until you own your own home to get started. The energy you put in will not be wasted; it comes back to you as joy.

on the ground: the beauty we love

Jane lives alone in a 1,025-square-foot house, on a 5,700-square-foot lot within walking distance to downtown. She has been living and gardening there for more than thirty-two years, so her garden is beautiful, mature, and complex. Her orchard—like her garden—is an altar to her life, as well as the entrance to her pantry. Her garden has spread from the creek that marks the boundary at the back of her property to the front of the curb, which is planted in fava beans, strawberries, and potatoes. In recent years her garden has overflowed to a small plot in the community garden across the street. "My motivation to live this way is that absolutely any, any, anything I can do to help the earth, I'm going to do it," Jane says. "We all live badly in the States, and there's one person—me—in this enormous house, so every single thing I can do that might be helpful, I'm going to do it."

When Jane and her husband first purchased the home, they were swayed by the size of the lot (50' x 100') bordered in the back by a creek. They installed a woodstove for heat, and it remains the only heat source. Over the years, the garden has spread, but when her children were small, large parts of the yard were places for them to play and dig and do whatever they wished. At one time, a labyrinth took up space in the yard, and at another time, a giant eucalyptus tree was felled. This wood was bucked and stacked, and became the sole source of heat for the house for over fifteen years.

Jane is able to grow all of her produce—she buys no fruits or vegetables from the store. She buys grains in bulk and stores them in plastic bins. She cans and freezes her extra produce, dries some of the rest, and has a giant stash of squash sitting on her dryer in the laundry room. Her yard houses ducks and chickens and three beehives. She raises the chickens for eggs and for meat. Her ducks sometimes fall prey to animals that enter the yard from the creek—raccoons, deer, and possum are visitors, and occasionally, a four-foot otter has come and whisked off a duck or two. The drama of such nocturnal visits aside, Jane appreciates living in a place where wildlife sometimes comes to call.

Jane lives an amazingly frugal lifestyle in close proximity to her work, local stores, and public transportation. Jane's frugality extends to her minimal car use (she not only walks to work, but when she gathers coffee grounds from the local coffee shop for her compost pile, she walks the empty wheelbarrow to the café, and wheels it home quite full). She diverts grey-water from the bathtub and washing machine into the garden, and until recently, her systems were rudimentary—the bucket method prevailed. Garbage cans gather water under her downspouts, and she acknowledges that other than beekeeping and learning to harvest her chickens for food, her learning edge is how to catch and store more water for her garden.

Jane's homesteading successes and failures all relate to the garden—different crops succeed in different years, and fail in others. She's had a hard time learning to tend to the bees. Like many bee colonies in recent years, the hives in the yard are not surviving the winter. She can't grow enough tomatoes, or zucchini, despite their reputation for rapacious growth. Her cabbages often "suck," though this year they are "gorgeous and beautiful and plentiful," and the grapes get eaten every

Jane and a giant cabbage in the community garden.

year by the raccoons. But her three apple trees yield enough fruit to can 81 quarts of sauce each year, which supply her with the fruit she eats all winter (as well as some gifts to her children).

Living alone enables Jane to make choices that would be challenging for a family—her near-zero-waste lifestyle, as well as her ability to provide for all her fruit and vegetable needs, would be nearly impossible to achieve for a group of people, especially if they don't have access to this much space. Additionally, because she is making decisions solely for herself, she's not involved in the ongoing negotiations implicit in all family relationships. We recognize that this is not how everyone does (or even wants to) live.

Without trying, Jane is an inspiration to the neighbors—her extensive strawberry patch and lavish apricot tree in front always draw hungry children, and her yearly fava bean patch is epic. The sunflowers are resplendent in summer. Puttering around her yard in her straw hat and old blue jeans, she offers produce and advice to dog walkers and neighborhood residents who wander by. "Just do it and you'll learn," she advises. "There's no right way to do it. It doesn't matter if you use raised beds. It doesn't matter how you do the compost. Just do it. You'll learn what works. So you plant lettuce here this year and it doesn't work. Next year you plant it somewhere else and you see what happens. And in another year, it could be different. It's just one big fat experiment and that's how you need to approach it. Just have fun. Start where you are and do what you can."

Where Can I Grow in the City?

People in cities often say, "I have no space for a garden. Where could I grow food?" Once you start looking around for it, you'll notice plenty of unused space that will work. The median in front of your house or the narrow strip between your place and your neighbors, empty lots, alleyways, rooftops, backyards, front yards, against a wall, and decks are all places where you can grow good food. Value the edges and the marginal: once your eyes are open to all the space that could be cultivated, you'll be amazed you didn't see it before.

If you don't have enough space at home, consider the options of community gardens, your neighbor's yard, and the wide variety of school gardening projects blossoming under the hands of children. The front lawns of municipal buildings, office parks, and other businesses might be ready for an upgrade, especially if you do it on the cheap and commit to taking care of it; the flat roof of your neighborhood movie theater or grocery store might be ready and waiting for a hive of bees and some raised beds.

Community Gardens

Community gardening and the victory garden have a long and venerable tradition in our country. When the United States entered World War I, Americans were encouraged to express their patriotism by growing food. During the Depression, victory gardens provided people with affordable food and a dignified means of support in a challenging time. Approximately one million people belong to the estimated 18,000 to 20,000 community gardens in the United States. Community gardens remain one of the most expressive pieces of the commons,

Digging in the school garden—an opportunity to use the head, the heart and the hands to make something beautiful and useful.   *Photo by Lauren Elder*

places where people from different cultures and backgrounds come together to share the practice of growing food.

Community gardens are popular—some have long wait lists for people to get a spot. If that's the case in your city or town, use it as an opportunity to start a new one. Often, municipalities hold unused lands that might be donated to people willing to tend it. Some are placed on available bits of city land, and some are placed on private lots that people are willing to lend or offer to the project. San Francisco recently passed legislation encouraging landowners who do not have the resources to build on their empty lots to open them to community gardening projects. Gardens, orchards, and small-scale animal husbandry projects are springing up in the empty places of the city—an excellent development strategy for a local homegrown economy.

FIRST STEPS IN STARTING A COMMUNITY GARDEN

A little bit of land, a group of interested people, dirt, tools, and a willingness to work is all it takes to start a community garden in your neighborhood.

- Gather a group of people interested in the project.
- Identify possible sites for a garden.
- Contact the owner of the land and negotiate a lease for the garden.
- Through dialogue with members of the project, forge an understanding about how the garden will run: land and water use, pesticide use, safety, responsibility for common areas, design issues, financial commitment, relationship to neighborhood and neighbors, dealing with vandalism, distributing extra produce, and managing conflict.
- Once land is secured, create a garden plan, including design of space, and a schedule for starting garden.
- Organize labor, organic materials, and tools, and schedule garden workdays.
- Dig in!

School Gardens

School gardens are another way urban children and their families are learning to grow their own food and connect to the earth. Many schools have gardening programs where kids are taught how to grow all the parts to their pizza, or make a salad, or which flowers are edible and how they taste. At the

A whole generation of urban farmers is being trained, getting ready to head out into the world, hoe in hand, to regenerate the world.
Photo by Lauren Elder

Front lawns are a great place to plant fruit trees, sprawling squash, and tomato plants, inspiring your neighbors to indulge in their own food-growing fantasy.

Closest to home—the backyard garden.

Photo by Jenifer Kent

Up on the roof: the Brooklyn Grange, a one-acre rooftop farm in New York.

Photo by www.brooklyngrangefarm.com

Vertical gardening in the city makes sense.
These planters grow a prodigious amount of
herbs, greens, squash, and other trailing vines.

Use sunlight where you find it. Lots of vegetable plants are happy
and productive in a climbing habitat: beans, tomatoes, cucumbers,
squashes, and vines like grapes and kiwis.

Once a home for
cars, this driveway
is prime growing
area on this small
city lot. Notice the
use of vertical
gardens, raised
beds, and small
pots and containers
along the edges.

*Photo by Kitty
Sharkey*

luckier schools, these edible gardens become central to the curriculum, sharing valuable hands-on knowledge about the cycle of the seasons, the relationship of our labor to our food, and the value and joy of shared work.

on the ground: spiral gardens

Daniel Miller runs Spiral Gardens Community Food Security Project in Berkeley, California. Growing out of his earlier social and environmental justice activism, Spiral Gardens supports itself in part by running a nursery, a farm stand, a community garden, and an educational project to teach people the skills they need for home food growing. Its central mission is to help local residents gain more control over their food supply by learning basic skills for gardening and food production, getting to be self-sufficient. Berries, trees, and vegetable starts are some of the urban farming supplies Spiral Gardens generates for local gardeners. It's also pushing for social change: "People who garden on their land are less likely to dump oil from their crankcase on it, because gardening changes consciousness about how to use land," Daniel points out. "People get healthier and start making better choices about upgrading their lives in all kinds of ways. It catalyzes people to all these beneficial effects."

Spiral Gardens is located on the edge of Berkeley, where residents often do not have access to fresh, healthy food, nor the ability to pay for it. "The idea for this project came because the census tract for the area around the garden has ten years earlier mortality than the rest of Berkeley, especially among people of color," Daniel says. "The number-one cause of death for a black man is heart attack and heart disease. All indicators leading to earlier projected mortality have to do with diet and exercise. If you overlay a map of food availability on the city grid, it exactly corresponds to health outcomes. . . . The folks in this demographic with the most poverty, the worst health outcomes, and the highest rate of unemployment are the folks who could most benefit from the skills and resources of urban homesteading."

SPIRAL GARDENS
COMMUNITY FOOD SECURITY PROJECT

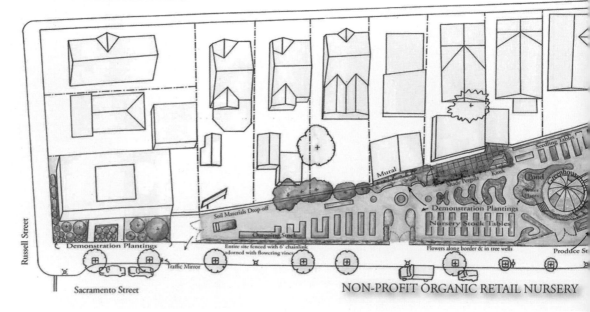

There are no good sources of fresh food in the areas where people are dying earlier, but there are lots of liquor stores. Produce at Pak-n-Save is subsidized and still way cheaper than organic food. Projects like Spiral Gardens are up against the deadly leviathan of television and print advertising, but as more information becomes available about chronic disease and its relationship to food, more people come to Spiral Gardens. For Daniel, "This work is about being one small part of the reparations that are needed, because reparations toward people of color haven't happened in this country. Activist urban gardening where you are dealing with food and ecology also deals with social justice, wise city growth, better urban planning, job creation, waste recycling. Working like this inspires me and keeps me happy and proactive. When you're gardening and growing food, you're not avoiding pressing issues, you're addressing them directly in a way that feels good."

The social enterprise of the nursery relates directly to the garden's mission to provide food to the food insecure. "We have the nursery, the produce stand, the community farm, and the fourth component of education. The produce stand has been going on since we've been on the site, and for better or worse, it's the most popular thing we do. We order food at wholesale beforehand from farmers who attend the farmers' market, pick it up at the beginning of the market, and take it to the corner and sell it at cost. It's a volunteer-supported model of the lowest-priced organic food around."

The community farm runs on a model that isn't quite like a community garden which is usually "divided into plots, and people work side-by-side but not together. This is a social experiment, where everyone works on the same plot, together. You can't come in and just do your thing. You have to work with other people to make decisions and get things done. This makes the whole project harder. There's a whole skill set in how people work and organize together that's hard to learn. I'd like the farm to be our most popular offering, but it turns out to be the most challenging thing for people to do."

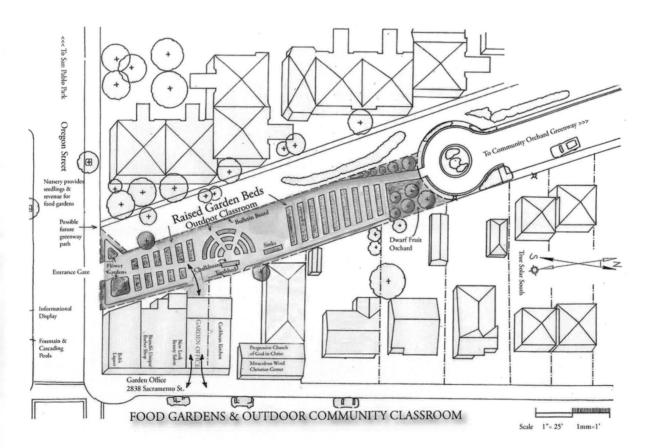

FOOD GARDENS & OUTDOOR COMMUNITY CLASSROOM

Scale   1" = 25'   1mm=1'

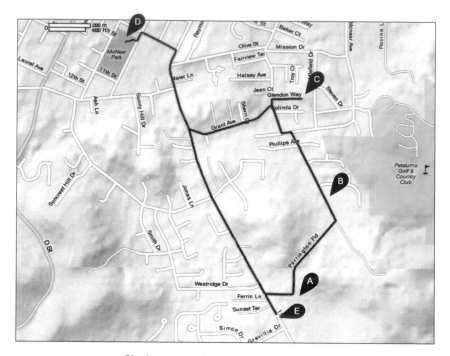

Sharing spaces: Rachel's Walking Gardens

A. Home Garden B. Chickens & Bees C. Garden Shared with Friend D. Community Garden E. Bees

More Creative Land Use Sharecropping

One way around the space dilemma is to borrow some from your neighbors and garden in their backyard. I was lucky enough to be offered a good-size bed in the backyard of an acquaintance. She was looking for someone to garden there in exchange for teaching her a few things. What began as a bare yard is now a flourishing garden with every available inch covered in green and growing things. I've been gardening there for three years, and that neighbor is now a friend. She went from "knowing nothing" to having lots of ideas about what we should plant, expanding the garden wherever she could, and eating more kale than anyone I've ever known. I might not have befriended her under ordinary circumstances, but her generosity of space and my willingness to teach her a few simple basics became the ground for a sweet, reciprocal friendship. Our family also keeps our bees in other people's yards, and we share our chickens with friends around the corner. This all makes for fine and friendly relations with the neighbors.

If you find yourself looking over the fence and lusting over your neighbor's unkempt yard, get down off that ladder, walk next door, and offer to turn it into a productive garden. You might get more than space for gardening.

on the ground: sharing the land

Patricia Algara is a landscape architect who practices urban agriculture on a double lot in Berkeley, California, that she borrows from a neighbor. Having come from a family of farmers in Mexico, working the land is instinctive to Patricia but when she came to this country, she felt overwhelmed by the difficulties of finding a place to farm. During a winter storm a fence was blown away, revealing an empty lot with perfect sun exposure two blocks from where she works. She met Giancarlo Muscardini, the neighbor to the right of the unused lot, and asked him about the empty space that was a huge contrast to his own lush, edible garden and orchard. He knew the neighbors and had dreamed of doing something on the next-door site for a long time.

"It drove him crazy to see the neighbors mowing the weeds every other week, so together we wrote a formal proposal for the use of the land," Patricia recalls. "We met with the neighbors. They were excited but had some issues. I had a lawyer draft an agreement so they wouldn't be liable if anyone got hurt. They didn't want an open community garden, so we fenced the property in a way that keeps it private for them and usable for us. I have keys and can bring people there; we have work parties and events, but always in collaboration with the owners' needs."

Together, Patricia and Giancarlo designed and built and farmed the garden. During the first year, work was done on setting up infrastructure—putting in a greenhouse and garden beds, but mostly amending the hard clay soil so that things could grow. Water was piped into the garden from Giancarlo's place next door. "When we first started, it was all grass so we had a lot of work to do," Patricia says. "It's been two years and it's finally a working garden. We sheet mulched, and did compost, and used a chicken tractor and did what we could to build the soil. Everything here is done through permaculture principles and the soils are much better now. They can still improve, but now they are productive and there's a lot of food."

Patricia and Giancarlo get food from the garden, as do the owners of the land and people who live in the units below Giancarlo's house. To share the excess produce, Patricia hosts an informal Friday lunch where all the food is prepared using the fruits and vegetables from the garden. The food gets harvested the night before to assure the freshest and most ripe flavors. The menu changes every week depending on what's in season. The lunch is also an opportunity to show people how the vegetable plants look while they are growing, their benefits to the garden, and their nutritional value. The lunches are mostly a way to expose people to local, organic, fresh flavors and get them excited to grow their own food.

As this demonstration urban farm project keeps growing, Giancarlo is inspired to deepen the permaculture design of the garden, as well as his own garden on the other side of the fence. He's looking toward planting more fruit trees, trees that can be coppiced (cut back to encourage the growth of straight suckers which can be used for building and garden projects), a more evolved food forest, and planting biomass-producing trees that add further nutrients to the soil.

Patricia's success with this garden leads her to work on food security projects in Richmond, one of the most impoverished communities in the Bay Area. "The more I learn, the more I see how this needs to happen everywhere," she says. "It is so daunting at an environmental level to understand what is going on, and sometimes it feels like a huge thing I can't do anything about. But growing food is something I *can* do.

"The more research I do, the more I see what a big impact food production has on so many aspects of the environment. We can have an impact every time we eat, breaking away from industrial agriculture, the fuel to transport the food, and the toxins that are used to grow it. If everyone grew something, one little small thing, it would matter. Food is basic. We need it three times a day. At least! Even if you just have a little window, you can grow mint. And that's one less thing to buy at the store.

"Food is the gateway drug to a more sustainable lifestyle. You start to become aware of the cycles of nature, the cycle of the moon, what's happening with the seasons and the climate, and you start to pay attention to the world. And it has a trickle-down effect—just doing this changes your behavior. It happened to me. Seeing this garden and the changes it's brought me makes me want to work so that more people can see this and do this. And it's not that hard. It's this basic human thing. We should all be doing it."

Guerrilla Gardening

Sometimes a garden springs up illegally in an abandoned piece of land. This happens a lot in places like West Oakland, Detroit, Philadelphia, New York, and Austin, where there's a lot of abandoned land and not a lot of places for local people to get good healthy food. Guerrilla gardening—squatting a piece of abandoned land—is more unsteady than gardening in a sanctioned place. But there are benefits—access to unused land, the potential for community energy to catalyze around urban use issues, and the sheer joy of trumping the laws of private property with the laws of human need and love. Sometimes guerrilla gardening can be as simple as lobbing a seed ball over a fence in the autumn, and surreptitiously watching the flowers bloom in the spring. Guerrilla gardens can evolve into more legal land holdings, or remain as temporary installations in empty urban spaces.

HOW TO MAKE AND PLANT A SEED BALL

You'll need some wildflower, clover, or other annual seeds you want to sow. Terra cotta clay powder, a little compost, and water complete the simple recipe.

- Mix 1 part terra cotta clay powder with 1 part compost. Add seeds to the mixture.
- Add the water slowly and mix while adding water. The consistency should be easy-to-mold clay—not too wet, not too dry.
- Roll the clay mixture into a small ball.
- Let dry for 24 to 48 hours, and store them in a cool and dry place.

Seed balls can be deployed in empty lots, backyards, highway medians, and other sites for guerrilla gardening in the beginning of your growing season. Surprise your friends and your neighbors with instant wildflower infusions.

Liberate the Streets—Or at Least the Driveway

When looking for space for gardening, you may find it necessary to take up the asphalt in a parking lot, driveway, backyard patio, or front yard area. De-paving your neighborhood is a great way to reclaim land for growing—it's liberation for the soil and land for you. Cities are covered in pavement, from sidewalks to back and front yards, to roads and streets exclusively for cars. Asphalt is impervious to water and organic matter, which is great for cars, but not so great for people. De-paving helps avoid toxic runoff from rainfalls—rather than running down the road and into the sewer, water stays where it lands, and seeps in to the ground. The more people localize their food sources in urban areas, the more necessary it will be to de-pave the city. And not a moment too soon!

It's easier than you think to jack up asphalt or concrete—using a pickax and a rock bar (a heavy iron rod with one pointed end and one chisel-shaped end), a few people can take up a driveway in an afternoon. Asphalt tends to be about two inches thick and concrete is quite a bit thicker.

Start by making a hole with the pickax in the concrete or asphalt. With concrete, you may need to begin this project with a jackhammer to make a hole you can leverage as you begin to pry up the concrete. One person pries up the edge of the impervious material while the other pushes the rock bar as far under it as possible. Jack up the asphalt or concrete using the leverage of the rock bar. Keep using the rock bar to pry it up until you've liberated the street.

Underneath, you'll find road base, which is broken up rocks, crushed stone, and granite. Some of this material can be reused in natural building projects, but first remove any road base contaminated by petroleum spills. Asphalt and road base should be landfilled, as they have no reliable reuse. Asphalt retains petroleum residues that leak out when heated and tends to get sticky when hot. It shouldn't be used for projects involving growing plants, and certainly not for plants you intend to eat.

If you are removing concrete, you can reuse that for different building projects, including the edges of raised beds and the foundations of natural building projects. Broken up concrete like this is called urbanite—so be an urbanite who reuses urbanite. Whatever you are jacking up, the soil underneath it will need plenty of amendments before it will be suitable for growing. You can aid the process by adding compost, mulch, and other organic matter to the area.

Containers, Raised Beds, and Barrels

Wherever you live in the city, you'll only have so much space, so much sunlight, and so much soil. Container gardening is a good way to maximize space when you don't have too much of it and containers can be placed in parts of the yard or deck or patio that have the most sun. Most fruiting plants, like tomatoes, cucumbers,

Once a driveway, the space has been de-paved, sheet mulched, and liberated to grow herbs, flowers, trees, and vegetables.

or peppers, need at least six hours of direct sunlight to thrive and produce well, but many greens and herbs will get along with less. As well as the amount of sun, pot size, soil quality, and proper watering will determine the success of your container garden.

A raised bed is a wooden box placed on top of the soil itself and filled with compost and fresh soil. You can also put a raised bed that's at least twelve inches deep right on top of your driveway and garden without any problem. In places where irritants like gophers are a problem, line your raised bed with gopher wire, or else all your vegetables will go straight to the rodent population. A raised bed also gives you the opportunity to start out with good soil, which is something many city gardens don't have a lot of in the beginning. This will change as you begin to make compost and add it to the soil. In the beginning, use the best potting soil you can afford, and include 10 to 20 percent worm compost in your initial mix. For the most successful container gardening, tend to your container soil as you would your ground soil. Container soil can get compacted and tapped of its nutrients; loosen the soil between plantings, top dress with worm compost, and change out the soil completely every year or two.

Many vegetables and herbs will grow in small containers and don't need to be placed directly into the ground. You can grow a great amount of carrots, or leeks, or potatoes in a five-gallon bucket. Lettuce can spend its whole life in small pots. "Cut and come again" when you harvest your lettuce: trim the tops of the lettuces down once they are large enough to eat (don't pull out the whole plant) and let them grow out again for a few weeks. This will keep you supplied with lettuce for a long time without having to reseed and start another plant. Tomatoes and squash grow in pots, as do many of the leafy greens. Beets, beans, peas, lemon cucumber, tomato, lettuce, chard, zucchini, radish, spinach, kale, mint, oregano, thyme, basil, dill, and rosemary are all easily grown in pots.

Most vegetables need at least twelve to eighteen inches of root space, so a five-gallon pot is the minimum size for growing food successfully, though a dainty compact lettuce needs less depth than a wandering five-foot-high tomato. Herbs and some other plants may succeed in smaller containers. Plastic pots will retain the moisture better than clay pots, though they can deteriorate in the sunlight over time. Glazed clay pots are more expensive, especially the large ones, but they last and retain moisture well. Wood planters are also an option, but they may rot over time. Whatever container you choose, be sure there is plenty of drainage so the water can escape from the bottom.

Container gardens need more regular watering than in-ground gardens, especially if you live in a dry climate. If the soil in the container gets too dry, the water will run out along the sides and never get to the roots of your plants. One way to deal with this is to mulch around the plants with straw or bark. If the soil is kept from getting a hard, dry crust, watering will be more successful. Finally, container gardens do very well on drip irrigation systems.

A hot spot alongside the house is a great place to grow sun-loving plants like basil, peppers, and tomatoes in large containers.
Photo by Rachel Kaplan

A series of raised beds placed right on the asphalt turns this area into a garden. Benches alongside the raised beds provide places to sit. The arbors will eventually hold vining plants, using vertical space and creating shade.
Photo by Lauren Elder

Self-Watering Containers

Self-watering containers save on watering time and are useful throughout even the driest summers. Constructed out of stacking plastic boxes or buckets and a few other, easily sourced a materials, they swiftly turn a perch, deck, or back patio into an easy-to-maintain garden. Their water-saving capacities make it simple to grow a diversity of plants in small spaces.

Materials Needed

2 5-gallon buckets (or any other container that can stack)

1 lid

1 plastic tub OR drain gate approximately the same height as the gap between the two buckets when stacked

1 2-foot-long, 1-inch-diameter plastic pipe (longer than the height of the buckets when stacked)

1 mesh baggie (find them as packaging for fruit or veggies)

Drill with 1-inch bit and 1-inch masonry bit

Utility knife with extra blades

Rounded file

Saw

Permanent marker

1. Mark the buckets.

   Hole for wicking basket: on the bottom of the first bucket, trace your drain grate or plastic tub and mark a circle on the bottom of the first bucket. Be sure your circle is smaller than the lip of the container.

   Hole for pipe: on the same bucket, mark a hole for the pipe, also one-half inch from the wall of the bucket.

   Side drainage holes: measure and mark drainage holes on the side of the second bucket. Just place the buckets one next to the other and figure out how much of a gap there is between them when they stack together. Mark two drainage holes, one on each side, just below that line.

   Second hole for pipe: on the lid, mark a hole for the pipe (½ inch from the edge).

   Holes for plants: next mark holes for the seedlings on the lid, or one big hole for an established plant.

2. Cut the holes in the buckets. Cutting plastic kicks up a lot of little plastic bits. Protect your eyes and nose and mouth accordingly. For the big holes on the first bucket and the lid, start them with a drill, using a 1-inch masonry bit. Use the utility knife to widen the holes. Cut drainage holes in the bottom of your first bucket, using a ¼ inch-diameter drill bit. Next, cut the side drainage holes on the second bucket. Do not cut the side drainage holes in the bucket with the holes in the bottom.

3. Prepare the pipe. Cut an angled segment from the bottom of the pipe, using your hacksaw. The reason you're doing this is so that water can flow out of the pipe when it's at the bottom of the buckets.

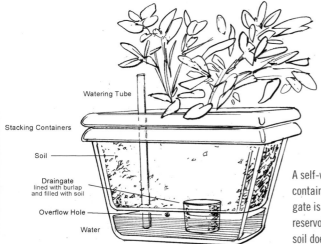

Watering Tube

Stacking Containers

Soil

Draingate
lined with burlap
and filled with soil

Overflow Hole

Water

A self-watering container consists of two stacking containers of any size. The top container has holes cut into it for a watering tube and a drain gate. The drain gate is lined with mesh or burlap and filled with dirt to wick up the water in the reservoir below. An overflow hole near the top of the reservoir ensures that the soil doesn't get overly soggy.

4. Assemble the wicking basket. Either line the drain with mesh or cut holes in your solid plastic container. You can also use food containers, as long as there is enough of a lip and they are the right height. The drain cover, though more expensive, seems sturdier and better for this project.

5. Assemble the bucket. Place the assembled wicking basket in the bottom of the bucket. Push the pipe through the holes in the lid and the bottom of the inner bucket. Stack two buckets, with the basket hanging between the two. Place the top of the bucket underneath the whole set-up to catch extra water. Fill your buckets up with soil and you're ready to grow.[32]

More Projects for Maximum Production in Small Spaces

Potato in a Barrel—Compact and Productive

Cutaway drawing of potatoes growing in a barrel. Potato plants grow up from the potato eye. Potatoes grow well in containers on a patio or deck.

1. Start with a large wine barrel or planting pot or even a regular-size garbage can. Make sure there is a drainage hole in the bottom.

2. Fill the barrel with about 4 inches of dirt.

3. Place a few potato eyes in the dirt and cover them up. (Potato eyes—those spots on the potato that often sprout when you don't eat your potatoes quickly enough—are the seeds of the potato. Take a potato and cut it in many pieces, each with its own eye, to start a new potato plant.)

4. Once the potato starts to grow, cover the plant with more soil, up to the leaves of the plant. The buried plant sends out spuds from its central shaft, and rather than fruiting from the top of the plant, fruits from its roots and stalk. You can continue to cover the growing plant three or four times over the growing season to encourage the growth of additional potatoes along its buried stalk.

5. After the potato flowers die back, it's time to harvest. Dig the potatoes out of the bin, or just dump the entire bin on its side. You'll be amazed at how many potatoes you'll harvest. If you have the space to do a few barrels like this, you can produce most of the potatoes you'll need for a full season.

Keyhole Bed

Traditional gardens tend to fall into the pattern of rows, for the ease of cultivation and harvesting, and because the line defines our cultural relationship to the earth. But when you bend a line into a circle, you get more surface area in a smaller space. In a garden, bending the line of the bed gives you more planting space, and puts different plants into beneficial relationships with one another. The keyhole garden works on this principle. Also in a small garden, a keyhole form gives you both planting space and a garden path in between the rows.

Orient your keyhole bed to face south for maximum sunlight, and try planting larger plants on the outside of the bed, and smaller ones in the middle. The outer plants form a windbreak for the smaller plants, and the shape will hold a lot more plants than an ordinary straight-row bed. A keyhole bed optimizes edge to encourage biodiversity and makes space for more plants, and more opportunities for pollinators and beneficial insects. This form is used a lot in permaculture design because it exemplifies numerous principles—biodiversity, efficient use of space, maximizing the edge, and using least effort to greatest effect.

The Herb Spiral

The herb spiral is a raised mound of earth that also maximizes space and can bring culinary and medicinal herbs closer to the kitchen where they are used. The mound creates more surface area than a flat garden bed, and

a denser, more diverse garden. It's easy to site in an urban setting, and harvesting and watering are simple. When creating the herb spiral, bank up the bottom with stones for good drainage, and mound dirt up into a peak. The pitch of the herb spiral will increase planting area and create a variety of microclimates (such as sun on the southern side, and cooler darker planting areas on the north "slope").

When planting, remember that some herbs spread disastrously—like some thymes and most mint—so be judicious when choosing them. As with the vegetable garden, there's no sense in filling your herb spiral with rare and difficult to grow herbs you never use; plant it with plants you like. You can also fill up your "herb" spiral with other leafy plants, like lettuce and chicory, especially if you like them more than dill and basil and parsley. Like the keyhole bed, the herb spiral is often chosen in permaculture designs because it uses space so well, maximizes diversity through the creation of microclimates, and makes something beautiful while expressing the principle of working from patterns to details.

Planting the herb spiral with mint, sage, artemesia, and nigella.

The herb spiral maximizes space and creates a diversity of microclimates for culinary and medicinal herbs.

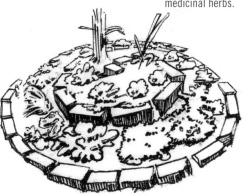

Start a Garden Wheel Project

Permaculture is not a back-to-the-land movement or a self-sufficiency movement. It's about permanent culture, which means creating community sufficiency and resilience through collaboration. If we're just applying these principles on an individual home-stead, you could live a nice life, but you wouldn't get to the higher aspirations of regenerative community that may be a key to human survival. In that spirit, we encourage you to apply all of these space-making strategies to your entire neighborhood by starting a garden wheel project.

A garden wheel is a great way to get neighbors and friends to participate in the creation of local gardens. It's similar to an ongoing work party where people give and get support for their food-growing projects, a gathering of people who are willing to help one another in exchange for getting help themselves. Your garden wheel group can be made up of neighbors, or it can be an affinity group—parents whose children are in the same school, or associates from work or from a church. We advocate for the close neighborhood association as a way to strengthen local relationships and webs of resources and sharing.

Laying out the keyhole bed on a former front lawn. The circular shape allows for more planting space and leaves a pathway in between beds. *Photos by Trathen Heckman*

The keyhole beds in the first season, filled with an abundance of leafy greens, kale, and early spring vegetables.

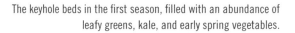

Helping out at a garden wheel event. *Photo by Trathen Heckman/Daily Acts*

When starting the garden wheel, neighbors or friends who want to get involved can share with one another their interests and desires for gardening and establish equitable ways to support one another in reaching their goals. A fun and useful way to structure meetings for the garden wheel project is to plan potluck dinners with all the participants. These become opportunities for community and relationship building, as well as garden planning.

A garden wheel group will be able to share not only labor but also space. Once you've talked to your neighbors or friends about working together, you may find that there are innovative ways to share space with one another. For example, some yards have great sun and are ideal places to grow tomatoes and other heat-loving plants. Other yards have a lot of shade, and not enough sun, so they'd be better places for growing tender greens, spinach, or lettuce, especially when it gets too hot out. Garden wheel members can become "specialists" based on their inclination, their growing conditions, the amount of time they have, and their ability to collaborate with their neighbors.

Planning the Potluck

If planning a potluck is as far as you get, you won't be disappointed. Getting to know your neighbors is a crucial first step in building community resilience.

1. Invite your friends, associates, and neighbors to a gathering at your house. Put a note in a mailbox or talk to neighbors on the street. Call your friends or send them e-mails. Direct contact is always best. People like to be personally invited.
2. Encourage everyone to come and bring something delicious to eat. Don't worry that there won't be enough or that everyone will bring chips. It always works out.
3. Cook something delicious you love to eat.
4. Before starting the meal, have everyone introduce him or herself, or, if they already know each other, have them share one small thing they are excited about.
5. Eat and be merry.

Meeting After Eating

1. Before you serve dessert, convene a short meeting with the following agenda:
   - Check-in—everyone goes around the circle and shares their name and the inspiration that brings them to the potluck.
   - Dialogue about projects that could be shared in the neighborhood. Sense where the collective interest is, what the opportunities and needs are, and make sure everyone gets a chance to speak.
   - Talk about schedules, available tools, and materials.
   - Set a date for the first garden wheel event.
   - Close meeting with a brief checkout.
2. Make sure you set up a time for the next potluck before people disperse. Have it at someone else's house.
3. Everyone help clean up.

A garden wheel project will help you get to know your neighbors, efficiently establish your garden, share space and produce and labor and friendship with the people around you, and in general, grow your neighborhood connectivity and resilience. These kinds of projects can evolve into dialogues about common concerns and additional projects that can foster community coherence and connections.

Urban Dirt Farmer

My whole life has been spent waiting for an epiphany, a manifestation of God's presence, the kind of transcendent, magical experience that lets you see your place in the big picture. And that is what I had with my first compost heap.

—Bette Midler[33]

O nce you've got your land (or deck or driveway) ready for growing, tending the soil is your next step. Growing healthy soil is essential for growing healthy food and herbs and medicine, recharging the urban earth with nutrients, organic matter, and the relationships of many organisms that bring life to the soil. Soil fertility is essential for the continuation of life; one of the hallmarks of all failed civilizations is their collective inability to maintain fertility in the soil.[34] Agribusiness has depleted and deadened the topsoil through toxic overfertilization and massive monocropping, fundamental instabilities in our current agricultural system that endanger our capacity to provide for our daily need for healthy food. While a reformation of industrial agriculture is surely in order, waiting for the system to change is not necessary. While much of the urban earth has been trashed by pollution runoff, toxic waste, lead, residue from unsavory building practices, or general neglect, we can rebuild the soil one backyard at a time.

Every patch of reclaimed dirt, every scrap of kitchen waste composted and returned to the soil, is a step toward renewing the earth. In the words of Wendell Berry, "The care of the earth is our most ancient and most worthy, and after all, our most pleasing responsibility. To cherish what remains of it, and to foster its renewal, is our only legitimate hope."[35] Even if you can't find the way to the garden, we hope you'll join the army of urban dirt farmers who are tending to the urban earth first. This is one of the single best ways city people can participate in revitalizing the city. Caring for the soil recreates conditions for life, and is an action each one of us can take. To commit to a piece of dirt and bring it to life even at the small microbial level is ultimately a passionate act of love.

Soil Ecology

All life on earth comes from the soil and returns to it, an endless cycle of death and rebirth. The miraculous process of decomposition incorporates all the nutrients that plants require. Good gardening respects this cycle and ensures that the soil in the garden is healthy and life giving. Soil is made up of three layers—topsoil (which, when healthy, contains decayed organic matter), subsoil, which is rockier than topsoil, and a deeper layer of rock. Plants send their roots into the different layers of soil, drawing nourishment from the soil and the water within it. Animals burrow through the soil, aerating it, and return nutrients to the soil in the form of excrement.

Many beneficial organisms flourish in a healthy garden—worms, bacteria, yeasts, insects, mushrooms, and algae among them. Worms drag organic matter into the soil and convert it into nutrient-rich worm castings. Bacteria perform vital functions in the soil, including taking nitrogen from the air and putting it back in the soil. Yeasts convert sugar into alcohol and carbon dioxide that lives in the soil. Insects aerate the soil, which permits oxygen to penetrate. This creates conditions for healthy soil and flourishing plant roots, and the decay processes that create humus. Fungi and algae, the smallest particles in the soil, continue to break down humus so that it can release its nutrients into the soil.[36] The more you can support this complex community of organisms in the soil, feeding them what they need to grow, and not interrupting their important work, the better.

Urban Earth

Most soil found in our backyards is a mixture of soil types and the residue of urban toxins and miscellaneous particles (debris, glass, plastic, etc.) Cleaning up the soil is a necessary first step in making our gardens good for growing. Plant growth is strongly influenced by soil structure that affects the movement of water, air, and roots through the soil. Soil structures include loamy sand, loam, clay loam, and clay. Each of these soil types is distinguished by the density of soil particles, their capacity to absorb or deflect moisture, and their ability to hold together when wet. Soil is made of small particles of clay, larger particles of sand, and medium-sized particles of silt, plus organic matter and air.

Simple Soil Tests to Do at Home

The Smell Test

To get a general sense of your soil health, take a closer look. While you're there, take a good whiff. Does your soil have a nice earthy smell? Unpleasant odors may indicate poor drainage or water-clogged soils.

The Squeeze Test

When your soil is still damp from recent watering or a rainstorm, take a loose ping-pong–size ball of it into your hand. Squeeze the ball of soil and then open your hand. If it falls apart, the soil is more sand. If it keeps its shape, the soil is more clay. Then take some of the soil and rub it between your thumb and forefinger. Grittiness indicates sand. If it feels slippery, the soil is more silt. Stickiness means clay. Again, take some soil and roll it into a log in the palm of your hand. A long log indicates more clay in the soil.

The Separation Test

This test can give you a general idea of the proportions of sand, silt, and clay in your soil. Collect one cup of soil from the garden. Place the soil in a one-pint canning jar with a lid (or a 16-ounce jar with a lid). Measure one cup of water and pour just enough in the jar to cover the top of the soil. Let the soil settle by rapping the jar gently on the counter and waiting a few minutes. Mark the soil level with a piece of tape. Add the remaining water. Put the lid on the jar and shake the jar vigorously.

Let the jar sit for 24 hours. The largest particles—sand—will settle to the bottom, the intermediate-sized particles—silt—will be in the middle, and the smallest particles—clay—will be on top. (If there is a gap between the top of the clay level and the tape marker, it is because all the clay has not fully settled out. It can take a week for the clay to completely settle.)

Look at the soil layers in your jar. It may be hard to determine clear edges, but try to get an idea of the proportions of each. Is there more sand, or clay, or silt? Can you estimate the percentage of each?

Soil Drainage Test

Collect a 46-ounce can and remove both ends of it. Dig a hole in your garden about 4 inches deep and push the can firmly into the bottom of the hole so that water will not leak out around the bottom of the can. Fill the can with water and wait an hour.

If the water level has drained by:

About 2 inches: the drainage is fine for plants that require good drainage.

About 4 inches or more: the drainage is fast, indicating you have soil that contains a good proportion of sand.

Less than 2 inches: the drainage is slow and most common garden plants will have difficulty growing in your soil.

Soil is different in different places—the soil in your yard will have its own composition, and can be amended by the addition of organic matter. If you are concerned about lead or live in an area that has been heavily impacted by toxic residue, garden in raised beds, buckets, barrels or other containers while you begin the process of renewing the soil. The addition of living compost, soil amendments, compost tea, mulch and mycelium to any compromised soil will heal the earth over time. There is very little soil in the world that would grow good healthy crops year after year without some help. The good news is that whatever your soil type, you can improve it by adding compost.

Two Methods of Garden Bed Prep: Sheet Mulching and Double Digging

There are many ways to start a garden, even on a spot that has never been gardened before. If you have a patch of bare land, or a bit of lawn, you can reclaim and redesign it as a food, herb or flower growing landscape. Sheet mulching and double digging are two methods for reclaiming the land for growing. They tend to the soil in different ways but don't think of them as an either-or proposition. Sheet mulching can be used to quickly transform a lawn or bare yard into a garden. Double digging can be used subsequently to increase the yield and productivity of the garden. Using both methods will be the most thorough way to prepare whatever soil you've got.

Sheet Mulching

The front lawn must go the way of the dodo. No longer will we spend our time in submission to the manicured lawn, wasting water and energy. Our rallying cry is: Turn Your Lawn into Your Lunch! Sheet mulching reclaims the lawn with an organic "lasagna" of cardboard, compost, and mulch. This is a low-effort way to create a garden bed while adding fertility to the soil. You can do this project in an afternoon with some friends.

Mulch
Compost
Cardboard
Compost

Sheet mulching is a layering of compost, cardboard, and mulch over a lawn or other area in preparation for making a garden. Sheet mulching harnesses the power of worms and composting microorganisms in the soil. Cardboard keeps the light out to kill the lawn or weeds underneath; worms eat through the cardboard to get to the compost, turning and conditioning the soil as they go, thus doing most of the hard work for you.

Easy Sheet Mulch Recipe

After a rain, or in the morning after one hour of watering, lay down 2 to 3 overlapping layers of cardboard, 3 to 5 inches of compost, and 8 to 12 inches of tree mulch. For every 100 square feet that you are covering, you will need 1 to 2 cubic yards of compost, and 3 to 4 cubic yards of mulch.

You can sheet mulch any time, but the best time to do it is in the fall, or before your biggest rainy season. You want your sheet mulch to get nice and wet so the worms will come up and do the rest of the work for you.

Materials for Sheet-Mulching

Cardboard: The cardboard smothers out all the weeds and provides a foundation for your sheet mulch. Get the largest pieces possible. Try looking behind furniture stores or bike shops and driving around to the more industrial parts of town to scavenge cardboard. Remove all the staples and tape and use brown cardboard if you can find it. Layer your cardboard, making sure to overlap the edges. This is important if you are trying to get rid of grass or other noxious weeds. Two layers will work, but for particularly tenacious or noxious weeds use three.

Compost: In some places, the best resource for compost is the dump, where the green garbage materials collected at the curb are turned into compost and sold back to the consumer. If you don't live in an area that has this kind of program, see if you can find a nearby farm that will sell (or give) you well-composted horse manure. As a last resort, you can buy compost in bags at the garden supply store, but it will cost you a lot to do this project that way. A cubic yard is a lot of compost. A regular pick-up truck with normal suspension can probably handle half a yard at a time.

Getting ready to sheet mulch the lawn at a public park.
Photo by Trathen Heckman/Daily Acts

Piling on the cardboard over the first layer of mulch goes quickly when done in community. This front lawn was sheet mulched in less than three hours by this homegrown guild. *Photo by Trathen Heckman/Daily Acts*

Tree Mulch: You can often access this by calling a tree trimming company and having them dump the contents of their truck at your house. They often have to pay to dispose of their waste, so they are happy to give it away for free, but beware: some places insist you take a full truckload, which can be between 10 and 20 cubic yards of material. Share it with your neighbors who are going to be so inspired by your sheet-mulching project that they're going to do one too. Check the yellow pages for arborists and tree doctors. Try to get mulch that is a mixture of wood and leaves, and make sure the mulch comes from a disease-free tree. Avoid highly acidic tree mulch (like redwood) or trees that are full of seeds and will sprout throughout the garden.

When Is it Ready?

Sheet mulch needs at least one full rainy season to do its magic. This is when the worms come out and eat through the cardboard to get to that delicious compost. If you are starting in the beginning of spring, you can plant in the same season. Try planting simple veggies like zucchini or potatoes in your sheet mulch. For zucchini, scoop out a hole in the compost and mulch about the size of a soccer ball and fill it with rich potting soil. Plant your starts into this and re-cover with a bit of mulch. Water thoroughly and often. For potatoes do the same, using a bit less dirt around each piece of seed potato.

If you are patient or your soil is really bad, try a two-year program. Sheet mulch in the fall, and rather than turning your mulch in the spring, plant zucchinis or potatoes or fava beans, which bring nitrogen to the soil. At the end of that growing season, water deeply or wait for the first rain; then use a fork to lightly mix the layers, but don't turn under or completely cultivate. Add another inch or two of compost and another 3 to 4 inches of tree mulch. The following spring, cultivate your soil as you normally would; you will have created 6 to 8 inches of gorgeous topsoil. Be sure to get under the lowest layer of what you added and mix the native soil in before you start planting your garden.

This kind of project is best done with friends and neighbors. You can do it in rounds: first my house, then yours, then yours, until the entire neighborhood has forsaken its lawns for the possibility of growing its own lunch.

Double Digging

Double digging is the energetic opposite of sheet mulching, a much more labor intensive but equally useful gardening technique that enhances soil fertility by intensively digging the area where you will be planting. Developed by British horticulturist Alan Chadwick, double digging has been used in gardens throughout the

Once you've sheet mulched, shape and form your beds as needed. This entire yard was sheet mulched after cement was removed and soil was amended. Bringing fresh compost, mulch, and straw to the site created these garden beds.
Photo by Erik Ohlsen

world to increase soil fertility and garden yield to an incredible degree.[37] John Jeavons extended this practice at his research and development site in Willits, California, where microfarming in largely inhospitable terrain was refined, and small double-dug beds became highly productive. Double digging maximizes production yield in small spaces and has been used to great effect all around the world, especially in Third World countries where people are more dependent on their small home gardens.

Part of the bio-intensive method of gardening, a double-dug deep bed of 100 square feet can produce from 200 to 400 pounds of vegetables a year. According to USDA statistics, the average American eats 322 pounds of veggies a year, so just one bed—20 feet by 5 feet—can keep an adult supplied with vegetables for an entire year. A backyard microfarm located on as little as an eighth of an acre can produce a large portion of a family's vegetable needs for the year.[38] Bio-intensive methods use less water per unit of land than commercial agriculture and still less per pound of food produced. With the double-digging method, plant roots can go down as far as ten feet if the soil is loose enough, rather than spreading horizontally, so there is less competition between plants for nutrients at the surface.

1.   Begin by spreading a layer of manure over the top of the bed.
2.   Starting at one end of the bed, dig a trench about a foot deep.
3.   Put the removed soil into a wheelbarrow.
4.   Loosen the subsoil by using a garden fork and plunging it into the soil and moving it around.
5.   Dig a second trench next to the first, throwing the topsoil and manure into the first trench. Be like a gopher and work the subsoil in the bottom of the second trench.
6.   Dig a third trench and repeat the process.
7.   To make a path, throw all the pebbles and rocks you dig up to the side of the bed.
8.   Dump the soil in the wheelbarrow (from the first trench) into the last trench.
9.   Once you've finished, throw the topsoil from the path-to-be on the top of the bed, and line the path with the stones and rocks you've uncovered.[39]

You'll want to avoid stepping on this bed because that will undo all the hard digging work you've done, and recompress the soil. Make your path spacious enough to accommodate your walking, and design your bed to be the right dimensions so that you can reach across it from either side without having to step upon it. Four feet across is about as big as you'll want it to be if you want to be able to reach across the bed from either side.

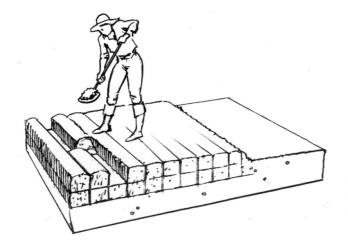

The double digging process enhances soil tilth and productivity by creating open space in the soil for downward root growth. Once your bed is dug, avoid stepping directly on it, as that compacts the newly aerated soil.

on the ground: wildheart gardens

Christopher Shein runs Wildheart Gardens, a sustainable landscaping company in Berkeley, California, and has taught permaculture at Merritt College in Oakland for the last nine years. His classes have touched more than 600 students who learn at the one-acre food forest planted with 200 fruit trees on the campus. Christopher got his start early with inspiration from his urban farming grandmother and his family of political activists.

"I got my feet wet with community gardening and food growing at a program called Detroit Summer, which was a youth leadership-training program where we learned about community gardening, community based murals, and toxic cleanup assessment," Christopher recalls. "We'd fix up homes for seniors and do other community projects like that. We'd meet with old-time organizers and activists and that's how I got the idea that urban agriculture, farming, and community gardening all had something to do with social justice. It seemed like a good way to join in the struggle. Teaching people to grow food is essential; it's the entry point for so many people. Once I was older, I saw that if we could get enough people growing their own food, and feeling empowered, the culture might change. I've seen how gardening is a good tool for helping get people organized around issues they care about and things they need."

His homestead—approximately 6,500 square feet of lush growing productivity—is skillfully designed, diversified, and beautiful. He grows bamboo for sustainable building projects, has extensive greywater and rainwater catchment systems, uses a simple outdoor composting toilet, and grows some of the food his family needs with a minimal time commitment of one to two hours per week, including raising a large flock of chickens and four ducks. "Raising food is easy and everyone can do it," Christopher points out. "When there are no supermarkets, then we all will be growing our own food. Until then, we are preserving the knowledge and building a decentralized network of do-it-yourselfers who are trying to teach people the old skills for an uncertain future." He shares his home space with his daughter and his wife, a doctor who opens the homestead to her clients on a daily basis so they can heal from the garden as well as from her integrative medical practices.

Christopher propagates plants for his landscape company on site. "I focus my nursery on perennial crops, like oca, yacon, and mashua. The benefit of perennials is that you grow them and they keep coming back. It saves you time, it saves you money, and it helps your landscape mature pretty quickly. A tree collard can keep you in greens season after season, year after year. They are a great source of food for humans and forage for chickens, and they look nice in the garden too. I'm not a perennial purist—I like cabbage too. But a garden works well when you mix them together. In the end, the determining factor is whether or not your plants are delicious, not deciduous!

"I was trained in 1994 as a 'master composter' through a county program. We have a pretty active composting setup here. Once a week, I go to a local restaurant and get fifty gallons of kitchen scraps because in our home we don't generate enough volume of green to make good compost. I take theirs, which lowers their garbage load, and it all goes back into the earth. The chickens eat the compost for a day or two, and we put straw and sawdust down for the chickens and then that goes into the compost too. Our chickens are a regular compost-making factory. I harvest more than ten yards a year of finished compost from the poultry for landscape projects. Surplus in abundance is our theme around here.

"We've been recycling our humanure since 1996. We have one composting toilet in the yard. This fertility really

Propagating plants, rare and common, at Wildheart Gardens.

adds to the garden. We're such a fecal-phobic culture—for some good reasons—but we can so easily deal with our shit better than we do. It makes sense not to poop in our water system, and waste all the nutrients we are taking from the soil. We're depleting our farmland and are flushing the nutrients away. It's not necessary or sustainable. We like trying to do it differently around here.

"This is another great green job! Recycling poop would create a lot of jobs. You can't get much more green than being a gardener, growing your own soil, mushrooms, and food, and making your own fertility in the way we're doing it back here. When I think about a future with more people doing this out of necessity, I see how our culture's going to be so much more local and sustainable, and beautiful."

Growing food and teaching permaculture is a good life, but it's the "community building and all the sharing" that really feeds Chris. "It's good enough for me to have enough and to share it. But for me, giving back all this information I've gathered is where it's at. I like working with these young folks who want to learn. They give me their time and I give them plants and food from the garden and experience. If I can teach people, I know that I'm living in a circle of giving. And that makes it worth doing. I'm really into teaching people that anyone can grow some of their own food. It connects us to the web of life."

Amending the Soil, One Backyard at a Time

Once you've prepared the garden beds or containers, you can enhance fertility by adding humus to the soil, creating environments where small and large organisms can flourish. The simplest way to grow humus in your own backyard is to compost your kitchen scraps and return them to the soil. Compost is one of the magical transformations of gardening. It helps us understand waste as part of a closed system, and the cycle of life as a companion to death and decay.

Backyard Compost Bin

You can build a compost bin in your backyard in an afternoon. You need a three-sided box ideally 3 x 3 feet in diameter. One of the simplest designs involves three pallets nailed together into the shape of a *C*. If you have the room, make a compost bin with two sections, in the shape of an *E*. Make a gate with some chicken wire in the front of it, but latch it in such a way that you can move it for the times when you turn over the compost in the pile.

Another simple option is to use four fence posts, held together with chicken wire. This will create a box with enough stability to hold the compost in place. Some people just pile the compost in a corner of the yard without building a box, and that works fine, but if rodents are an issue, you'll want more protection than that. A circle constructed of chicken wire also works to keep the compost in one place.

Once you've got the bin built, it's easy to get started composting your kitchen and yard wastes.

Place a layer of dry organic matter at the bottom (leaves, hay, straw, even wood chips will do). After the layer of carbon matter, add your kitchen scraps.

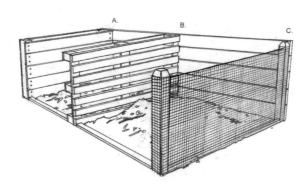

This drawing shows three options for creating a compost bin from simple materials.
A. Plank and wooden posts
B. Pallets
C. Metal fence posts and hardware cloth

When one bin is filled, start on the next. By the time the first bin is composted, the second and third are already on their way. Three compost areas maximize space and yield compost in succession throughout the season.

You can also add the trimmings from the garden or any other organic matter into the compost bin, but do not add meat or dairy products, as they draw animals and do not compost well in a small-scale backyard composting system.

Each time you add your wet compost of kitchen scraps, cover it with a bit of dry organic matter (leaves, straw, etc.).

Water is the final ingredient for getting the compost pile going. If you live in a particularly dry climate, water your compost bin well to get started. In wetter climates, there is usually enough water for the compost pile to do its magic. You'll know the decomposition process is working when your compost pile literally gets hot. In some climates, you will see it steaming. In dryer climates, you probably won't get the steamy drama, but if you put your hand into the pile, it will feel hot. Bless the bugs in the pile, and all the worms you see there. They are doing the hard work of transforming food scraps into dirt. Compost usually takes about four to six months to make right. Once the materials in your bin have composted down, take them out of the bottom of the bin and dig them into the garden before planting new plants. Turn the contents of the compost bin over when you are digging dirt out of the bottom to mix up the partially composted parts with the newer additions to the pile. This will speed the process along.

Forget the Gym: The Berkeley Compost Method

If you want a quick composting method combined with a 10-minute daily workout, try the Berkeley method. A hot, aerobic process with a high carbon-nitrogen ratio, this process takes about 21 days from start to finish.

To start, collect as many coffee grounds as you can. You can get them from your local café. If you keep chickens or ducks, make sure you incorporate their manure into the pile as well. The next ingredient is organic matter, usually clippings from the garden and food scraps. The key to the Berkeley method is to chop up the organic matter very fine—nothing longer than your pinkie finger. Use a machete or clippers to chop it up. Layer the manure and coffee grounds, kitchen scraps, and grass clippings or other organic materials. The pile should be a 3 x 3-foot cube, or better yet, a 4 x 4-foot cube, and will need to be very wet to get it going well. Let it sit for one day.

For the next eight days, turn the pile every day. Turn the whole pile inside out, taking the middle to the sides, and collapsing the sides to the middle. Float the bottom on top of the pile, and let the rest spill back over. On top of getting some aerobic exercise, use this as a metaphysical exercise to compost your personal patterns. It's your 10-minute meditation on change. What habit of mine would I like to compost? What habit would I prefer? What can I turn it into? That should get you sweating.

The hard working earthworm turns kitchen scraps into vermicompost.
Photo by Daniel Miller/Spiral Garden

After eight days of turning the pile, do it every other day. By the end of another 10 to 12 days, you'll have a finished compost pile. Screen the dirt before you use it, but expect to get about 37,000 cubic inches of good rich soil from this process. Whatever doesn't get thoroughly composted can go back into another pile, or spread out under a tree.[40]

Worm Bins

If you don't have enough space to host a compost bin or energy for the Berkeley method, a worm bin will be the best way to compost your kitchen scraps. Building and maintaining a worm bin is a great urban gardening project—it doesn't take up a lot of space, uses available kitchen scraps and turns them into soil, and can be easily placed on a deck, back porch, or small yard setting. Children love watching worms work their magic, and it's the kind of project they can lend a hand in building and maintaining. Once the worms eat your garbage, you'll end up with vermicompost, one of the best soil amendments available.

Materials Needed

Two 8- to 10-gallon dark colored plastic storage boxes
Drill (with 1/4-inch and 1/16-inch bits)
Newspaper
One pound of red worms

1. Drill about 20 evenly spaced ¼-inch holes in the bottom of each bin. These holes will provide drainage and allow the worms to crawl into the second bin when you are ready to harvest the castings.
2. Drill ventilation holes about 1 ½ inches apart on each side of the bin near the top edge using the 1/16-inch bit. Also drill about 30 small holes in the top of one of the lids.
3. Prepare bedding for the worms by shredding newspaper into 1-inch strips. Moisten the newspaper by soaking it in water and then squeezing out the excess. Worms need moist bedding. Cover the bottom of the bin with 3 to 4 inches of fluffed up moist newspaper. Old leaves or leaf litter can also be added. Throw in a handful of dirt for "grit" to help the worms digest their food.
4. Add your worms to the bedding. One way to gather red worms is to put out a large piece of wet cardboard on your lawn or garden at night. The red worms live in the top three inches of organic material, and like to come up and feast on the wet cardboard. Lift up cardboard to gather the worms. An earthworm can

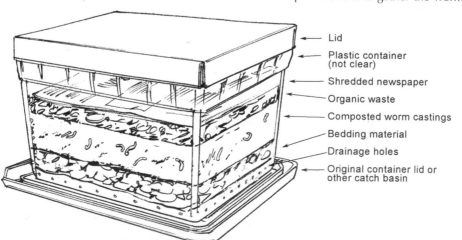

— Lid
— Plastic container (not clear)
— Shredded newspaper
— Organic waste
— Composted worm castings
— Bedding material
— Drainage holes
— Original container lid or other catch basin

In this worm bed, organic kitchen wastes are easily transformed into "black gold;" a superior fertilizer for the garden.

consume about half of its weight each day. If your food waste averages ½ pound per day, you will need 1 pound of worms or a 2:1 ratio. There are roughly 500 worms in one pound. Don't worry if you start out with less than one pound—they multiply very quickly. Just adjust the amount that you feed them to fit your worm population.

5. Cut a piece of cardboard to fit over the bedding, and get it wet. Then cover the bedding and the worms with the cardboard. (Worms love cardboard, and it breaks down within months.)

Place your bin in a well-ventilated area, preferably in the shade. Place the bin on top of blocks or bricks or upside down plastic containers to allow for drainage. Use the lid of the second bin as a tray to catch any moisture that drains from the bin. This "worm tea" is a great liquid fertilizer. Feed your worms slowly at first; as the worms multiply, begin to add more food. Gently bury the food in a different section of the bin each week, under the cardboard. The worms will follow the food scraps around the bin, and burying the food scraps will help to keep fruit flies away. Feed your worms a vegetarian diet. When the first bin is full and there are no recognizable food scraps, place new bedding material in the second bin and place the bin directly on the compost surface of the first bin. Bury your food scraps in the bedding of the second bin. In one to two months, most of the worms will have moved to the second bin in search of food. Now the first bin will contain (almost) worm free vermicompost.

If the worms are dying or trying to escape, it's probably too wet or too dry in the worm bed. Try adding more bedding, moistening the bedding, or harvesting the compost from the bin. If the bin stinks, there is probably not enough air, too much food, or it is too wet. Try drilling more ventilation holes, not feeding the worms for a few weeks, or adding more bedding. If you've got fruit flies, you probably have exposed food scraps. Try burying them under the bedding material and see if the fruit flies disappear.[41]

Another excellent design that literally stacks functions is to place your rabbit hutch on top of the worm bin and let them drop their poop into the bin. It feeds the worms in a balanced way, the urine keeps the worm bin moist, and the yield is a fantastic compost of worm castings and rabbit poop combined, some of the richest fertilizer you can use for your garden.

Compost Tea

Compost tea can help boost microorganism populations in your soil, increasing beneficial interactions within the soil and between soil and plants. Applied to soil around plants, it can both add nutrients to the soil and increase plant uptake of nutrients. It can be sprayed onto sheet mulch to aid the breaking down process or used as a foliar spray to protect and fertilize the plants. It also boosts microbial interactions in container gardens.

Materials Needed

5-gallon bucket
1 gallon mature compost or worm compost
1 aquarium pump
1 long waterstone
3 feet aquarium hose
Non-chloraminated water
Unsulphured molasses

1. Set up your pump and waterstone in the five-gallon bucket. Try it with just water first to see that you get some good bubbling. The microorganisms need air to proliferate.

2. Add compost to the bucket.

3. Fill with non-chloraminated water to within six inches of the top. Chloramine will wipe out the beneficial bacteria you are trying to grow. Use spring water or rebalance tap water by using a chloramine removal product (in the same section where you get the aquarium pump) or a teaspoon of powdered vitamin C.

4. Add 1/4 cup molasses.

5. Start the pump.

6. Let it brew for 2 to 3 days. Stir occasionally.

7. Strain the tea. It should smell sweet and earthy. Do not use it if it smells rotten; put it back in your compost pile.

8. Use immediately. Spray directly onto plants and water around the base of plants. This process can be repeated every two weeks or less regularly as desired.

9. If you continue to add compost and compost tea to your garden throughout the growing season, your soil will become balanced over time. The original composition of clay, loam, or sand will be amended by the addition of organic matter to the garden beds.

Mulch and Manure

Mulching your garden—adding organic material that has not fully broken down, such as wood chips, straw, or leaves—adds additional fertility to your soil. This enhances soil tilth, the ability to support plant growth. Mulch holds water in the soil, preventing evaporation and keeping your plants well watered. It suppresses weeds, moderates soil temperatures, improves soil structure, encourages earthworm activity, stimulates beneficial microbial activity, acts as a substrate for fungi growth, and beautifies and improves our environment. It's also a great closed-loop practice: rather than sending mulch material to the landfill, spreading it on our gardens and paths keeps it on site where it provides benefits.

Mulch can be used on garden pathways, and in a garden bed to hold soil in place and keep it healthy. Use mulch in your vegetable garden, around annuals and perennials, and under fruit and ornamental trees. Remove weeds, and water thoroughly before applying mulch. You'll get the best weed control when you weed first and then spread the mulch. Mulch is most effective when applied to soil that's already moist. If the ground is too dry, it will be difficult to get moisture to the soil below, so use smaller amounts on poorly draining soil, especially if you are laying down mulch during a dry season. Add additional layers when the ground is moist. Leave a ring of mulch-free soil at least 6 to 12 inches wide around the base of trees and shrubs. A 2- to 4-inch layer of mulch works for most landscape or agricultural applications. A thick layer will be most effective for moisture conservation and weed control. A plant's roots need to breathe, so don't mulch so heavily that you interfere with this. Any woody material that is incorporated into the soil will temporarily inhibit the soil's ability to supply nitrogen to the plant. Keep mulch on top of the soil to prevent this.

You can make mulch at home using the "chop and drop" method—simply cut plant trimmings into smaller pieces with pruning shears and place them as mulch around the base of the same plants. Weeds—except for invasive weeds like ivy, bindweed, or crab grass—can be placed on top of the soil surface as mulch. Grass cycling (mowing the lawn and leaving the grass clippings behind to dry out rather than gathering them up) is a good way to mulch a lawn area. This keeps moisture in the ground and lessens the need to water. The only downside to mulching your entire garden is that some pollinating insects need bare soil so they can make their nests. If you have space, and in order to increase the number of beneficial insects in your garden, make sure to leave a small mulch-free area to encourage the ground nesting of pollinating insects. This will also increase the productivity of your garden.

Another excellent addition to the soil is manure—animal and human. Chickens and rabbits offer excellent manure for the soil. Rabbit manure can be put right on the garden bed; chicken manure should be composted in the compost bin for a few months to reduce its nitrogen load. Horse manure that has been modestly composted can be brought on-site from a local stable and used directly in the garden. Human manure, when properly composted, can also be used to enhance soil fertility. (More on this in Chapter 17, Zero Waste.)

The Magic Mushroom

Fungi live throughout the soil of our planet in an interconnected mat of electric energy—the "Internet of the forest"—a grand recycling, fruiting, decaying organism that pervades the biological chain and brings health and continuity to our forests, rivers and gardens.[42] Mushrooms may be an important key to mending parts of our broken environment in a swift, nontoxic way. Mycelium, the connective tissues of fungi, which look like a thin white mesh or net when you dig them up in the garden, have the capacity to transform toxic oil spills into neutralized soils (myco-remediation); begin the renewal process after a forest fire (myco-forestry); clean the waters of toxic spills (myco-filtration); and offer nontoxic pesticides to control insect infestations (myco-pesticides). This is an exciting area of earth renewal going on in forests, riverine systems, and laboratories around the world.

Mushrooms are important keys to human and planetary health, living in a symbiotic relationship with trees and other plants of the forest. Most ecologists now recognize that a forest's vitality is directly related to the presence, abundance, and variety of mycelial associates.[43] The biologic intelligence of mycelium binds the soil to itself and encourages successive plant growth. Many trees cannot grow to maturity without a symbiotic association with mycelium. This relationship can be utilized in our gardens as we rebuild the soil in our yards and enhance the productivity of our gardens, orchards, and trees. We can grow mushrooms for these remediation projects as well as growing edible mushrooms, adding another taste treat to the long list of foods easily grown in small urban spaces.

Building the Mother Bed

Mycelium "runs" or spreads throughout the soil in a fibrous film-like network. To generate connectivity and soil health, we want to spread mycelium through the garden. A simple way to do this is to build a propagation bed that will contain and grow mycelium, as well as sprouting edible mushrooms. Some simple-to-grow edible mushrooms are the oyster mushroom, the garden giant (wine cap stropharia), shiitakes, and the shaggy mane. Do a little taste test before starting this project to decide which mushrooms you want to cultivate in your yard and start with the one you like best.

Edible mushrooms cultivated in a propagation bed are saprophytic, which means they live on dead things like mulch, wood chips, or wood shavings. A propagation bed will hold a hardwood substrate and provide an environment for the mycelium to spread and come to fruit (the mushroom). Hardwood is the best medium for growing saprophytic mushrooms because it gives the mycelium more to eat as it grows. Pet stores often sell alder wood shavings that are perfectly serviceable for this project. If you get wood chips with twigs and leaves in them, clean them out a bit so they're not so twiggy. This limits the growth of other kind of fungi that find their way into the substrate, and protects the strain you are cultivating.

Materials Needed

¼ inch hardware cloth, at least 2 foot square

Hardwood chips, mulch or wood shavings (alder or oak).

(If hardwood is not available, use a soft wood like pine or eucalyptus).

Mycelium or spores for mushroom you wish to cultivate.

A common mushroom to cultivate in this manner is the wine cap stropharia.

1.   Build a square box between 10 and 24 inches wide, 4 to 5 inches deep. You can also use pre-made plastic boxes with ventilation holes on the bottom.

2.   Staple ¼-inch hardware cloth on the bottom.

3.   Soak the substrate of hardwood chips, mulch, or wood shavings for about 30 minutes before you put them in the propagation bed. If you're using wood shavings like pine or eucalyptus with volatile oils in them, they will need to be soaked overnight.

4.   Drain the substrate, and fill the box with the materials.

5.   "Seed" the substrate with mycelium spores. You can get them from a friendly mycologist (usually an eccentric, inspired sort, like the local beekeeper) or online at one of these sites: www.fungiperfecti.com or www.fieldforest.net.

6.   Divide the mycelium spawn into five sections. If you are using a package of inoculated wooden dowels, use five per propagation bed.

7.   Place them in the box with some distance between them to encourage the spreading of mycelium through the bed.

8.   Put the box in a shady area of the garden. Keep this box well watered, but not drenched.

Within a month, the mycelium will have spread throughout the box. When the first mushroom sprouts, leave it to refruit into the propagation bed. Wait for the next mushroom to appear before you start to harvest your food. If you want to generate more mushroom patches throughout the garden or spread the mycelium around to support plant life in the garden, take a portion of the mycelium-infused substrate from the propagation bed and place it under an apple tree, or even into a garden bed underneath a shady plant, where it will get enough moisture and humidity to continue to fruit and spread. This mycelium and its fruiting body will enrich your growing soil and can be left to grow and decompose in the garden, or be eaten, whichever you prefer.

Portable mother beds house mycelium in a hardwood substrate.

A formerly "useless" alley alongside the house becomes a growing zone for mushrooms. Greywater from the bathtub upstairs is used to water the bed in the dry months, and in the wet months, the bed proliferates with edibles for dinner.

Construct a cylinder of chicken wire and fill it with substrate to feed mushroom spores. Add water. Oyster mushrooms are fast and simple to grow and offer a plentiful yield.

Growing Mushrooms in Straw Bales

You can also grow mushrooms in straw bales placed in cool, shady places in the garden. Oyster mushrooms are easily cultivated in this manner. Wet the straw bales well, and stuff the spawn deeply throughout the straw bale. Keep the bale watered and wet. It is best to inoculate just before the major rainy season in your area so you don't have to do much of the watering yourself. It should take a few months for the bale to start sprouting mushrooms. Once you have harvested the mushrooms and the straw bale begins to decompose, you can mulch it into the garden, which will also spread the beneficial mycelium.

Inoculating Logs or Stumps for Mushroom Growing

Some mushrooms grow better in denser substrates such as logs rather than wood chips. In order to inoculate a log or a tree stump for mushroom cultivation, you will need to order inoculated wooden plugs online. You can grow mushroom varieties such as shiitake, reishi, the Oregon polypore, maitake, the conifer coral, lion's mane, pearl and phoenix oyster, chicken of the woods, and turkey tail in this way.

Generally the best time of year to inoculate logs and stumps is the spring, after the last hard frost. In temperate climates, fall is best, before the rains. You can inoculate logs any time up to 2 to 3 weeks before consistent freezing temperatures set in for winter. Mushroom mycelium will go into wintertime dormancy, and needs some time to establish itself in its new home before winter begins in earnest. Logs can be left outdoors over the winter, under a layer of straw or a burlap tarp, shade cloth, or other vapor permeable cover. In areas where winter is harsh, logs can be stored in a shed or garage.

1. Harvest a hardwood log no more than 14 inches in diameter, and about 4 feet long. Harvest this wood a few months before inoculating it with mushroom spawn to provide a more viable substrate for the spawn.
2. Use a $5/16$-inch drill bit in a high-speed drill to make 2-inch-deep holes no more than 4 inches apart, evenly spaced in a "diamond" pattern along the length and around the full circumference of the logs.
3. Stumps should be inoculated along the circumference of their face, in the border between the bark and the heartwood. Insert 1 inoculated plug per hole and whack it in with a hammer. A 3- to 4-foot log can take 50 or more plugs.
4. Hammer the plugs into the substrate and cap them with beeswax. This will keep other fungi from growing in the log.
5. Keep these logs stored in a shady place, and keep them moist.

This is a long-term mushroom cultivation method. It will take anywhere from 9 to 12 months for the log or stump to be fully colonized by the spawn and begin fruiting, but once properly inoculated, can fruit for 5 to 10 years.

Oak log inoculated with shitake mushroom spawn plugs. Once the mycelium has spread through the log, this cultivation method can produce food for 5-7 years.
Photo by Trathen Heckman/Daily Acts.

Seeds to Stem

Don't judge each day by the harvest you reap, but by the seeds you plant.

—*Robert Louis Stevenson*[44]

We became homesteaders because we love plants. We love to grow them, we love to look at them, and we love to harvest them, eat them, and turn them into medicine. We love to save their seeds, and sprout their sprouts, and participate in the cycle of life they so swiftly and repeatedly demonstrate. What follows are some much-loved practices with plants, as well as information about plant families, plant relationships, and plant functions. The first rule in the garden is to plant what grows. Notice the plants that thrive where you live. It makes no sense to plant a cactus in cold and rainy New England, or water-loving annuals in the middle of the desert. Notice which plants you love, and also the ones you aren't so attracted to. Plant the ones you love.

Seasons

Plants live and die by the seasons, so it is important to understand something about your location when choosing plants. Planting zones, also called hardiness zones, divide the United States, Mexico, and Canada into 11 different areas, each based on the annual minimum temperature. This gives us information about what you can plant when and where throughout the country. Check the USDA online for planting zones in the continental United States (www.usna.usda.gov/hardzone/hrdzon4.html).[45] Learn the seasonal cycles in your area as you make choices for your garden. Some places are great for growing corn or tomatoes, and in other places, you can't grow a tomato to save your life. Don't waste your time planting things that won't grow. Focus on what works, rather than lamenting about what doesn't. The Master Gardener Program established throughout the country trains avid gardeners in horticulture; in return, master gardeners give back to the community in the form of lectures, demonstrations, school and community garden projects, and other projects. Find one in your area to get you started.

Align your energies with the season, and enjoy each one for the energy it brings—the first flush of life in spring, the abundance of summer, the harvest of autumn, the rest of winter. Living with the seasons will help you make decisions about what kind of work needs doing and what kind of food needs eating. It's another primary practice for rooting yourself in place.

Fig tree leafing out in early spring. *Photo by Dafna Kory*

Seed Starting

The practices of seed starting and seed saving are a pleasing part of a homesteading life. While both practices can become detailed and complex, it is simple to participate in the ancient rituals of seed saving and sowing without a lot of expertise and get excellent results. It's the task of winter to look through the catalogs that come in the mail, plan your summer garden, and salivate over the beautiful seeds you might purchase. The benefits of starting your own plants include earlier harvests, greater variety (you can grow that incredible tomato you saw in the catalog but not in the nursery), stronger and healthier seedlings, cheaper plants, and, of course, the delight in doing it yourself.

Seeds are self-contained embryonic plants, surrounded by a supply of stored food, awaiting the right moment to emerge from dormancy and become a plant. Different seeds need different conditions to move from dormancy to a more active stage of growth. Cold can break dormancy (some seeds prefer to be chilled before growing), as well as light (almost all seeds need light to begin growing). Seeds also need water and a medium in which to grow. The planting medium can be a combination of vermiculite, perlite, and potting soil. A simple potting soil includes one part compost to one part vermiculite. Make more than you will use at one time, so that next time you want to plant some seedlings, you'll have the mixture you need. If you don't want to mix it yourself, buy a bag of the best quality organic potting soil you can find.

Once you have your seeds and your soil mixture and a place where your seeds can sit while they begin to germinate, you're ready to begin. Make sure your setup has enough heat and light for the seeds to grow.

Be creative with containers for your seeds. Use milk cartons, plastic gallon jugs, clay pots, or other plastic containers. Reuse old six-packs that housed the seedlings you couldn't resist at the nursery last spring, or build a flat to grow seeds in. The most important thing about the containers you plant in is that they are very clean when you begin. Contaminated containers are the single biggest reason why seed starts don't thrive. If you are using any recycled or reused containers, wash them in hot soapy water and sterilize them with a mild white vinegar solution.

Read your seed packs and note the days from germination to harvest. This will help you recognize which seeds are good for planting at which times of the year.

Prepare your flats by making sure the soil mixture is in the container, but not firmly tamped down, and that the flats have a tray underneath them to catch water. Prepare the seeds by giving them the cold or heat they require. If the packet says "cold stratify," put them in the refrigerator

COOL WEATHER CROPS

Broccoli, cauliflower, kale, lettuce, Brussels sprouts, cabbage, kohlrabi, spinach, beets, peas, carrots.

WARM WEATHER CROPS

Tomatoes, eggplant, peppers, beans, corn, squash, melons, cucumbers, tomatillos, ground cherries.

Early spring.

High spring.

High summer.

Autumn.

Early summer.

Winter.

for a few days to simulate a chill. Other seeds will need heat; place them near a heater or heat light for a day or so.

Cover the seeds. Seeds should be planted as deep as they are large. A tiny seed needs to be barely dusted over with soil, and a half-inch seed needs to be pressed a half-inch down into the soil. Make sure to label the containers with ink or pencil that won't run when you water your seeds.

Your seeds need warmth and sun in order to grow. This is very important, and can be challenging during the winter. You can place an electric heating pad underneath the flats to stimulate germination. You can buy these affordably at nurseries or hardware stores. Direct sunlight for some part of the day is essential. If you do not have this crucial ingredient, your seeds will have a hard time germinating, and they will spend most of their energy reaching for whatever sun they can find. This produces leggy plants that lean in the direction of the sun. Better to install a simple grow light to simulate the sunlight the seed needs. These grow lights are affordable and easy to install.

Create a germination spot from already available spaces—like underneath a piano bench, or a bookshelf area, or a box. Hang the grow light on the underside of the bench or the shelf and place your seeds there. Keep the light no more than 4 inches away from the sprouting seeds, as seeds require proximity to light in order to germinate. Once they have sprouted, move the light a bit higher so as not to burn their leaves. Use the artificial light until the sprouts are strong enough to be transplanted to a larger pot, or to a cold frame. Seeds need about 12 hours of light per day, and a period of darkness as well.

If you encounter problems with your sprouts—too spindly, or discolored, or falling over at the soil level—you may have used unclean containers, or excess or deficiency in soil, water, air, or light. Clean your containers well and make sure your seedlings have sufficient air, water drainage, proper temperature, and soil, and you'll soon be happily growing seeds on your own.

Transplanting and Hardening Off

Once your seedlings have two real leaves (the leaves that come out after the first set of round cotyledon leaves), they are ready to be transplanted into bigger containers. As you transplant, move slowly and with care, keeping as much of the developing root structure as possible connected to the plant. These transplants will still need a protected environment of light and heat, but after a few

Sprouting seeds in a variety of containers—some new, some recycled. *Photo by Dafna Kory*

Sprouting seeds in handmade flats, recycled from packing crates.

When you have more than you need, make sure you spread the wealth.

This simple "greenhouse"—easily found at many hardware stores—works for starting sprouts and hardening off seedlings. Plastic cover keeps a moist and warm environment when sprouts are just starting, and can be opened up to let in some air on a warm day,   *Photo by Dafna Kory*

weeks, if the weather is warm enough, they will be ready to go outside to get accustomed to the temperature fluctuations, wind, and sun of the great outdoors.

This process of slowly exposing your sprouts to the outside world is called "hardening off." It makes the plants strong enough to live outside in the garden. Find a spot in dappled sunlight to provide enough light, air, and heat, but not too much. Don't leave starts in hot direct sun for too long because they will wither and burn. Also, they will need a lot of water as the sun will dry out the soil.

Once your seedlings are strong enough to be planted out, you'll have to figure out where to put them in the garden and how to protect them from wind, weather, and pests until they are strong enough to stand on their own. You can cover small starts with Reemay, a thin translucent cloth placed over the plants (use twigs or bamboo sticks to make sure it floats above the plants) to keep away the strongest light and wind.

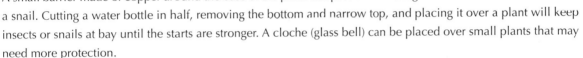

A small barrier made of copper around the base of the plant can prevent the ravages of a snail. Cutting a water bottle in half, removing the bottom and narrow top, and placing it over a plant will keep insects or snails at bay until the starts are stronger. A cloche (glass bell) can be placed over small plants that may need more protection.

If you are unlucky enough to be battling an army of gophers, a basket made of gopher wire would be the planting option of choice (put the whole plant in the basket so that the central roots are not accessible to the rodent). You can buy gopher baskets premade, or make them yourself out of gopher wire. If you are feeling intrepid and lucky, you can trap and kill gophers, but make sure you protect your house pets from accidentally nosing into a trap by covering it with an overturned plant pot.

Buying Starts

A lot of urban gardeners buy starts from a nursery to get their gardens going in the spring. This is a great way to go—it supports a local business and gets you good quality seedlings. If you're going to buy starts from the local nursery, here are some suggestions for buying starts with confidence.

1. Buy things when it is appropriate for them to go in the ground. Do the research to know if it is the right season to start them. Peppers and tomatoes need nights above 50 degrees.

2. Read the label. Know how much space, sun, and water the plant will need and purchase appropriately for your space and watering capabilities.

3. The plant should fit the container. It should neither look too small—swimming in a sea of soil—nor too large or tall, as if it has already outgrown the pot.

4. Avoid plants that are leggy (stems overly long so that they are falling over or trailing on the ground), wilting, or tortured-looking in any way. If the plant looks stressed out, it probably is and won't transplant well.

5. Avoid plants that are root-bound or flowering. If they are already flowering, it is likely they are also already root-bound. Plants need to establish roots when they are put into the ground to give a good foundation for later fruit production. A plant that is already flowering is two-thirds of the way through its life cycle and has

already turned from root growth to fruit production. It may still produce but far less than if you had bought a younger plant.

6.   Finally, don't be seduced into buying more than will fit in your garden. Think through where each plant will fit as you select for purchase. Do you really need six packs of five varieties of tomatoes?

Building a Cold Frame

Building a cold frame is a simple way to create a growing spot that gets heat and sun early in the growing season, so you can start seedlings when it's still cold out. A cold frame can buy you as much as a month of extra growing time in the beginning of the spring, and depending on your climate, your cold frame can be used all year round to grow vegetables like lettuces and member of the brassica family. In this way, we expand our smaller spaces by expanding the amount of time we have for growing. A cold frame is basically a box with a window on top of it to keep the heat and light inside. It's easily made out of salvaged materials like wood and windows you can find on the street, a salvage yard, or a used building supply store.

Materials Needed

Salvaged window, any dimension
Scrap wood, at least 1" x 6"
Screws
Drill
Hinge (optional)

1.   Measure the window frame.
2.   Use scrap wood at least 1" x 6" to construct a box that fits the measurement of the window. Construct it so the window can rest comfortably on this wooden frame.
3.   Screw together your frame into the shape of the window. Set the window over the frame.
4.   Optional: Attach the window to the frame with a hinge. (This will make it possible to prop the cold frame up on hot days to provide ventilation for plants, but the cold frame will function with or without the hinge.)
5.   Put it in a sunny location and you're ready to go.

You can skip steps 1–3 if you have an already constructed box—an old cabinet or a set of shelves you find at a garage sale or thrift store can work well. Simply turn this into a cold frame by filling it with dirt and covering it with a properly sized glass window.

Start plants in separate flats underneath the glass, or fill the cold frame with dirt and grow them right in there. The dirt will need to be periodically replenished as you lift the seedlings out and transplant them into the garden. Attach the window to the frame with a hinge, and use a stick to

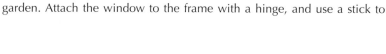

The cold frame is a simple constructed box with a glass cover. Useful for starting seeds in spring and extending the growing season, especially in four-season climates.

prop open the cold frame on warm spring days. Remember to prop it open when the days warm up, or you'll fry your starts. Site your cold frame so it gets maximum sunlight.

Easy Mini Hoop House

One of the best ways to stretch your growing season especially in four-season climates is by using a greenhouse. In most urban settings, you just won't have room for a true greenhouse structure, but you can simulate greenhouse conditions easily and provide another zone for starting sprouts in a small space. If you do have space for a greenhouse, we envy you, and encourage you to go online to find a design that suits you. Some of the best ones are narrow structures placed against the sides of buildings and constructed out of PVC pipe and greenhouse plastic. For the rest of us, simulating greenhouse conditions in a smaller space will have to suffice.

To make your mini hoop house, you're going to be building a tent made out of bamboo, rattan, PVC pipe or some other flexible material and covering it with plastic sheeting. You can make a free-standing setup that you place on an outside deck, patio, or yard space, or you can attach the greenhouse tent to a table and place it in a sunny place in your yard.

1. Find a long table that won't mind living outside (something metal or plastic will work).
2. Create a frame around the table by attaching bamboo or wire in an arched shape to the edges of the table. If you are using PVC, get 3-way fittings to make the corners. This will create a square base around the table and allow you to fit the hoops into the corner pieces. If you opt for material like bamboo or rattan, attach the corners of the tent with strong wire to hold them in place.
3. Make the form high enough to provide space for starter flats and roomy enough above the flats for the starts to grow.
4. Cover the frame with greenhouse-grade plastic or Reemay. You'll have to purchase this new, and it may cost a bit, but it lasts for years and provides the protection and sunlight plants need. Attach it with clamps, or clothespins if the framing material is narrow enough.
5. Attach the cover to the greenhouse on three sides, and leave a fourth side moveable. This way your plants have protection, and you have access to your plants.
6. Place your flats with seeds in them on the table and cover with the plastic. Make sure you situate your "greenhouse" in a sunny part of your yard. Check frequently to make sure moisture, heat, and light needs are being met.

You can take this contraption apart in the cold of the winter when you won't use it. If you set it up on a folding table, it should be easy to stash somewhere until the next growing season.

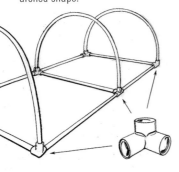

3-way plastic corners hold the PVC pipe in an arched shape.

Hoop house "tent" can also be constructed as a square for smaller areas. Reemay or greenhouse plastic is draped over PVC pipe.

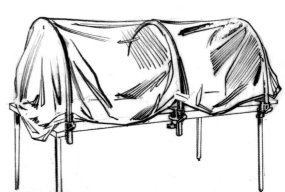

PVC, willow branches, or flexible reed can be attached directly to a table and covered with plastic sheeting for an instant greenhouse.

SEEDS OR STARTS? SUN OR SHADE?

Direct Seed in Ground

Carrots, beans, corn, garlic, turnips, beets, chard, cucumbers, squash, parsnips and other root crops, calendula, zinnia, sunflower, cosmos.

From Seeds or Starts

All greens, squash, broccoli, cauliflower, kohlrabi, kale, tomatoes, peppers, eggplants.

From Starts

Artichoke, asparagus, leeks, onions, many medicinal and some kitchen herbs, berries (bare root).

Full Sun (6 to 10 hours)

Sunflower, tomato, melon, pumpkin, pepper, eggplant, beans.

Partial shade (3 to 4 hours)

Cool weather crops in summer, lettuce, snap peas, chard, kale, broccoli, zucchini.

Happy in the Shade

Mints and mushrooms.

Winter Sowing (outdoors in temperate climates, in greenhouses in four season climates)

Onions; lettuce; all the brassicas including broccoli, kohlrabi, kale, and cauliflower; Asian greens like mustard, tatsoi, and pac choi; leeks; Swiss chard; and peas.

Borage, a self-seeding annual herb. Bees love the pollen and the flowers are good in salads.

Green cauliflower.

The scarlet runner bean in flower.

Squash flower unfurling.

Urban
artichoke.

Tomatoes
on the
vine, high
summer.

Early
lettuce.

Pea
shoots
in early
spring.

Early
mustard,
late
winter
garden.

Broccoli raab.

Kitchen Botany: Important Food Families

Knowing the food families helps you balance diversity in the garden and understand plant relationships to one another, much the way you know someone a little bit better when you meet her mother, her father, or her brother. Plant families also trace the evolution of seeds, seed saving, and seed starting needs. Once you know about how to save a broccoli seed, for example, it's easy to apply the process to its cousin, the cauliflower. Seed starting, growing, and harvesting needs are consistent within family lines.

***Asteraceae*—Sunflower Family.** (*Compositae* is an old name for the family.) Crop examples: endive, radicchio, artichoke, sunflower, lettuce, Jerusalem artichoke, and cardoon. Pollinated by self and insects. Seeds are viable for three to five years. These plants are relatively easy to grow—artichokes are high yielding perennials and also beautiful to look at. All the lettuces are simple to grow and harvest.

***Apiaceae*—Carrot and Parsley Family.** (*Umbelliferae* is an old name for the family.) Crop examples: Celery, celery root, dill, chervil, cilantro, carrot, fennel, parsnip, and parsley. Pollinated by honeybees and hairy insects. Seeds are viable for three to five years. This family includes herbs and vegetables. They are a little more challenging to grow than the *Asteraceae* family, but are favorites throughout the year. The starts of this family of plants need more attention in the beginning (consistent watering is essential for starting carrots), and they require good, loamy soil for proper growth.

***Brassicaceae*—Mustard Family.** (*Cruciferae* is an old name for the family.) Crop examples: mustard, kale, broccoli, Brussels sprouts, cauliflower, collards, kohlrabi, turnips, cress, and radish. Pollinated by insects. Seeds are viable for three to five years. This family provides stable crops in the fall, winter, and spring, but do not grow well in the heat of summer. Kale is best for the small garden—one or two plants will provide most families with all the greens they can eat. They also reseed freely if you let them go to seed. Broccoli is also suited for a small garden, as it has been bred to send up a single large head in the beginning of its growth, and then many off shoots of smaller heads throughout the season.

***Lamiaceae*—Mint Family.** (*Labiatae* is an old name for the family.) Crop examples: basil, mint, lavender, rosemary, thyme, shiso, hyssop, and sage. Pollinated by insects. Seeds are viable for four years. This family brings us a wide range of important medicinal and culinary herbs. These plants are simple to cultivate and grow, bring bees and other pollinators to the garden, and can be used in cooking, herbal medicines, sachets, and other sweet smelling potions.

***Amaryllidaceae*—Onion Family.** Crop examples: onions, leeks, scallions, chives, garlic, and shallots. Pollinated by flies and bees. Seeds are viable for two to three years. These are important staple foods, some of which grow perennially, and some of which can overwinter in milder climates. Garlic is planted in the fall for an early summer harvest.

Rosaceae—**Rose Family.** Crop examples: rose, apple, pear, raspberry, strawberry, and plum. Pollinated by bees and hairy insects. These plants are mostly propagated from bare-roots, roots, suckers, and grafts. The flowers, scents, and pollination opportunities this family offers are a welcome addition to any garden.

Solanaceae—**Nightshade Family.** Crop examples: sweet and chili peppers, tomatoes, tomatillos, eggplant, potato. Pollinated by self and insects. Seeds are viable for 3 to 7 years. Some people can't digest the fruits in this family, which have been shown to have some effect on autoimmune conditions, including rheumatoid arthritis. If you have a tendency toward these imbalances in your body, don't grow too many tomato plants! Potatoes are the single most important staple crop to grow in a small garden plot. Inch for inch, they deliver more carbohydrates and nutrients than any other plant we grow. And because it's almost impossible to harvest every single one once you've planted them in a garden bed, expect to see them year after year in the same place.

Cucurbitaceae—**Squash Family.** Crop examples: gourds, watermelon, cucumber, muskmelon, summer and winter squash. Pollinated by insects. Seeds are viable for three to five years. Hard-working, heat-loving vines bring plentiful bounty to our gardens during summer and early fall. These plants can be grown in vertical spaces as well as along the ground when we are compensating for lack of space.

Gramineae—**Grass Family.** Crop examples: Corn and sorghum. Pollinated by wind. Seeds are viable for four to six years. Some of the world's most successful plants are in this family, most especially King Corn and wheat. They require a lot of nutrients, so be sure to replenish the soil both before and after, and companion plant with beans. Corn's great in the garden, dramatic and delicious, and provides great stalks for beans and other vining plants.

Chenopodiaceae—**Goosefoot Family.** Crop examples: Swiss chard, beet, orach, quinoa, and spinach. Pollinated by wind. Seeds are viable for three to five years. This family of greens and grains provides good staples for the garden that are simple to grow and eat. In some climates, they grow all year long; in other climates, they are among the longest lasting crops, growing from early spring through late fall.

Leguminosae—**Bean Family.** Crop examples: common beans, peas, fava beans, lima beans, runner beans, peanuts, soybeans, lentils, jicama. Pollinated by self and insects. Seeds are viable for three to five years. Beans are essential for the self-sustaining gardener. You can plant them in rounds every three weeks or so throughout the season, and have fresh beans, and beans for drying all year long. They are simple to grow, delicious, and come in an astonishing variety.

on the ground: tiny town farm—rachel's story

For a lot of years, I wanted to escape the city and create a land-based community called the Art Farm. The vision was to have friends and family living together on a rural property, raising children, making art, and growing food, together. We'd live in separate little houses and have a communal kitchen and make lots of art. I spent years driving around California with a friend, looking at properties. Some were beautiful, and some were horrible, but none of them felt generative to us or were so far outside our price range as to be completely unrealistic.

Driving around the California countryside gave me insight into the land and its uses, and what it might mean to live outside the city. While it was beautiful and felt calming to my system, I also imagined that living rurally could be quite isolating and would, I suspected, include a fair amount of driving. In short, it never seemed sustainable, and for lots of reasons—like the fact that starting a land-based community when your tribe is a bunch of low-budget anarchist city dwellers is like herding cats—the Art Farm never happened.

I got pregnant and had a baby and my partner and I did leave the city for a more rural town. I started gardening more and learning more about the earth at that time when I was largely housebound with my infant daughter. This was in the years after 9/11, when George Bush II was president and I was becoming more and more alarmed about the path our country was taking, the insanity of our leaders and what seemed to me an inevitable collapse of the systems we need to sustain life.

So I started growing things. Experimenting. Learning about plants. I taught myself to can (and proved that even someone who doesn't follow the rules will have a hard time poisoning her family with canning mistakes). I began to study permaculture, agricultural land use, and sustainable development. I watched my daughter and our family grow.

On top of the general political insanity that characterizes our time and the direness of our ecological situation, having a child is the whole context for my sense of urgency about caring for the world. When you take on the responsibility of a child, your connection to the future isn't tenuous anymore—it is embodied, made real in the life of your child. What will the world be like for her when she is 30 or 60? It didn't look good, and I knew that while I had no ultimate power over the future, I wanted her to grow up with the values of the earth and the power of life, and I wanted to live them with her. Everything that's come to our family—our gardens, our beehives, our backyard chicken flock, and our posse of earth-loving homesteading friends—comes from this impulse to care for the future that our daughter will live.

I think it's funny when people are surprised that their children start repeating their own fantasies (and idiosyncrasies and challenges). What else do children learn, other than what surrounds them? When she was five, my daughter invented a fantasy world—a land-based community where everyone lived in tiny houses around a central building with a shared kitchen and a beautiful working farm. There were horses to ride down to the river for water and salmon in the stream and tree houses and outdoor ovens and bugs and bees and birds and paint supplies everywhere. She called it "Tiny Town Farm." So we named our little 5,000-square-foot farm Tiny Town Farm, and that's where we live now. If you're going to teach your children how to live, the happiest way is to live it with them, even if it doesn't meet the highest image of your dreams.

Saving Seeds

Another great practice, especially after you have come to befriend your plants, is to save your own seeds for the next garden year. Seed saving is an ancient co-creative practice that evolved the enormous variety of fruits, herbs, and vegetables we love today. Over millennia, different seed traits have been selected out for flavor, beauty, productivity, and pleasure until we have the cornucopia of today. Due to the limiting practices of industrialization, we are on the verge of losing much of the agricultural diversity it took humankind 10,000 years to create; seed saving is an important way to support the diversity of our food sources.

Saving radish seeds by hanging the plant up to dry. Once dried, seeds are easily shaken out of the pods. *Photo by Rachel Kaplan*

Until the beginning of the twentieth century, farmers had available to them as many as 1,500 different plants, each with thousands of different cultivated varieties. In the last 100 years, 75 percent of genetic plant diversity has been lost and 30 percent of livestock breeds are at the risk of extinction. Industrial agriculture virtually controls our entire food system. At this point, six companies control 98 percent of the world's seed sales, and only four (wheat, rice, corn, and soybeans), provide 75 percent of humanity's caloric intake.[46] Seventy-five percent of the world's food comes from 12 plants and only five animal species, making our global food supply highly vulnerable to disease and famine. The selection and species limitation practices of agribusiness have placed the diversity of the food chain at risk. Super-hybrids (which bear no viable seed) monocrop the globe, replacing the diversity of our heritage breeds. More ominously still, transnational corporations like Monsanto are arrogantly patenting the gene pool of different food crops, thereby limiting people's rights to food choice and diversity, as well as profoundly impacting the livelihood of small subsistence farmers around the world.

Diversity is one of the strengths of the unfettered natural world. In a healthy ecosystem, different species live in proximate relationship to each other, with pollinators and animals benefiting from the unique flowering of each plant. In nature, planned redundancy in the system provides security in the event of a change in the environment (such as pests or disease). We have effectively stripped this "security system" from our food chain through monocropping and the narrowing of the gene pool. This puts our culture's sustenance in a dangerously dependent and unhealthy position.

Fortunately, seed savers around the world have committed to the ancient practice of seed saving, and are generating global seed-saving networks, and international seed banks. Home farmers and gardeners can play an important role in supporting the rich diversity of plant life by learning to save their own seeds from varieties that perform well in their ecosystems. While seed saving is somewhat compromised by gardening in small

HEIRLOOM VEGGIES—SIX REASONS TO PLANT THEM

1. Better taste.
2. More nutritious than new varieties—industrial farming yields industrial food. With its eye on the bottom line, nutrition and earth care are lost. You can taste this in your food.
3. Heirlooms are open-pollinated, which means you can collect their seeds. This reduces your reliance on seed companies and helps create a more closed system at your home site.
4. Heirlooms are less uniform than hybrids, which means they don't ripen all at once.
5. The seeds of heirlooms are more affordable than the seeds of hybrids.
6. Heirloom vegetables hold within themselves a long history of their ongoing relationship to the earth. Planting them keeps us in touch with our lineage as farmers and eaters within a community. Hybrid seeds, on the other hand, reinforce a lineage of dispossession.

Sunflower seeds dry on the stalk, but if you leave them out in the sun too long, the birds will eat them all.

spaces and the impact of cross-pollination, if you are willing to let some of your plants go to seed rather than harvesting them for their fruit, you will be able to gather seeds for next year's garden, breed seeds from your healthiest and most productive plants, and participate in this ancient agricultural ritual. Seeds are generally saved from annual and biennial plants. Start with plants whose seeds are easy to recognize: beans, lettuce, sunflowers, and calendula flowers are very simple seeds to save, and easy to grow as well.

Allow the seeds to dry naturally on the plant. Corn, beans, and garlic are good examples of this method. Pull the cornhusks when the corn is fully ripened and allow it to continue drying on racks (protected from birds and squirrels) or in paper bags indoors until they are thoroughly dried. Once dry, twist them in your hands to get the kernels to fall off. Package, label with name of the variety and date or year of harvest, and store.

For garlic, the same drying method applies. Pull the garlic in late spring, and let the plants dry (with their leaves still attached) in a cool, indoor place. Do not leave them to dry out in the sun. Garlic can be braided and hung from nails, or stored in open woven bags while they are drying. Part of your garlic harvest will provide next year's seed garlic. Save the best heads of garlic for next year's garden. Each fat clove will grow into an entire head.

Lettuce and brassicas such as broccoli can be collected directly from the plant. When the seeds are dry and about ready to fall off, directly pull the seeds off by hand into a waiting paper bag. Similarly, for beans or peas, just wait until the pods dry on the vine, pull them off and shell. Let the seeds dry out completely before you store them. Most of the harvest will be for eating, but set enough aside to start the coming spring.

The most common method of vegetable seed storage is to remove the seeds and allow them to air-dry. Cucumbers and squash plants dry this way. Allow the fruits to fully ripen even to the point of the fruit starting to turn yellow so that the seeds inside fully develop. Then cut open the vegetable and scoop out the seeds. You can also remove the seeds from vegetables like this when you are eating them—take the seeds out of the watermelon, or cucumber, or squash before you eat them, place them on a paper towel until dry, then store them in well-marked envelopes for next year.

Fermenting is needed for tomatoes, as the viscous pulp inhibits germination. The easiest way to achieve this is to slice open your tomato, squeeze the contents into a glass jar, add water to the top of the jar, stir, and set aside for a few days. You will notice a residue collecting on the top of the water, as well as some seeds (the dead ones). The water will then clear and the good seeds will sink to the bottom of the jar. After about 4 to 5 days this process will come to an end, so carefully scoop out the stuff from the top and throw it away, pour off the water from the jar, and lastly, pour out the seeds from the bottom of the jar onto newspaper or paper towel for final drying. When the seeds have dried, they can be removed from the paper and stored.

Home-saved seeds will retain their vigor if thoroughly dried and saved in airtight containers in the freezer for extended storage or in a cool place for next season. While some vegetable seeds can remain viable in storage for as long as 15 years or more, and grains may remain viable much longer under stable environmental conditions, every year in storage will decrease the amount of seed that will germinate.

SEED SAVING TIPS

What are some things I should consider before beginning to save seeds?

Learn about the plant's method of pollination. Will it cross with other plants? When will it bear seeds? Does it have any special requirements to be a viable seed? Start with beans, peas, tomatoes, lettuce, and peppers. These self-pollinated crops are the easiest for the small gardener. More experienced seed savers can work with corn, cucumber, muskmelon, radish, spinach, and squash or pumpkin. These vegetables produce seed the season they are planted but require separation to keep unwanted cross-pollination from taking place. An expert seed saver can work with beets and Swiss chard, the brassica family, carrot, escarole, onion, radicchio, endive, turnip, and Chinese cabbage. These vegetables normally require more than one year for seed production and mandate separation to prevent cross-pollination.

Are there plants that shouldn't be saved?

Seed from hybrid plants will not reproduce true to the parent plants. These seeds should not be saved.

What are the methods of pollination for plants?

Pollen can be airborne or insect-borne, or can be self-pollinated. Some varieties need to be separated by large distances or care must be taken to ensure that plants are not cross-pollinated (by using alternate day covering or staggering of plants). Remember that plants that are related to one another can pollinate each other too.

How do you pick which plants to save from?

Observe your plants through the growing season. Choose disease-free plants that have the qualities you like. Some qualities include: flavor, plant size, harvest time, bolting time, fruiting abundance, yield, resistance to pests, and early bearing.

Which seeds are good to save?

Seeds must be mature if they are going to germinate. They are mature when flowers or pods are faded and dry. In the case of fruiting bodies, a good rule of thumb is that when they are ripe or overripe the seeds are ready to be collected.

Are there other things I should do?

Take notes so that you can remember why you saved those particular seeds. Record the dates of harvest, source of seeds, and variety traits.

Urban Seed Bank

Creating an urban seed bank is a valuable community project. Due to space issues and the fact that seeds cross-pollinate in small spaces, the best way to save seeds in the city may be as a collective endeavor, with different gardeners saving the seeds of different plants over time. Seed saving for each species does not have to be done yearly; one kale plant, for example, can yield hundreds of seeds. These can easily be saved and shared with your fellow gardeners. But because of the cross-pollination that happens in small spaces, having one person save the kale seeds while another saves the spinach makes good sense.

If you are not part of a seed-saving collective, this is something you could create in your neighborhood or among homesteading friends. Harboring safe spots for seeds is important and the best seeds are ones that produce well locally, so saving, storing, and sharing them is a job that can best be done locally. Doing this as a collective

Baker Creek Seed Bank, a premier seed-saving organization with branches in Missouri and California. The old bank vaults are filled with heirloom seeds from around the world. *Photo by Rachel Kaplan*

endeavor maximizes diversity. Finding a safe, cool, and dry place to store seeds is important, and following guidelines for each variety also ensures having a collection of strong, healthy seeds. Once you've tried your hand at seed saving, try hosting a yearly seed swap to share the bounty of your seeds. Plan it as a simple potluck, sometime in February when gardeners get to longing for the earth. It will remind you of the pleasures of spring to come.

In Richmond, California, a seed-lending library has been established, functioning much like a regular book library, with a little twist. Citizens can "check out" seeds in the beginning of spring, use what they need in their gardens, and then follow the seed-saving guidelines that come with the seeds. The unused seeds, plus a newly grown supply, is returned to the library when the season is done. The library encourages people to borrow seeds at their seed-saving level of experience, and offers an orientation to the entire process. Check out www.richmondgrows.org for more information on how this crafty system works.

A variety of heirloom drying beans, including Calypso, Hidatsa Shield, Trail of Tears, Tiger's Eye, and Good Mother Stallard.

From Patterns to Details

Nurture all that is authentic by acknowledging three simple realities: nothing lasts, nothing is finished, and nothing is perfect. To accept these realities is to accept contentment as the maturation of happiness, and to acknowledge that clarity and grace can be found in genuine unvarnished existence. Filled with subtlety and depth, this way is a river flowing toward and away from you, and always within you.

—Richard Powell[47]

Once you've started (and saved and stored) all your seeds, you'll need to figure out where to plant them in the garden. And because spaces are small in the city, wise strategies for design and for growing different plants in close proximity to one another are important. Working within the permaculture framework, we design our gardens to use the least effort for greatest effect, maximizing the strength and intelligence of nature so we don't have to do so much heavy lifting.

Selecting perennial plants that grow year after year creates successional growth and change in the garden without a lot of redesigning each year. Planting beneficial "weeds" that flourish happily in your location, plants that encourage bees and other pollinating insects to visit the garden, and artfully combining different plants in small spaces are strategies we can employ to great effect in the urban garden. The design process prioritizes ongoing relationships between the different elements in the garden to maximize yield, create biodiverse ecosystems, and oases of beauty in the urban landscape.

Some Thoughts About Ecological Garden Design

Designing your garden is an ever-evolving process of observation and inspiration. Experiment. It's going to take time to figure out what grows well and what you can and want to grow. It will probably change from year to year to fit your needs. We've found though that once we started it was hard to stop and that one garden bed leads to another, until we're awash in food and fruit and herbs for many months of the year, and nearly every inch of our yards is filled with plants.

When you first start your garden, look around and see what resources are available to you. Plan your garden *after* you have observed your site. Enthusiasm for the task at hand is great, but intelligent design is better. Ask yourself questions about sunlight throughout the seasons, as this will change all year long. Do you have more vertical space than horizontal space? What kind of vegetables will be happy growing up, rather than out? Tomatoes can grow hanging upside down and pole beans can be 12 feet tall by the end of the season. How is your soil? Do you get visits from cats, dogs, or predators? Are you surrounded on all sides by neighbors? Where's your water source and how close is it to where you want to garden? Can your neighbor grow a great tomato and you grow a great peppermint and cucumber bed? Can you share resources and space with one another? What do you want to do with your space other than garden—outside sitting areas, places for children to play? How much time do you have to tend your garden?

Each of these questions will yield different solutions—some will be easy to answer, and others may take the trial and error of some seasons before you know what you can grow best and where your energies are best spent. Remember to look carefully at the resources of soil, water, light, and air before you start planting. Sowing is the easy part. Tending and bringing the fruit to harvest is where the true art is.

As you practice, you'll learn which plants you can grow easily in your location, and which will be more challenging for you. Some areas are ideal for heat-loving plants like cucumbers, melons, and peppers, and in some places, you'll be hard pressed to grow a tomato because you just don't get enough heat. Some plants, like squash, corn, and melon, simply take up too much space for an urban garden, but others, like lettuce, beets, carrots, and beans, are easily grown in pots on a back deck, or in a small backyard bed. Some berries, like blueberries and strawberries and raspberries, can be grown in pots, and there are many varieties of fruit trees that can be grown for many years in containers, and offer prolific yields. If light is an issue where you garden, you will need to work with plants that need less of it—greens, lettuces, peas, and beans all do well in low-light conditions, but you will probably have to forego the homegrown pepper, tomato, eggplant, or cucumber.

Where Will Everything Fit?

Zonation helps us ask and answer questions about location: How often do I need to go to the herb bed? The vegetable garden? The compost heap? What will be the benefit of placing this plant, or the chicken coop, or

the beehive, close to the house? The answers will help us make decisions about where to locate different elements of our homestead. A compost heap is best placed near the garden for ease of use once the compost is done. The chicken coop can be close to the garden too, so that chickens can peck in the garden when allowed, and the manure can be easily spread on garden beds. Your bees should be at a distance far enough from the house so

A diversity of flowers in the garden brings pollinators and beauty to the garden.

After a few seasons, the bare lawn has been transformed into a gorgeous garden filled with flowers, trees, and food.
Photos by Trathen Heckman/Daily Acts

you aren't walking into their flight path as they go about gathering pollen. Your orchard can be planted near your vegetables for greater integration of plants and trees, and herbs and vegetables can be integrated as well for pest management, plant strength, and biodiversity.

How Do I Know How Much to Grow?

This is a hard question for beginning gardeners. When I first started, I grew 13 kale plants for my family of three (one of whom was under two.) I could have fed the entire block. Now, three kale plants per season more than keep us happy. How many tomatoes will you eat? Do you even like peppers? The answer to how much to grow always depends on how much you will eat, preserve, and want to share. Some people grow extra because they have the space, and then they can give it away. Understanding how many plants you need helps maximize space and yield and diversity in the garden. It'll take a few years to figure this out—expect to make some mistakes in the beginning, and if they are mistakes of abundance, make sure to spread the wealth around.

Permaculture Design Superstars

The following design features are classic to a permaculture garden. They exemplify different principles of design and have been used repeatedly in permaculture landscapes for many decades. All of them can be adapted even to very small spaces.

The Food Forest and Perennial Plantings

A food forest is a seven-level planting structure, beginning underground and rising up to the tree canopy. This includes the rhizomatic (or root level), spreading ground cover, the herbaceous level, shrubs, lower trees,

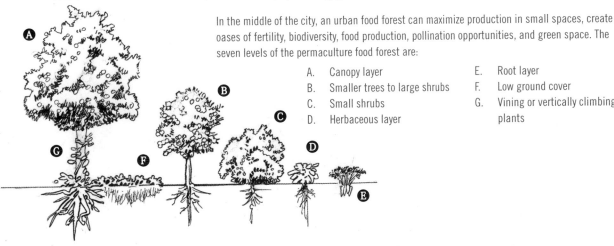

In the middle of the city, an urban food forest can maximize production in small spaces, create oases of fertility, biodiversity, food production, pollination opportunities, and green space. The seven levels of the permaculture food forest are:

A. Canopy layer
B. Smaller trees to large shrubs
C. Small shrubs
D. Herbaceous layer
E. Root layer
F. Low ground cover
G. Vining or vertically climbing plants

vines, and canopy trees. Permaculture design employs the food forest strategy to mimic the forest ecology, which teaches us the benefits of plants sharing space with one another, using every level available for growing, the natural succession of plants over time, planned redundancy within a system, and the use of multiple plants to serve multiple functions. Within a small area, the amount of biodiversity and biomass and food productivity in the forest ecology is astonishing.

Urban food forest with apple tree, lemon tree, perennial kale, and other herbaceous and flowering perennials.

While we are unable to replicate the forest ecology exactly—due to our lack of understanding as well as our limitations of space—this model is worthy of study and emulation. Designing our urban gardens on the food forest model allows us to grow more food in smaller spaces, generate positive relationships between plants, and experiment with these relationships over time. In a small garden space and a busy urban life, mixing perennial and annual plants in the food forest design makes sense. Perennials live and yield for years—herbs, trees, and some vegetables are perennials, and many flowers are perennials as well. They give a garden shape and food year after year while reducing our workload. Planting perennials in layers is part of creating a sturdy food forest design. It creates ecosystems of depth and duration that rely on the succession of evolution to create diverse, water-wise environments.

Some perennial vegetables include rhubarb, artichoke, and asparagus, as well as some more unique plants like oca, mashua, and yacon. Lovage is a perennial herb that tastes a bit like celery and is flavorful in soups and stews. The tree collard is a perennial brassica that's easily propagated from a branch of an existing plant, or from seed. You can eat long and well from a tree collard planted in a shady part of the garden. All fruit trees are perennials, as are berry crops, including blackberries, raspberries, and blueberries. Some vines are perennially fruit bearing, such as grapes, passion fruit, and kiwi (although these are happiest in temperate climates). Perennial herbs are the norm, rather than the exception, and some herbs, while not truly perennial, are self-seeding and so act like perennials in our garden design. We are particularly fond of those perennials that yield food *and* beauty year after year.

Homage to the Weed

Weeds are plants no one wants in their garden. But one gardener's weed is another gardener's dinner. Homeowners around the country have made war on the lowly, lovely dandelion, but some of us cultivate the dandelion for its leaves, which are excellent in salads and

Urban food forest—trees, perennial and annual vegetables, herbs, flowers, and plenty of room for a child to play.

Art in the garden and the budding of spring.

provide a cleansing tonic for the liver. In some gardens, a dark green ground cover with small leaves trails across an empty bed. A gardener may pull it out without recognizing it as purslane, another excellent salad green, tonic, and great supplemental food for chickens. Stinging nettle is considered a pernicious weed, but it too is an excellent spring green, provides tonic to the kidneys, support for the respiratory system, enhances the fertility of the compost bin, and provides benefit to pollinators like bees and butterflies. Nettle is the kind of plant to search out and dig from someone's garden or ditch, just to have your own patch in the backyard. Just pay attention to the stinging barbs when you harvest them.

We often say "It grows like a weed" in a disparaging tone. What we mean is: that plant is extremely successful and productive. Don't overlook the plants that are happiest right where you live. Integrate "weeds" that provide benefits to your life into your garden design.

Plant Guilds Ancient and Modern

One of the best relationship-based designs in our gardens is the plant guild. This design combines different kinds of plants in the same small space to generate synergistic relationships, maximize space, and increase yield. Plant guilds mimic the natural order of things and tend to the different needs of a garden simultaneously (yield, mulch or biomass production, and the development of root systems) by using multiple planting layers and plants with different functions in a small space. When we are trying to enhance biodiversity, soil fertility, and habitat for a variety of pollinators and other creatures, as well as grow food or medicine for ourselves in a small space, guild plantings are the way to go.

The classic example of a guild is the Three Sisters planting—corn, beans, and squash planted together in hills. These plants have been grown in proximity to one another on the American continent for many millennia. The corn grows in mounds that are covered by squash. The beans trellis up the corn, using available vertical space. The beans are nitrogen fixing and provide nutrients for the corn and soil. The corn plants reach up toward the sun. The squash leaves cover the roots of corn and beans, providing a natural mulch that conserves moisture and lessens the need for water. Together, these three plants produce a lot of food in a very small area.

To plant the Three Sisters in your yard, create four-foot-wide mounds about two feet high, and about three feet from one another. Flatten the top of the mound. Plant four corn seeds, one in each corner of the top of the mound. When the corn has grown to five inches high, plant pole bean seeds around the corn. This way, the beans will not overwhelm the corn as it grows, but will use it as a trellis. Once the beans have sprouted,

The Three Sisters—corn, beans, and squash—is the most tried and true guild planting on the North American continent.

FROM PATTERNS TO DETAILS

95

An experimental guild combination—olive tree, delphinium, euphorbia, and lavender.

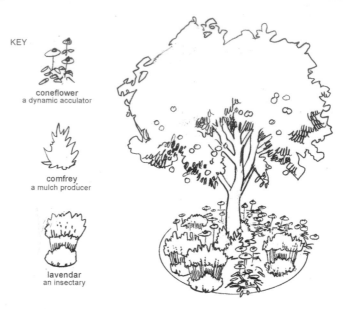

KEY

coneflower
a dynamic acculator

comfrey
a mulch producer

lavendar
an insectary

Apple tree guild. The tree is the centerpiece of this sample guild.

plant your squash seeds on the edges of the mound, no more than four plants per mound. Eventually you will want to thin this out to no more than two squash plants. As the summer wears on, you will lose sight of your soil because it will be covered by foliage. The three plants can be grown until they die back and start to dry, which makes harvesting these easy-to-store-vegetables a simple task.

In a less time-tested guild, we combine different *kinds* of plants to produce the same beneficial results as the Three Sisters plantings. The different kinds of plants are *dynamic accumulators*, which pull nutrients from deep in the soil up to the topsoil; *nitrogen fixers*, which replenish the soil with nitrogen; *mulch producers*, which are often herbaceous and easily compostable; and *insectary plants* that provide forage for beneficial insects. Low-growing grass-suppressing plants or ground covers complete the guild. In a plum tree guild, the plum tree is in the center, surrounded by garlic chives and low-lying herbs; insectaries like dill or fennel; accumulators like chicory or yarrow; and a mulch producer, like artichoke or comfrey. This guild offers plants for food or forage for numerous species, herbs for healing and eating, pollination opportunities, and a water and energy conservation strategy. Experiment with the guild strategy in your own yard and see which plants work best together in the place where you live.

Biodiversity and Beauty

Wherever you garden, remember to plant a diversity of plants that have both similar and different functions. This is like planning an ideal community, or business—different people bring different skills and talents to the table. Biodiversity is maximally threatened by the devastation of ecosystems currently underway during this reign of corporate terror; the rate of species extinction is rapid and alarming, and shows no sign of reversing. Valuable pollinator species like the honeybee are at risk of habitat loss and pollution-induced decimation. While our backyard gardens cannot overcome the large-scale devastation of the earth, small steps toward large goals is the name of the game. Do what you can.

Within an urban setting, gardeners create biodiversity by selecting a variety of plants, bushes, and trees that offer opportunities for important pollinators (bees, birds, and bats), habitat niches for small animals, and fertile zones like compost or worm bins or mushroom habitat for microorganisms that replenish and remediate the soil. This is a good example of stacking functions. When the plants we choose feed more than one need—food, fodder,

Growing a wide diversity of plants in a small space on a back deck. At the height of spring, this 27 square foot container garden contained 26 different kinds of herbs, vegetables, and flowers.
Photo by Rachel Kaplan

pollination opportunities, medicine, pest management, fiber, and beauty—we create complementary and variegated ecosystems that support many beings.

You can work toward biodiversity even in a small container on your back deck. Try planting different kinds of vegetables in the same place, making sure you choose veggies that harvest at different times, and use different levels of soil in the barrel. For example, you could plant kale, carrots, beets, and parsley together. The kale will yield food for much of the growing season; the beets and carrots will come later, using the lower layer of soil the kale doesn't need; and the parsley will peep its leaves around the other plants for many months. Try a different combination of compatible plants the following season.

In an urgent time, beauty is often sacrificed for expediency, but beauty is an essential design feature, not an expendable one. Beauty cultivates humanity's progress. It creates reverence and feeds the soul. Remember to include beauty as part of your design. Especially in a time of contraction and destruction, the way beauty feeds us cannot be underestimated. It lifts our spirits, signals our creativity, and gestures toward the world we are growing into being. Post-industrial capitalism is ugly. Our lives don't have to be. As artists, we always include handmade objects in our gardens, or along our fences, or in small details against the house. We plant herbs because they make medicine and bring bees, but also because they are lovely to look at and to smell. Some homesteaders only plant edible plants. We like to be able to eat our gardens, but we also like to be able to go outside and see

Let the beauty we love be what we do.

something beautiful, like the honeybee in the center of a bright red zinnia. This feeds us in another important way.

Microclimates

One of the interesting things about gardening in a city is that once you begin to observe where you are and the forces that are at work in your garden, you may also notice that across the street, your neighbors are dealing with a whole different set of issues. They have more sun, or more shade, or more snails, or more wind. This difference from house to house is a called a "microclimate"—it means that one site will have one set of elemental impacts, and another will have a different set. This can be seen yard to yard, but also within one yard there can be different microclimates. Observing them helps us make decisions about where to place different elements in the garden.

For example, the wall in the front of my house is sunny all year long, even in winter, and heat radiates off it from the house's thermal mass. Because of the reflective surface, the

Mandarin orange tree placed close to a staircase, where it gets protection from winds and plenty of sun during the day.

heat of the sun and the house and the protection provided by the wall, this is a great place to grow citrus trees, which flourish in these particular conditions. This would not be a good place for plants that need a lot of water or shade. In another spot in the backyard is an area where water pools in the winter, and is fairly windy during the spring. This is not a great place for those citrus trees, but works for native grasses, sages, and other plants that can tolerate water and wind.

You can create microclimates in your own garden through your selection of plants, water features, and trees. A cluster of trees will create a cool, shady area. A water feature will bring pollinators and birds into a corner of the garden. A ground cover in one part of the garden not in full sun will create a moist habitat for plants and animals. Bare ground will provide habitat for nesting bees. A patch of lavender will invite bees and other pollinators; the proximity of fruit trees to insectary plants will also draw pollinators.

Observing the environment reveals microclimates all over. And learning about plants and what they like

(which happens more by trial and error than by reading about a plant's habits), informs our choices about where the citrus tree belongs, where the bunnies will be happiest, where the tomato plants will thrive. Patience and a spirit of experimentation will serve you well. If you don't get it right one year, you're bound to get it right the next. In the meantime, you'll be learning about the habits of your plants or the creatures you are tending, and the relationships between them.

A water garden creates a habitat for fish and frogs, and when properly planted, brings other pollinators into the garden. The plants in this newly planted water feature will fill the tub by the end of the first growing season.

on the ground: dog island farm

Rachel and Tom farm on a 12,196-square-foot lot in Vallejo, California, an economically depressed suburb of San Francisco. They live in a 750-square-foot house with a 215-square-foot detached water tower. "Our front yard is approximately 2,400 square feet, and while it's not currently under cultivation, we plan to start it next spring," Rachel says. "We have an approximately 1,000-square-foot patio (bigger than the house) with surrounding landscape that equals another 3,696 square feet. Our goat and chicken yard is an additional 700 square feet. The plans for the front yard are specifically for our neighbors, since we live in a depressed area with no real grocery stores in walking distance. We wanted a part of our homestead to be a place we could share with the people around us."

Though largely self-taught, both Rachel and Tom had some teachers in their lineage. "My grandmother grew up on a homestead at the base of the Grand Tetons," Rachel recalls. "Her family lived without any of today's modern conveniences. That lifestyle has always called to me at a really deep level. However, I love living around a lot of people, so I'm not sure I would be okay with the isolation. As for my grandfather, he was an avid gardener. I spent a lot of my early childhood with him teaching me how to garden. During the summer he would take me to visit his brother, who still was a farmer in Iowa. I just remember running around his farm barefoot. Those are some of my best memories.

"My father's side of the family is Mormon. Self-sufficiency and family history are a huge part of their lives. I think part of me is still trying to connect with that side of my heritage and this is the best way I know how. I know that my grandfather's influence is what really pushed me to choosing my career in landscape design."

Tom's paternal grandfather was a big influence on him. He spent his summers on his grandfather's farm helping him grow, harvest, raise, and slaughter. At Dog Island Farm, he is the "muscle," doing the physical work of digging and slaughtering and building. Rachel does most of the canning and preserving, except for the tomatoes, which are in Tom's domain. They share animal-related chores and the pleasure of harvesting.

Rachel and Tom's blog (www.dogislandfarm.com) calculates the amount they spend compared to the amount they save by growing their own. Within their first year of farming, they've moved from spending more to grow their own, to growing more than they spend. Each month the numbers change. In the beginning of the spring, their expenses outstripped their income; by midsummer, their savings hit a mere $37.41; by the end of August, savings were approximately $580. By the beginning of harvest season, the savings easily cleared the $1,000 mark. As the homestead grows to include meat birds and honey from the hive and the garden gets more established, Dog Island Farm's economy will continue to grow.

Dog Island Farm, before planting, 2009.
Photos by Rachel Brinkerhoff

Dog Island Farm after a complete season, 2010.
Let the harvest begin!

Although their blog reads "Learning How to Be Self-Sufficient in Today's Urban Environment," neither of them truly imagines reaching self-sufficiency. "I think it's a great goal, but naive to think you can reach self-sufficiency," Rachel points out. "We need others along the way, whether it's the company that builds our solar panels, or the guy we sell goods to so we can pay property taxes (you can pay off your mortgage, but the taxes will always be there). No one person can do everything. That's why we live in a society. We work towards being as self-sufficient as possible, but we will never get there."

Rachel adds, "Whether or not we can be self-sufficient isn't the thing—but everyone doing something is important. If a person lives in an apartment, they can grow a window farm to provide some of their food. They can use pots on a balcony, join a community garden, start a community garden, buy locally grown food, reduce consumption or take alternative transportation. There are a lot of things people can do, no matter how much space they have. It should never be an 'all or nothing' kind of thing."

Pollination Ecology

Insect life is an essential part of a biodiverse, healthy organic garden. Plants that attract and support beneficial insects (hover flies, native and honeybees, lacewings, ladybugs, wasps, flies, and other garden friends) are welcome in the garden. A well-tended garden will eventually bring the bug population into balance with itself, and the numbers of pests will be greatly reduced. In a healthy ecosystem, each insect species is kept in balance by the other insects and small animals that live around it. We can see this life cycle in the compost bin, as well as in the garden itself.

Most people know that plants reproduce through pollination and setting seed but are less aware that flowering plants depend on insect or animal pollinators to transport the male genetic material (pollen) to the female reproductive organs of another plant of the same species. There are many thousands of insect pollinators such as beetles, butterflies, and moths, and animal pollinators such as hummingbirds and bats, but bees (*Apoidea*) are the most prolific and efficient of the pollinators. Their bodies are covered with fine hairs and physical structures designed specifically for pollen collection. There are over 25,000 species of bees worldwide and 4,000 native species counted in the U.S.

The honeybee is one of our most important pollinators, pollinating nearly 70 percent of our food supply. In recent years, bee colonies have been decimated by a "mysterious" colony collapse disorder. Like other sensitive indicator species, bees are reacting to environmental degradation and pollution, non-organic, chemical-driven practices, and are threatened by our abuse of the planet. But the honeybee is only one of many thousands of species of bees, most of whom live solitary lives (not in a colony) and many of whom partner with specific native plants to

Three honeybees working on echinops flower, gathering pollen for the hive.

Bumblebee pollinating a tomato plant.

create a vibrant ecosystem full of pollination dances. The value of the bee as a pollinator is crucial to our survival; providing forage for bees does service to the bees, the health of the food chain, and the vitality of our ecosystem.

Bees are generally divided into several categories. There are social and solitary bees, generalist and specialist bees. Honeybees are the only truly social bees, living year-round within the social structure of the hive, with a clear division of labor. The honeybee is native to Europe, but has naturalized worldwide. Bumblebees, some species of which are native to the U.S., are in the same genus as the honeybee and are considered semi-social. The queen bee builds a seasonal colony, which works and lives together during the warm months of food abundance and dies out over the winter. Mature bees overwinter in a dormant state waiting to hatch in the next warm season. Bumblebees are important pollinators and have a special relationship to certain plants that will only release their pollen through the strong vibrations of the bee's buzz. In particular, bumblebees pollinate plants of the *Solanaceae* family and are often used commercially as pollinators in tomato hothouses.

All the other bees are solitary bees. They may share nesting areas, but they build their nests and live alone and do not tend their young once the egg is laid. Among these bees are the mason bees, carpenter bees, plasterer bees, and digger bees. They range widely in size and shape and in the manner of building their nests. Nests are commonly built in snags (dead wood) or directly in the ground and are lined with gathered material such as leaves or mud. Nests can range from one to sixty cells, depending on the species. Once the egg hatches, the larva eats, poops, and pupates. Depending on the species and the time of year the mature bee will then emerge and start the cycle over, or overwinter in a semi-dormant state, waiting for the right weather conditions. They will often emerge just at the moment their preferred host plant is coming into bloom.

Bees offer their pollination services for plant rewards of pollen and nectar. Some bees are generalists and can gather and claim rewards from a variety of plants. Others are specialists who have developed an evolutionary relationship with a particular plant. Many times their life cycle and physical structures are attuned to this plant in such a way that the two are dependent on each other for their further existence. If one or the other is endangered both are endangered. Thus if a plant habitat is diminished in such a way that fewer bees visit it, pollination does not occur and their seed set is also diminished. The population may dwindle to the point that bees no longer come to offer their services or die out completely along with their preferred plant. Conversely, if the bee habitat is continually disrupted, by clearing of dead wood and brush or through seasonal tilling, there may not be enough bees for effective pollination.

To entice native pollinators to your garden, provide a square yard or more of diverse forage plants with flowers in whites, yellows and purples. Bees will find places to nest in any diverse garden; providing nesting boxes adds additional incentive.

A. Mason bee/leaf cutter bee block made of a block of wood with holes cut into it. They can be drilled in $3/16$" and $5/16$" diameters and should be at least 4" deep.
B. Mason bee/leaf cutter bee block made from round reeds that provide nesting space for the bees.
C. A bumblebee habitat with a "lobby" area and a brood area. This habitat should be buried in the ground and protected from excess water. Proper ventilation and entrance/exits should be considered in design.

Making a Home for the Bees

Fortunately, native bees can be quite enduring. Even specialist bees can learn to survive on plants other than their host plant. Bees will return to an area that has been decimated given the right conditions and can exist quite well even in an urban setting given enough of the right kind of habitat. Native bees need much the same things as we do: shelter, food, and water. Bee habitat can be provided by creatively arranging stumps or other deadwood in your garden and by leaving exposed earth in perennial plant beds unturned. You can create bee habitat by drilling holes into scrap lumber. Native bees prefer a range of diameters between $^3/_{16}$" and $^5/_{16}$" and between 4 and 8 inches deep.

Native bees are a vital link in the cycle of life and their role in the continuance of these cycles is often underestimated. Every time a species becomes extinct we lose access to valuable information and wisdom contained in the gene sequence. Once that information is lost, we can no longer get it back and can only wait and see how this loss may affect the entire web of life on the planet. We have a choice to support life or destroy it. Creating habitat for bees and plants is a small part of a bigger solution.

Planting for Pollinators

Important pollinators often thrive in cities because of the wide diversity of cultivated plants in people's yards, on their decks, and in the cracks between the sidewalks. Providing nectar, pollen, water, nesting materials, and open ground in your garden for pollinating insects, birds, and bats is a collaboration with nature that results in higher yields in your garden, and gives support for the much-needed pollinators. Small patches of pesticide-free safe havens for all pollinators play a small but vital role in reversing the dramatic pollinator declines we are seeing in the bee population. Magnified across your city, your state, and the country, these gardens form a patch-work quilt of pollinator-friendly places and serve a vital role by feeding and protecting many threatened species. This is especially true for other migratory pollinators (animals like nectar-feeding bats or monarch butterflies) that travel long distances across state and international boundaries. Along these "nectar corridors" the migratory pollinators can take a much needed nectar break within your newly constructed pollinator garden.

Think about diversity when planting a garden for beneficial insects. Companion and interplanting strategies are good for the insects—calendula next to carrots, and borage next to beets. Plant as many different kinds of plants as you can. Most beneficial insects live on nectar and pollen for at least part of their life cycle. Choose species that flower throughout the season, and include plants with blossoms of different hues and shapes. Most insects prefer single blossoms (flowers with one row of petals) as opposed to double blossoms (flowers with many rows of petals).

Due to selective color vision, bees like flowers in the white, yellow, blue, and purple ranges. There are many native and exotic plants that will draw bees to your yard. Among them are mints, lavenders, yarrow, clarkia, gaillardia, delphinium, poppies, penstamon, milkweed, ceanothus, grindelia, fireweed, verbena, and dusty miller. To provide and maintain a regular visiting spot, one to two square yards of plant material seem to be a good minimum. Bees prefer flowers in clumps, rather than different individual flowers. If you provide water in a small pond or birdbath, be sure that there

Water features in the garden provide opportunities for pollinators like the hummingbird. *Photo by Erik Ohlsen*

Coated in pollen, this bee returns to the hive covered with the sweetness of summer.

are plenty of landing areas for the bees. Flat rocks or sticks that angle gradually into the water are excellent for this purpose and will keep the bees from slipping into the water and drowning.

Plant lots of herbs and let them flower. Provide shelter from drying winds—layer the garden with trees, shrubs, vines, peren-

BEE HAVEN

Spring is the glory time for beekeepers, as one swarm after another races across the sky in search of a new home. My partner Adam is called to the bees, and holds a vision of a beehive in every yard. Last weekend as he was getting the mail, Adam heard a loud buzzing sound. He looked up and noticed the cloud of bees overhead. He came running, gleefully shouting, "A swarm, a swarm!" We all piled onto the street to watch the bees until they settled on the branch of a Monterey pine across the road. The neighbors who own the tree weren't home so Adam hesitated to start gathering the bees, but another long-time neighbor said, "Go ahead. They won't mind." Adam took out his ladder, his bee veil, and brush, and began.

Beekeepers generate interest wherever they go, because of current concern about the plight of the bee and because bees are fascinating creatures. Most of us never get to watch a bee colony in action, but if you do, you'll notice that the bees form an integrated, non-neurotic, cooperative mass. Every bee knows its purpose, and wastes no moment wondering about its career choice. We can look to bees as a great model for rising to our purpose (be it worker, drone, or queen) to get the job done. The bees remind us that each role is important, every task part of the whole.

As Adam methodically moved about his beekeeping tasks, neighbors from up and down the road began clustering like a human swarm. Some were nervous, but most were curious to watch the bees. The air was full of their buzzing, and everyone stepped closer to see them hanging in the early evening twilight. It was dinnertime, and soon a picnic table had been dragged to the edge of the road, a bottle of wine opened, and people began to share an impromptu dinner. All the while, Adam was climbing slowly up the ladder and carefully brushing the bees into his swarm box. Eventually, the bees gathered together, Adam came down from his perch, gratefully accepted a glass of wine, and the adventure was complete. After dinner, everyone drifted back home, satisfied and newly connected.

When we answer to the call of the earth, surprising conjunctions of love and possibility happen. A spontaneous conjunction of bees and people came together to bring us a moment of true conviviality and friendship. The bees teach us that we have a purpose, a role, in making what is already sweet—life—even sweeter.

Who knows what kind of collective solutions to our current dilemmas we might find as we set ourselves with a good will to the task? All I know for sure is: it's time to find out. Look around your own neighborhood—what can you grow with the people you share it with? Like the bees seeking the right place for their colony, it's time for us to claim our own place. Maybe that means swarming to a new spot where there's room to spread out, but it's more likely that it means settling into where you already are and learning to work with your neighbors, your own human swarm.

nials, and annuals. Accept a few "bad guys" (aphids for example) so that the good guys (ladybugs) have something to eat. Don't buy insects—instead, plant things that will attract them. If you have the space, leave a small area in your garden unmulched and available for ground-nesting bees and other insects. It can take a few years to convert to a successful low- or no-pesticide garden. While you're waiting, keep your interventions at a minimum.

Plant Families that Attract Pollinators

- The *mint family (Lamiaceae)* attracts native bees and honeybees, bumblebees, hover flies, and other beneficial insects that find mint family plants attractive. Most common Mediterranean herbs like rosemary, oregano, sage, lavender, and thyme are effective. Try agastache, nepeta, and monarda as well.
- The *carrot family (Apiaceae)* including dill, fennel, cilantro, parsley, and chervil, when allowed to flower, attract the tiny non-stinging wasps that control so many garden pests (including aphids). Many other pollen- and nectar-feeding beneficial insects use this plant family as well.
- The *sunflower family (Asteraceae)* is highly attractive to most beneficial insects because of their open centers. Erigeron (Santa Barbara daisy), asters (Michaelmas daisies), yarrow, feverfew, and goldenrod are among the best choices. Native coyote bush (*Baccharis*) is known to attract over 700 species of insects. Also try artemisia, coreopsis, echinacea, and rudbeckia. Almost anything in the thistle tribe is effective. And don't forget annuals like cosmos, sunflowers, chamomile, and zinnias.
- The *mustard family (Brassicacea)* includes wild radish and mustard, both good choices (and you can eat the leaves before the plant flowers). Any cabbage family veggie that is allowed to flower will attract beneficial insects. Good candidates are arugula and many of the quick-growing Asian greens, like bok choy and mizuna.

Companion Planting and Interplanting

There are a few tried and true strategies for growing multiple plants in the same places that are worth experimenting with in a small urban garden. Companion planting is a folkloric gardening technique that pairs different plants with one another depending on their mutual benefits. Some companion plantings discourage harmful pests without losing the beneficial alliance, while adding diversity to a garden bed. Some plants, when placed in proximity, maximize productivity in other plants. Some plants bring beneficial insects into the garden, which bolsters pollination and yield. The practice of companion planting helps create a balanced ecosystem in your garden, allowing nature to do its job without you having to work so hard.

Helpful companions are certain herbs, weeds, flowers, or vegetables that deter insects and encourage plant growth when planted near compatible plants. Borage, chamomile, and lovage enhance growth and flavor in nearby vegetables. Garlic planted near roses and raspberries should deter aphids. Horseradish repels potato bugs. Marjoram improves the flavor of nearby food plants. Mint, sage, and rosemary are the traditional enemies of cabbage moths. Catnip deters flea beetles.[48] And so on . . . a veritable cornucopia of plant relationships to enjoy.

Interplanting is another kind of companion planting, a way of mixing different plants in the same space according to their roots structures. Plant a bed by pairing shallow-rooted plants like chard and celery with onions and carrots, whose roots go more deeply into the soil. This utilizes some of the same thinking as the food forest and guild designs, which use the different level of planting space to maximize diversity and yield. Urban gardeners can rely on this "intensive intercropping" to make the most use of small spaces.

Polycultures

While companion planting gives us some information about the interactions between plants, and interplanting begins to imitate plant design in nature, the permaculture practice of polycultures blends the best of both strategies by integrating the habits of planting varieties that enhance each other, and combining crops that minimize competition for space and nutrients. Polyculture beds—where a variety of plants take up different space, have different root structures, and live and die on different calendars—relies on plant succession as a way to lengthen the growing season. Succession is time unfolding in the garden—early spring crops yield to later spring crops, which yield to heat-loving summer crops, which yield to the late summer and early fall crops (generally the same as early spring crops). Polyculture beds pair quickly growing vegetables (like radishes) with vegetables that take a longer time to come to fruit (like cabbages and parsnips), interspersed with vegetables like beans and lettuce that fruit somewhere in between.

Information about how long it takes a seed to come to fruit is always printed on a seed packet. For example, if the packet says 55 days, it will take 55 days from planting the seed to eating the food. Once you learn when in the season you should plant each vegetable, you can balance that with how long it will take to fruit, and maximize your available space. Some of the principles involved in creating polyculture beds include the rates of germination and fruiting of different plants, the habitat that plants provide for each other and for a variety of pollinators, and their value in attracting beneficial insects, or repelling pests. Thickly planting seeds creates a living mulch for the soil, which minimizes evaporation and the need for water. A polyculture bed can yield a steady harvest of food for several months in a very small space. Noted permaculturist Ianto Evans has evolved a polyculture bed that works in myriad climates to great effect.

Prepare a garden bed, allowing about 20 square feet of bed for each person who will be fed from it.

Two weeks before the last frost: Indoors, start about ten cabbage plants per 20 square feet of bed. The cabbages should be ready for transplanting a month or so after the seed mixture below is sown; to extend the season, choose both early- and fall-maturing cabbages.

Week one: In early spring, sow seeds of radish, dill, parsnip, calendula, and lettuce. For a lengthy harvest season, select several varieties of lettuce. A mix of loose leaf, romaine, butter, iceberg, and heat-tolerant varieties will stretch the lettuce season into summer. Broadcast all the seeds over the same area to create a mixed planting. Cover the entire bed with a light scattering of seed (sow each type separately—don't mix the seeds and toss them all onto the bed, because the heavy seeds will be flung the farthest, and you'll wind up with the radishes all on one end, and the parsnips all on the other). Sow at a density of about one seed every couple of square inches. Cover the seed with about ¼ inch of compost, and water gently.

Week four: Some of the radishes should be ready to pluck. In a few of the gaps left by the radishes, plant cabbage seedlings about 18 inches apart.

Week six: The young lettuce will be big enough to harvest. The dense sowing of lettuce will yield a lovely mesclun blend when the plants are young. With continued thinning, the remaining lettuce will grow up full-sized. If you've chosen varieties carefully, you'll be eating lettuce for up to four months.

Late spring/early summer: When the soil has warmed to above 60 degrees F, plant bush beans in the spaces left by the lettuce. If more openings develop in early summer, sow buckwheat and begin thinning their edible greens shortly after they appear. The next crops to harvest after the lettuce will be the dill and calendula (the blossoms are edible and make a beautiful addition to salads.) The early cabbages will be coming on at about this time too, followed in midsummer by the beans. Parsnips are slow growing and will be ready to eat in fall and winter. As gaps in the polyculture appear in early autumn, mild winter gardeners can plant fava beans. Others can poke garlic cloves into the open spaces to be harvested the following spring.[49]

on the ground: small space, high yield

Jenifer Kent and Michael Erlich and their daughter Zoe live on a small lot in Oakland where they grow food, raise chickens and bees, and practice a variety of homesteading arts. "We're interested in living as sustainably as possible and growing our food as much as possible," Jen explains. "It's something we're trying to put into effect every day, by gardening and farming, canning and preserving, thrifting, making our own stuff, mending and repairing all of what we have." Their desire to move to a larger rural property was thwarted by the economic downturn, so they find themselves tailoring their homesteading to the limits of the city, and making much of what they have in the process.

Their homestead is less than 8 percent of an acre with approximately 1,000 square feet for growing food. "The major limitation of our site is how small it is—we've really had to be creative with what we have," Michael says. "We've built raised beds, have trellising and hanging plants all around, and we are constantly trying to figure out other efficient ways to use the space. We've been able to grow a lot in a very small space using the square foot gardening method, limiting space between plants in the beds to increase the growing space we do have. In the future, we plan to espalier trees up against the fence and trellis berries everywhere. We're trying to maximize the space, and balance that with wanting it to look nice and be a good place to hang out and where our daughter can happily play."

They follow some strict rules for planting in the garden: "Only edible plants. We've got medicinal herbs and fruit trees and vegetables, chicken and bee forage plants. There's a little ravine—the wilds—in the back of the property, which creates diversity in animal and plant life. Raccoons, squirrels, possums, and other varmints come to visit, and there's some great wild plants like chicory, cleavers, arugula, cress, and claytonia (miner's lettuce) back there, as well as greens and berries planted by previous tenants."

Good dirt was an issue when they bought the property. The previous owner had used the site as a landfill and had dumped automotive waste on it. When Jen and Michael started digging a garden, they found car parts and other debris in the dirt. They dug about three feet down, renewed the garden with organic soil, and put the toxic dirt in a remediation pile in the back of their property. "We're interested in more aggressive remediation, maybe using mushrooms, but certainly using time," Jen says. "The chips we use in the garden have a lot of spore in them, so they grow massive amounts of mushrooms that also help remediate the soil. We also have chicken poop, which really helps the soil."

A typical "day off" is filled with homesteading tasks. "This Sunday, we woke up, ate breakfast (fresh eggs and home-baked bread), went outside to survey our plan, finished building new raised beds, filled them with dirt and compost, weeded and cleaned out areas to prepare for new seeds, put an apple tree in the ground to espalier on the trellis, put a lemon balm

in the ground, and moved rock to make a nice sitting space," Jen describes. "We cleaned out the chicken coop, gave them new wood shavings, collected eggs, and changed their water. We trimmed the bush around the chicken coop to make more sunlight on the space, and cleaned out areas around the coop to create a new run for them so that when we plant new seeds they won't be in garden so much. It's a pretty busy place."

Apricots in early spring.

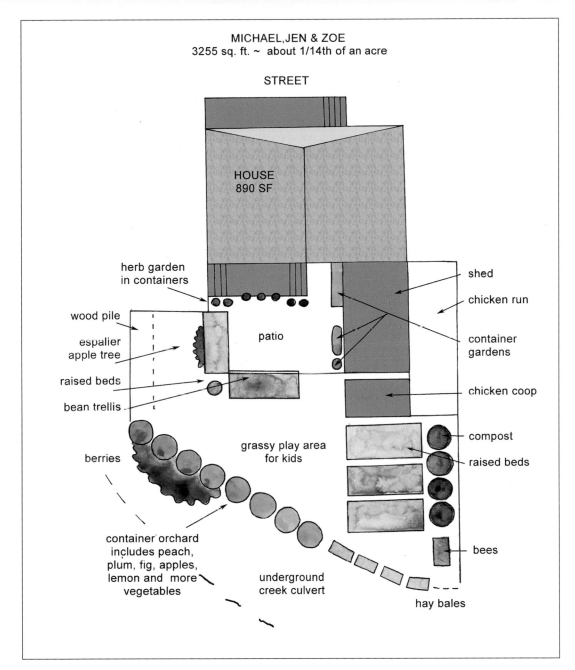

MICHAEL, JEN & ZOE
3255 sq. ft. ~ about 1/14th of an acre

STREET

HOUSE
890 SF

herb garden
in containers

shed

chicken run

wood pile

espalier
apple tree

patio

container
gardens

raised beds

chicken coop

bean trellis

berries

grassy play area
for kids

compost

raised beds

container orchard
includes peach,
plum, fig, apples,
lemon and more
vegetables

underground
creek culvert

bees

hay bales

After three years of being "not very successful," their garden is beautiful and full of food and trees and herbs. "We had an amazing garden last year for the first time," Jen recalls. "During our first couple of years we didn't know what to do and we had to figure it out. We made a lot of mistakes. To new people starting out, we say: Be patient. It takes a little while to figure out what you want and what you can do and how you can do it where you are. It's always a work in progress."

Urban Orchard: Growing Fruit in the City

Fruit trees are worth the commitment—they provide food, forage for beneficial insects, shade, and beauty in the garden. Fruit trees are perennial and provide a lot of food for many people over many years. An urban orchard is surprisingly easy to tend. Many fruit trees grow happily in pots for many years, or thrive in small, out of the way places. Columnar trees (pruned to grow straight, rather than branch out) are easy to set into small places at the top of a driveway or the corner of a deck. An espaliered tree won't take up much space, provides shade and privacy when flattened up against a fence, and yields a bounty of fruit. Planting fruit trees among other plants, in guilds or throughout the garden, is part of a perennial strategy that evolves into lush urban gardens. Trees need water, but not as much as an annual vegetable or herb, and are usually satisfied with a good drenching once or twice a month. Fruit trees planted in the ground can do with less water than trees planted in pots. Once established, a tree in the ground will live and produce for many years.

Some Basics for Tree Selection

1.   Pick the right tree and plant it well. If your tree doesn't have full sun or decent drainage or if you pick a tree with bad structure, taking care of the tree will be harder in the long term.

2.   Fruit trees should be planted in places where they will get regular water but not too much water; planted high so that the root flare (at the bottom the tree where the roots diverge) should be visible; and planted in full sun for most of the day. If possible, trees should be planted 10 feet away from buildings or other trees.

3.   Trees should be chosen which are appropriate for their location—check the chill hours, the pollination needs, and frost tolerance for your area before selecting a fruit tree.

4.   Good nutrition, good placement and good structure are the three most important things to think about in keeping your trees healthy. Are the trees getting the nutrients they need? Are they in a good place with plenty of sun and water? Do they have good branch structure? Fruiting problems always follow from a deficit in one of these areas.

Pruning Trees Planted in the Ground

Pruning fruit trees is important in an urban orchard where space is at a premium. Prune in mid-winter to support the shape of the tree, and in summer to minimize tree growth and encourage earlier and heavier fruiting. Pruning early in a tree's life for good structure makes it easier to deal with later on. For the task, you will need a one-sided pruning saw with a curved blade, a folding saw, and single-cut pruners, depending on the size of your tree.

1.   For the strength of the branches and the health of the tree, create 12-24 inch spaces between branches.

2.   The first branch should be at least 3 feet off the ground and should be trained to be as horizontal as possible. Shape a tree in its early years by hanging a weight from a branch to separate it from a nearby branch, or using a piece of wood between two branches to push them away from each other. This is a gentle way to give the tree direction as it grows.

3.   Most pruning should take place when the tree is dormant or after it has fruited and before it loses its leaves. If you prune while a tree is fruiting or losing leaves, it makes it susceptible to illness. If a branch breaks at any time of the year, remove it. Vigorous trees should be pruned in the summer, but a weaker tree should be pruned before it buds (between February and April, depending on your location).

The central stem of an espaliered tree is trained vertically against a wall, and lateral branches grow out at right angles to the central trunk.

The espalier, small pyramid, and fan shapes often work best in the city because they use the least amount of space. The espaliered tree is flattened against a wall by training the branches to grow on the vertical plane when tied to a preset trellis or fence. When pruning, remove branches that do not flatten against the surface where the espalier has been planted. Attach remaining branches to trellis or fence and encourage them to grow along these supporting structures.

The pyramid shape is a basic Christmas tree shape. Side shoots are kept very short, which make them less likely to break under the heavy weight of fruit. When pruning, imagine the tree larger and fully formed. Be ruthless in removing small branches that are not essential for the shape you want the tree to have, that are too close to other branches, or that point inward or at odd angles.

The fan shape is pruned to spread the branches wide and flat. Allow only two shoots of the young tree to grow at a 135° angle to the ground. In early spring each year, let a suitable number of branches shoot from these to form your fan. This shape is easy to train against a wall or a fence. Trees in a pyramid shape can be planted as part of a guild in the center of a garden bed. If you have enough space alongside a fence or property line, plant an espaliered Belgian fence (trees planted to cross at at right angles to one another) to form a living, fruiting garden fence.

Caring for Trees in Containers

Some fruit trees are grafted onto dwarf stock, which means they won't grow above a certain height—these are often ideal for smaller spaces, as well as for long lives in containers, barrels, or large pots. When keeping trees in pots, fertilize them and prune them each year to keep them happy in their small spaces. Trees in pots need

more fertilizer, more water, and should be kept smaller than trees planted in the ground. A tree in a container should not be allowed to grow out of proportion to the container itself. A 12-foot tree in a 3-foot container will not thrive. Be vigorous and regimented about fertilizing and cutting back your trees when they live in containers. You may need to work with their roots every few years or so, tipping them out of the pots, loosening the roots as much as possible, and replanting them in a bigger pot. When potting up is no longer feasible, the roots can be trimmed and the tree replanted in the same pot with fresh soil.

Another great urban project is a community orchard, managed by a few devoted arborists, but planted throughout a neighborhood or city. Sometimes it just takes a few motivated people to turn an empty yard, street median, or abandoned alley into a food-growing zone. "Borrowing" people's yards and tending them in exchange for some of the shared bounty is a strategy for

Urban orchard growing in containers along the edge of a backyard.

Petaluma Gage plum scion grafted to plum rootstock. Rootstock can accommodate numerous grafts of different cultivars, creating a "fruit salad" tree.

planting an urban orchard that could eventually feed many people locally and provide an urban arborist with a living.

Check out www.commonvision.org for orchard-planting resources. These folks will come to your town in their painted bus loaded down with fruit trees, and plant 50 to 80 trees in an afternoon in places where they will flourish and grow. These events are affordable and great community builders, drawing in children, adults, and elders in the work of creating the regenerative future today.

Grafting

The purpose of grafting is to get the fruiting spur (or scion) of one tree to grow into the rootstock of another by bringing the cambium layers of each into contact with one another. The cambium layer is the growing part of the tree, the whitish area just under the bark. For successful grafting to take place, the vascular cambium tissues of the stock and scion plants must be placed in contact with each other. Both tissues must be kept alive until the graft has taken, usually a period of a few weeks.

If you have a large fruiting tree already in the garden, you can graft scions of different varieties onto the rootstock of that tree. You can only graft a scion onto compatible rootstock, such as plum scion onto plum rootstock, and you can graft numerous scions onto one tree, creating a "fruit salad" with different varieties emerging at different times of the season. This is a good way to increase the diversity of plant varieties in a small garden. Grafting creates and supports new relationships and flow between different elements and forms—in this case, scion wood and rootstock.

Grafting's got a good metaphorical leaning too. As Trathen said, "You graft a new habit or practice. What is the range of fruits and flavors and scents you need to make your system more resilient and diverse? What new skills, ideas, visions, and mantras do we need to graft into our own lives? We need to have that good, clean, quick connection between our source and what we want to graft. It's all about joining the layers so that the energy flows."

Cleft grafting is a simple way to join scion to rootstock in an urban orchard.
1.   Saw through all the main branches of the rootstock a foot from where they join the trunk.
2.   Take a sharp cleaver and use it on one of the sawn-off stumps to make a cleft in the stump. Cut two scions to a chisel shape at one end.
3.   Force open the cleft and insert the two scions into this, lining up the cambium layer of scion and old wood. Let the cleft close up, thus clenching the scions into place. Pour hot beeswax all over to protect cut surfaces.[50]

Natural Pest Control

While insects can be beneficial, when out of balance they can also be hazards in the garden. Your best protection against pests is healthy soil and strong hearty plants. Interplanting and companion planting strategies also

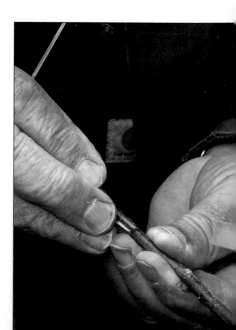

Cleft grafting.
Photo by Trathen Heckman/Daily Acts

provide natural pest control. While waiting for your pollinator and beneficial insect beds to mature, there are some other low-level, nontoxic ways of dealing with pests in the garden. Unless the pests are completely decimating your plants, sometimes the best option is just to share. Your greens might have a few holes in them, but that just proves they are organic.

1. Hand-picking insects and other pests off plants is an easy, front-line defense. Snails can be handpicked off plants every night for up to two weeks and exported to interrupt their breeding cycle. They can also be fed to the chickens, which snap them right up.

2. All-Purpose Spray (works against many pests, including slugs and Japanese beetles).

    1 garlic bulb

    1 small onion

    1 tsp cayenne pepper

    1 quart water

    1 tablespoon mild biodegradable liquid dish soap, or Dr. Bronner's soap.

    Chop garlic and onion in the blender. Add the cayenne pepper and water. Steep for one hour and strain. Add dish soap so that spray will stick to the leaves. Mix well. Spray mixture on both sides of leaves. Do not spray greens or plant parts you want to eat. Store mixture in the fridge or it will go bad. This also works with just garlic. You can also make a mixture of hot pepper and powdered garlic and store it until you need it. Mix with water, spray the leaves, and wash the leaves off before you eat them.

3. Hot pepper spray, a mixture of ¼ cup hot peppers and two cups of water, can be blended, strained, and applied to plants every day for five to seven days until pests are gone.

4. Watering in the morning or at soil level, instead of overhead watering, can also help with a number of pests and disease issues. It gives leaves time to dry, which helps prevent fungal diseases that thrive in damp conditions. It also gives things a chance to dry out before sundown, thus preventing slugs and snails from traveling (which they like to do at night, in moist conditions).

5. A jar lid, saucer, or other shallow container settled into the soil then filled with beer will attract and drown many pests, especially earwigs and slugs.

6. A short section of old hose or a rolled up newspaper will attract nighttime marauders like earwigs. These can be collected and moved well away from the garden in the morning.

7. A board laid on the soil with a little bit of crawl space is good for collecting slugs and snails. In the morning, they can be gathered and fed to the chickens.

8. Birds eat lots of insects. Providing a birdbath may attract them. A birdhouse or two will encourage birds to stay and pick off insects for you. (They will also eat your peas or beans when you plant them, so you'll have to cover those plants with netting until they sprout.)

9. Lizards, frogs, and toads are great insect catchers. Make them feel welcome. If they have a favorite place in a pile of pots, on a pile of rocks, or in a water trough, let them claim it as home and don't disturb them.

10. Plant a little extra for the bugs so you can share your bounty.

Maybe most importantly, we need to change our attitude about pests. We've lived at war with our environment for too long, and the results of that are staring us in the face. In whatever small ways we can unlearn our ill-conceived war on nature on our own homesteads and relearn ways to live in relationship with other creatures—pests, predators, or germs—the more generative and long-lasting our living systems will be.

Life Is with Creatures

The greatness of a nation and its moral progress can be judged by the way its animals are treated.

—*Mahatma Gandhi*[51]

A human being is part of the whole universe… We experience ourselves, our thoughts and feelings as something separate from the rest, a kind of optical delusion of consciousness. This delusion is a kind of prison for us, restricting us to our personal desires and to affection for a few persons nearest to us. Our task must be to free ourselves from this prison by widening our circle of compassion to embrace all living creatures and the whole of nature in its beauty.

—*Albert Einstein*[52]

Human/animal relationships are ancient stories of companionship and compatibility, and have evolved since the beginning of human agricultural practice. There are two contemporary stories about human/animal relationships: One is called "Capture and Enslave," and it's the story of industrial animal husbandry. The more we hear this story, the less we want to eat. The other story is called "Mutual Benefit," in which animals get shelter, food, health care, and love and in exchange offer healthy food for humans, fertilizer, and friendship. This story's way more fun, and way more delicious.

Animals turn a backyard garden into a farm. They bring a deeper level of complexity and relationship to our homesteads, along with their unique and incredible animal natures. Any animals you have will make a major contribution to the fertility of your garden. They swiftly turn backyard garden scraps into compost, and provide your homestead with some of the most nitrogen-rich fertilizer you won't ever have to buy again. And if you add rabbit and

Walking a goat in the neighborhood. Good for the goat, good for the girl.

chicken meat and eggs to the vegetables and fruits you produce, you are surely moving in the direction of greater self-sufficiency, while providing yourself with a familiar and varied diet. Many fear that animals will tie them down—no vacations or going away for the weekend. But if you keep animals cooperatively with neighbors, or agree to take turns caring for each other's livestock, these daily tasks are easily managed.

This chapter is the lowdown on the animals we live with and how to best house them, care for them, and forge reciprocal beneficial relationships with them. There are a few obvious choices for small animal husbandry in the city. Bees, chickens, ducks, rabbits, quail, and the occasional goat rise to the top of the list. This chapter also includes information on how to butcher animals. We weren't too keen at first on killing the animals we cared for. Now that we've tried it, we can safely say that if everyone looked their dinner in the eye before they ate it, this would be a different kind of world.

Considerations for Animal Care

When you're thinking about taking on any animal at your homestead, you'll first want to consider a few important things, since all animals require a certain quality of care, and it's not advisable to take them on if you can't offer it to them. Most barnyard animals need one another to thrive, so make sure you have space for more than one when you are planning to take them on.

Shelter/housing/protection. All backyard animals need predator-proof homes that are dry, clean, and spacious enough to meet their needs. Chickens need more space than quail or rabbits, and goats will need more than chickens. If you are breeding animals for food, you may have to provide separate housing for differently sexed animals until you bring them together to breed. Most animals prefer to have their houses out of direct sunlight, in close proximity to places where they can roam freely for part of the day, and lifted off the ground to provide protection from predators and rain.

Nutritional needs and food sources. On most urban homesteads, it will be unrealistic to imagine that you could produce all the food your animals will need, though some of them, like ducks and chickens, supplement their food with snails and slugs and bugs from the garden. Other animals, like goats, chickens, and rabbits, will eat kitchen and garden scraps and convert them into fertilizer for the garden, but this won't be enough nutrition for them overall. Some urban farmers feed their animals by dumpster diving. This may be a fine short-term solution for feeding an animal you are going to butcher at the end of a season, but maybe not the best choice for a long-term relationship. Make sure you have access to the proper kind of feed at a price you can afford before you take on an animal that's going to eat every day. Different animals have different nutritional needs—providing the animals the food they need to keep them healthy and productive is key.

Composting. With any animal, you're going to have to figure out what to do with their waste. Fortunately, the waste of many backyard animals provides excellent fertilizer for the garden. Chicken poop needs to be composted for some time before being added to the garden; rabbit poop can be integrated into the garden without a stint on the compost pile, and goat manure is also excellent for the garden as is. Ducks poop in their watery habitats; when the pond needs to be cleaned, this water can be added to the compost pile to add to its fertility.

LIFE IS WITH CREATURES

The ease of mucking out your animal's home should be considered. A rabbit hutch with a wire bottom is simple: just let the poop drop on the ground underneath the hutch, rake it up, and put it in the garden. A chicken coop can also be designed with wire flooring to serve the same function. You will be glad to limit the number of times per year that you have to walk into your chicken coop to clean it out. Pre-built structures can also be adapted to make cleaning and composting an easier task.

Legality Before you start keeping farm animals in the city, get familiar with city ordinances regarding small livestock in your area. Some places allow chickens, but not roosters; some places forbid bees. Many urban centers balk at goats or pigs but are happy enough to house the rabbit, inside or outside the house. Check with your local municipality to find out what's allowed in your area. Use your best judgment coupled with local ordinances to make decisions about how many animals to keep. More important than the law, though, are the needs of the animals themselves. Too many animals shoved into a small space makes for unhealthy conditions and unhappy animals, the exact opposite of kind of quality living we want to provide.

Neighbors. Keep your neighbors in mind when getting backyard animals. If your yard is small, chances are that a rooster under a neighbor's window won't go over too well. Animals that emit a strong smell aren't usually much appreciated either. But for the most part, neighbors are excited and interested in animals, and they are a great way to make good relations. Very few people say no to a fresh egg, or a jar of honey or even the chance to watch a chicken do her thing.

Alliances with other community members. Keeping animals is a pleasure, but it is also a daily chore. Some animals, like goats, require more care, and sometimes the best way to provide it is to share the work with others. To meet this need, in Oakland, California goat-sharing cooperatives are springing up and providing the people who own goats some help, while giving members the opportunity to learn how to take care of goats and enjoy their milk products.

Time and attention the animals will need. Once an animal's home is established, its care doesn't take much more than minutes a day. Make sure you assess your own time limitations before you commit to a new animal on your homestead. If you travel a lot, a good way to manage your animals is to share their care, feeding, and produce with neighbors and friends. Three rabbits really don't take much more time than one rabbit, but breeding rabbits and taking care of the babies takes extra time. The same with chickens—the size of flock doesn't add time to your day, but if you are taking care of small chicks that need more attention to survive, factor that in when you start your animal-keeping project.

Predators and pets. Some city neighborhoods have an abundance of predators that can be dangerous for your animals. A large gathering of feral cats will threaten small chickens, ducks, or other fowl. Dogs can also be a problem for smaller barnyard animals and some are wired to grab a chicken by the neck and shake it, even when they know you love the chicken too. As mentioned before, the design of the animal's shelter should reflect an awareness of opportunistic predators like raccoons and possums, as should your habits for bringing your animals in and securing them at night.

on the ground: green faerie farm

When Jim Montgomery and his roommates went looking for a home, they focused on houses with large lots because they planned to do a lot of growing. They bought Green Faerie Farm in 1995, a 50-by-200-foot lot, where they raise much of their own food, eggs, and meat. The collective household shares gardening tasks, but Jim is at the center of the animal husbandry at Green Faerie Farm, and he's been one of the inspirations for small animal husbandry in Berkeley and Oakland, CA.

GREEN FAERIE FARM
50' x 200' ~ 10,000 sq. ft. ~ just under 1/4 of an acre

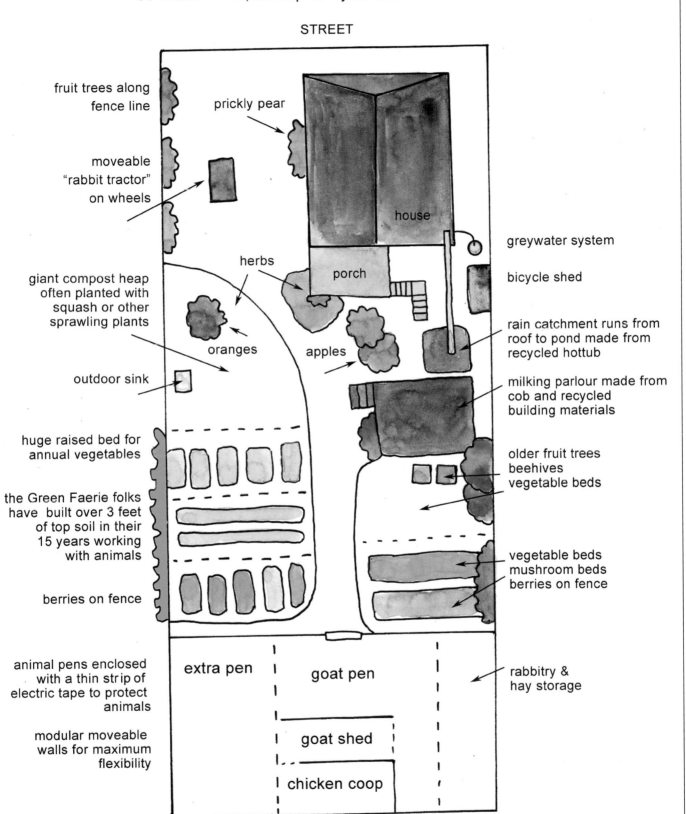

STREET

fruit trees along fence line

prickly pear

moveable "rabbit tractor" on wheels

house

greywater system

bicycle shed

herbs

porch

rain catchment runs from roof to pond made from recycled hottub

giant compost heap often planted with squash or other sprawling plants

oranges

apples

milking parlour made from cob and recycled building materials

outdoor sink

huge raised bed for annual vegetables

older fruit trees
beehives
vegetable beds

the Green Faerie folks have built over 3 feet of top soil in their 15 years working with animals

vegetable beds
mushroom beds
berries on fence

berries on fence

animal pens enclosed with a thin strip of electric tape to protect animals

extra pen

goat pen

rabbitry & hay storage

modular moveable walls for maximum flexibility

goat shed

chicken coop

Jim Montgomery (and baby goat). "Living with goats is about having a dialogue with them, a negotiation. You have to check in about what the goats need in a way you don't with a chicken or a rabbit. They are very smart and loving and funny."

Jim has lived and worked with animals since he was a teenager. "I've always had urban livestock and learned pretty quickly how to both raise and kill them. This gave me an early understanding of death as a part of the cycle of life, how no one lives without other things dying. There is a balance in nature, and it can be cruel or it can be kind. I prefer to make it as kind as possible." In addition to raising urban livestock, Jim also instructs other homesteaders in the art of conscientious animal harvesting. As a teacher, "I'm really into creating space where the animal is respected, so that's one of the things I teach to the interns, and my students—respect for life."

His attention is always primarily focused on the animals and what they need. "If you're going to raise animals in the city, you can do it in a small space. You can raise quail in a 3-by-3 closet with a good lamp. If you are growing sprouts in the window and sharing them with the quail and maybe even giving them some window time, that would be ideal. Chickens need more space than quail, and goats need more than chickens. It's important to make sure the animals have what they need."

Green Faerie Farm currently houses five humans, three adult and six baby goats, 28 chickens, and over 20 rabbits. Even so, "raising animals doesn't feel like a lot of work," Jim said. "Going to an office and earning someone else's profit for them, that's work. I get a lot of pleasure and restorative and mental health in working with the animals. Being in this yard, in a place where there's food growing, where there are happy animals, where we have all the things we need to live well, feels empowering and restorative to me. There's so much pleasure in it.

"Sometimes I say my garden is my altar," Jim says. "If I'm going to swoon, it's going to be with the earth, and the regenerative healing energy of the earth. My garden and my animals are a place where I commune and build relationships and understanding and independence."

The closed-loop system created by animals on a small urban homestead is working at Green Faerie Farm. The yard is lush and abundant, filled with food for people and animals. The animals, in addition to making compost for the garden, are also involved in balancing its ecosystem. "We find snails and put them in the pens with the chickens, and they eat them right up. Our excess food goes to the animals, and the animals provide richness for the soil, and the extra carbon materials—like straw bedding that we take from the waste stream and use and then recycle into the garden with animal poop in it—also help the garden. The animals bring efficiency to our site; we don't need to fertilize our gardens beyond what we get from them. They are part of rebuilding topsoil and making a really healthy garden."

No isolated plant or animal systems prosper the way they do in a closed system. Healthy plants need to be married to some sort of animal system. "A lot of people would turn those fava beans into green manure, but we will actually eat them. The crop residue provides nitrogen to our animals, so we derive benefits from growing the plants—for the animals, for our bodies, and for the garden." The animals also support productivity in the garden. "Tomatoes are hard to grow in this microclimate. One year we placed a cold frame over our tomato bed with rabbit cages integrated into this mini greenhouse. We gave the rabbits an escape hatch so they could get out if it got too hot in there. The rabbits provided heat and extra carbon dioxide for the plants. We harvested our first tomatoes in April, and our last ones in January that year. We were proud of those tomatoes!"

Is this sustainable living? "Sustainability has to do with a whole society closing the loops of waste and consumerism. There's no sustainability for one without sustainability for all. I think we are modeling more sustainable living, but *sustainable* means that you don't have any waste and all your actions are going toward building healthier, stronger, more resilient systems. Some people fantasize that you are only taking out what you put in, but I think you have to err on the side of building the system rather than just breaking even. You have to design toward increasing the resources in the system because if you try for an exact balance, we're still running a deficit. This is one way the animals are so important—they give back so much to the system. Green Faerie Farm is a step in the right direction. But in the end, it comes down to changing the culture."

Bees with full pollen sacs entering the hive.
Photo by Rhan Aladjim

Bee Kind

Beekeeping is thrilling—the opportunity to have contact with a hive, to catch a swarm, to gather honey and beeswax, and to engage with the wild busy entity of the hive mind is awesome. Standing in the middle of a bee yard or in the middle of a swarm of bees, you can feel the electric energy of the hive as it moves from place to place. Bees are important to us because of their pollination activities, but they are also an amazing life force we are privileged to encounter when we put on the veil and become beekeepers.

Bees are a true addition to the homestead because they offer pollination for plants and food for humans; in exchange, beekeepers and gardeners can offer healthy opportunities for home and forage to an endangered and essential pollinator. Bees forage within a three-mile radius of their hive, so they pollinate your neighbor's gardens as well. Bees are feral and untamed; they bring wildness into the garden. Bee venom has healing properties for people with autoimmune illnesses including arthritis, multiple sclerosis, Crohn's, lupus, and rheumatoid arthritis, and local honey can offer some relief from pollen allergies.

The hive is a complex, integrated entity that includes a queen bee, worker bees, and drone bees, each with its own specific task. In keeping with the principle of designing from patterns to details, every hive organizes around a certain set of tasks, but each hive develops in its own way. Every time my beekeeping partner comes back from a visit to a hive, he has a different story to tell about what's going on inside. As we try to puzzle it out, he invariably quotes one of his beekeeping mentors, "The bees just don't read the same books we do!"

While beekeeping is relatively easy to get started on, it is also an art and a lifelong practice. Next to humans, bees are the most studied and talked about creatures on earth. There is a lot of information out there, so we recommend that you read books and seek community to support you, as countless questions about what's going on in the hive will come up along the way. Beekeepers are a garrulous and opinionated lot who like sharing resources, swarms, information, and bee plants. Many places have local beekeeping associations that are well worth joining and that will hook you up with resources, information, and community. As with anything, learning from someone who knows is the best practice.

Site, Sun, and Water

Beehives can be sited in a backyard, on a roof, or in an empty lot. Bees use the sun to direct their flights into and out of the hive and thrive in a place with morning sun and afternoon shade. The entrance to the hive should be south facing, and set in such a way that the bees' flight path into and out of the hive is not in the middle of human pathways through the garden. Bees forage for pollen from a variety of plants and are happiest with patches of plants they prefer, rather than single plants. They appreciate a water source during the hot season, especially if you live in a region with long dry spells. If you place a bucket in the shade with some water in it, make sure you place a stick or some sort of ledge in the water where they can stand and drink without falling in. Bees don't present any kind of composting issues, though you will undoubtedly find yourself with some residue of beeswax once you start harvesting it from the hive.

31 Uses for Beeswax

(Besides making candles, of course!)

1.   Unstick a drawer. A thick coat of beeswax on wooden rails makes the wood drawers slide smoothly.
2.   Lubricate window sashes.

3. Free up rusted or stuck nuts by lubricating the bolt's threads with melted wax.

4. Wax wood. For structural elements that need to look good but take no wear (such as exposed ceiling beams), heat equal parts beeswax, linseed oil, and turpentine. Apply with a burlap rag while the mixture is still warm.

5. Preserve bronze. To ward against oxidation caused by moist air, brush on a solution of 1/3 pound beeswax melted in 1 quart turpentine. Buff it with a towel to create a thin, hard coat.

6. "Whip" frayed rope. Wrap a waxed length of string tightly around the rope's tip about a dozen times. Tie off the loose end and trim the excess.

7. Lubricate nails or screws to make them drive smoothly and resist corrosion.

8. Condition a wooden cutting board. Add half-teaspoon beeswax to a cup of mineral oil, heat on stove or microwave, and apply the mixture to the board with a soft cloth.

9. Polish concrete counters. Rub melted beeswax over the surface with a chamois cloth. Let it dry and then wipe.

10. Preserve a patina. Seal a copper sink by rubbing it with softened beeswax and polishing off the excess with a lint-free rag.

11. Waterproof leather. Combine equal parts beeswax, tallow, and neat's-foot oil. Warm the mixture and use a rag to rub it on your work boots or gloves.

12. Use as a resist for batik.

13. Use as sealing wax.

14. Seal and protect tree wounds. Melt and brush on.

15. Grafting wax. Melt equal portions of plant resin and beeswax in a double boiler. Allow the mixture to cool and roll it out into sticks. Wrap in wax paper and store in a cool, dry place.

16. Salve. Infuse olive oil with calendula or comfrey and combine by heating with beeswax, 1 part beeswax to 8 parts oil. Add a few drops of essential oil such as lavender.

17. Wax thread for easy threading of a needle.

18. Use as a resist for dying fancy Easter eggs.

19. Make oilcloth (see waterproofing leather above).

20. Finish and protect ironwork. Mix 1 part mineral spirits and 1 part linseed oil, then melt in beeswax. Add enough beeswax to give the mixture a consistency of motor oil.

21. Mix together equal parts of melted beeswax and honey for a good home remedy for the cracked hooves of animals. Clean and dry the crack before applying the mixture.

22. Make crayons. Melt equal parts grated soap and beeswax in a double boiler. Color with dry artist pigment or food coloring paste. Pour into molds.

23. Dental Floss. Coat thread or string with a mixture of beeswax, jojoba wax, carnauba wax, myrrh, and propolis.

24. Mouthpiece for didgeridoos.

25. To start and care for dreadlocks. Melt and work into dreads before the wax hardens.

26. Ear candles. Cut strips of muslin. Dip into beeswax and roll around a cone-shaped form.

27. Cheese wax. Mix with a small amount of olive oil so wax stays flexible when cold.

28. Moustache wax. Heat and mix equal parts beeswax and Vaseline. You can adjust the stiffness by increasing or decreasing the amount of Vaseline. Add essential oil such as wintergreen.

29. Earplugs. Mix beeswax with cotton and lanolin.

30. Lost wax casting process for jewelry or sculpture.

31. Fire starters. Dip small pieces of wood, pinecones, cloth, or cotton balls into beeswax.

The Law and the Neighbor

In most places, it is perfectly legal to keep bees, though in others, an absence of laws about beekeeping can be taken for consent. When people are concerned about bees, it is usually because they think of them as stinging hazards, and do not understand the benefits they provide. You do the bees a service by educating people about the gentle, non-aggressive nature of the bee, who only stings in defense of her babies and her food. All female bees have the capacity to sting, but no male bees can sting, and most bees, when left alone and unthreatened, will not sting. If a bee approaches you, it is possible that you smell good to her, like a flower. If you walk away calmly and avoid making abrupt movements, you are also unlikely to be stung.

If you have a neighbor who is truly allergic to bees, some caution may be warranted when placing your hive, but otherwise education and the sharing of local honey is often enough to help the neighbors relax and enjoy the hive. Bees have few predators, unless you live in a city with bears, and dogs and cats tend to stay away, especially if they've been stung once.

Equipment

A beekeeper needs a few pieces of equipment to maintain the hive successfully—a bee veil, a hive tool, a bee suit, gloves, a brush or a feather, and a smoker. Some beekeepers work without the gloves or the smoker; some eschew the veil. In the beginning of beekeeping practice, before you're really accustomed to the hive, do what makes you feel comfortable and unafraid of the bees. Your sense of confidence with the bees will help the bees feel confident of you.

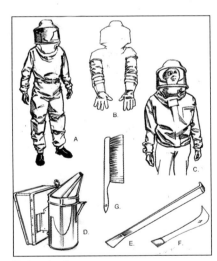

Beekeeping Gear
A. Full-body bee suit
B. Gloves
C. Half jacket & veil
D. Smoker
E. Top bar hive tool
F. Langstroth hive tool
G. Bee brush

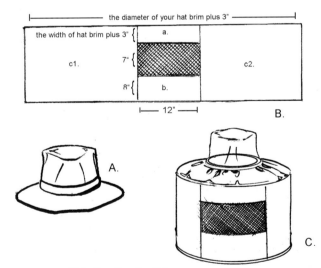

Make Your Own Bee Veil
A. Hat
B. Bee veil layout
C. Finished veil with elastic at top.

How to Make a Bee Veil

A bee veil is an important piece of beekeeping equipment. A veil will protect you and help you feel calmer around the bees. The bees will respond to your state of mind, so it's good to be as calm as you can when you approach the hive. You can buy a veil through a beekeeping supply store, but here's a simple way to make one yourself. (Please see illustrations on previous page.)

1. Start with a sturdy, breathable straw hat with a medium sized brim.
2. Measure and cut your cloth and mesh (see drawings for measurements). Cloth can be any light colored cloth or canvas. Mesh can be light plastic screen door mesh, available at a hardware store.
3. Sew section a & b on either side of the mesh to make center panel.
4. Sew side panels c1 & c2 to middle panel.
5. Sew together ends to make tube.
6. Fold over top and bottom edges twice and sew to make a tube approximately 1/2 to 3/4 inch wide, leaving a gap in your hem to feed in your string or elastic.
7. Feed string or elastic into the top and secure to fit tightly around brim of hat.
8. Feed string into bottom tube.
9. To wear, pull string tight around back. Pull under arms and tie around chest in front.

The Hive Box: Langstroth or Top Bar?

People have kept bees in different kinds of vessels for many thousands of years. About a hundred and fifty years ago, the discovery of "beespace" led to the development of hives with moveable frames. Even in nature bees leave a measurable and consistent amount of space between each paddle of comb. Moveable combs meant that the beekeeper no longer had to destroy the bee's work inside the vessel to manage the bees and gather the honey. Previous to this, harvesting honey or beeswax meant destroying all or part of the work of the hive.

While there are many types of beehives, the two most common ones are the Langstroth—those inimitable white boxes we are all familiar with--and the Top Bar Hive, which can be made in a variety of shapes, but is generally long and narrow. Both hives work with beespace. The Langstroth boxes are the dominant style in the U.S.and can be easily purchased from any beekeeping supply. In this system, the bees build their comb inside rectangular frames on a pressed wax or plastic foundation that shows the bees where to build their six-sided cells. Because it is the dominant system in the U.S. it is well understood and there is a lot of support available.

Langstroth hives meander down a San Francisco hillside.   *Photo by Philip Gerrie*

A Top Bar Hive apiary in an urban backyard.

The Top Bar Hive, used by the Peace Corps as a low start-up-cost cottage industry for African villagers, is gaining popularity among natural beekeepers in the U.S. Not readily available yet commercially, they are inexpensive and easy to construct with minimal carpentry skills. In this system, the bees build free-form comb off bars that sit across the top of the opening of the box.

As we have already said, beekeepers are an opinionated lot and there are strong opinions about which system is better, both for the beekeeper and for the bees. The Langstroth set-up is more traditional and familiar to the western beekeeper. Advocates insist that bees prefer building vertically and like being able to recycle the honeycomb on the frames when they harvest their honey (in the Top Bar system whole combs are crushed, in the Langstroth system just the caps are cut off and the honey is expelled with a centrifuge). Top Bar beekeepers like the low-tech, DIY nature of their system and believe it is healthier for the bees to manage things more naturally. Only the bees know for sure what feels best and they're too busy gathering pollen and turning it into honey to be bothered to report.

Whichever style you choose we recommend finding other beekeepers keeping bees in this style as there are some differences in managing the two systems. A good beekeeper in either system hones their awareness of the season and the cycles of the bees and ultimately this will determine your success in beekeeping more than the shape of the box. Hives more often die due to inappropriate human interference than to lack of human contact. The bees already know the way.

How to Get Some Bees

There are three main ways to get bees: buy them, split them from an existing colony, or catch a swarm. Many apiaries sell "packages" available in the early spring. A package is a queen bee along with three or four pounds of worker bees. Do your research ahead of time to find a supplier in your area and order ahead of time as well, as many apiaries sell out by the time bee season begins. The cost to purchase bees is currently between $50 and $100 per package. If you are in contact with other beekeepers, it is also possible to get a "split" from an existing colony. A split is a quantity of fresh brood and nurse bees along with some pollen and honey in combs. The bees will raise a new queen when isolated from the old queen. Once you become more experienced at beekeeping, you can catch a swarm of bees if you find them clumped somewhere as they search for a new home.

Managing Your Hive

The bee colony has a natural ebb and flow through the seasons, and managing the hive should both reflect and support this flow. During the summer the beehive can have as many as 50,000 individual bees, and in the winter, when food resources are scarce, they downsize to about 15,000. Ninety-nine percent of the bees are female worker bees. In the summer there may be about 1,500 male drone bees, and generally speaking, there is just one queen.

Pulling frame out of Top Bar hive to check the brood, honey, and health of the hive.

How to Build a Top Bar Hive

One of the appeals of the Top Bar Hive is its affordability and the fact that with some basic carpentry skills, you can build one yourself. The Langstroth hive has more pieces, some of which are a bit challenging to build without some evolved carpentry skills. Here's a description of how to the build the TBH yourself at home. The total cost of building a Top Bar Hive should be less than $100. Design by K. Ruby Blume.

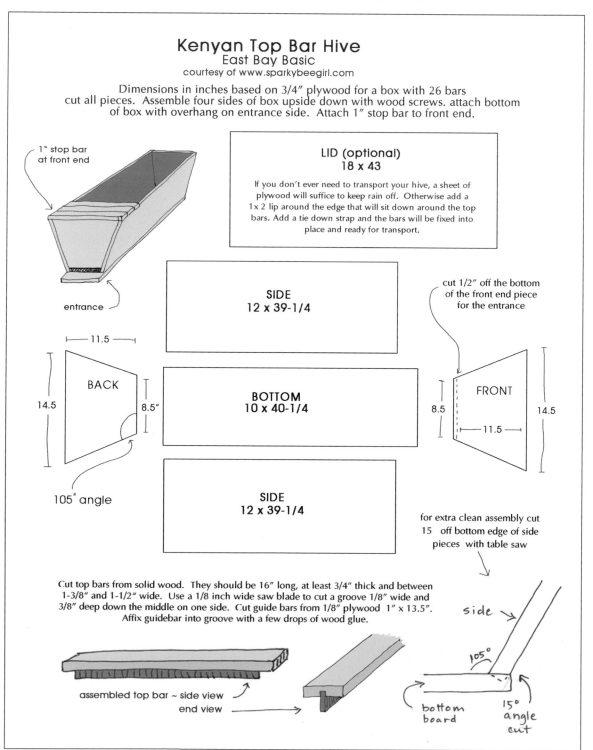

Kenyan Top Bar Hive
East Bay Basic
courtesy of www.sparkybeegirl.com

Dimensions in inches based on 3/4" plywood for a box with 26 bars
cut all pieces. Assemble four sides of box upside down with wood screws. attach bottom
of box with overhang on entrance side. Attach 1" stop bar to front end.

1" stop bar
at front end

entrance

LID (optional)
18 x 43

If you don't ever need to transport your hive, a sheet of plywood will suffice to keep rain off. Otherwise add a 1 x 2 lip around the edge that will sit down around the top bars. Add a tie down strap and the bars will be fixed into place and ready for transport.

SIDE
12 x 39-1/4

cut 1/2" off the bottom
of the front end piece
for the entrance

11.5

BACK

14.5

8.5"

BOTTOM
10 x 40-1/4

8.5

FRONT

11.5

14.5

105° angle

SIDE
12 x 39-1/4

for extra clean assembly cut
15 off bottom edge of side
pieces with table saw

side

Cut top bars from solid wood. They should be 16" long, at least 3/4" thick and between 1-3/8" and 1-1/2" wide. Use a 1/8 inch wide saw blade to cut a groove 1/8" wide and 3/8" deep down the middle on one side. Cut guide bars from 1/8" plywood 1" x 13.5". Affix guidebar into groove with a few drops of wood glue.

105°

assembled top bar ~ side view
end view

bottom
board

15°
angle
cut

To harvest honey from a TBH, comb is cut from the frame and mashed. The mashed-up comb is then strained to separate wax from honey. Traditional harvesting of Langstroth hive requires cutting the caps off the comb, then using a centrifugal force extractor, which "spins" the honey from the comb.

Your hive should be managed every two to six weeks during the warm season. For top bar hives, every three weeks is a good idea. For Langstroth boxes, every four to six weeks will suffice. Manage your hive on a warm sunny day in the middle of the day. Bees return to the hive at night and they should not be disturbed at night or during inclement weather. When looking though your hive, assess the strength and quality of the hive. Does it feel active, full of bees, and full of honey? Make sure there is a queen or evidence of a queen (eggs, larvae, or capped larvae, collectively called "brood").

In the spring you will be checking for the possibility of reproductive swarming and trying to prevent this, so as not to lose your entire population right when the nectar begins to flow. A swarm happens when the bees decide the conditions are right to send out the old queen with a good percentage of the hive population to form a new colony. The bees gorge on honey so that they can barely fly. Following the old queen, they leave the hive and land on an intermediate location, sometimes the branch of a tree, or a telephone pole, or a hole in a tree. When they are clumped in this intermediate location, you can catch the swarm by brushing them or knocking them into a box (either a transport box or the actual hive you want them to move into). When catching the swarm, the most important thing is to make sure the queen is in the new location—the rest of the hive will always follow its queen. If they are not captured in this way, the bees send out scouts to find a new home. With luck, this usually occurs within 24 to 48 hours.

One way to prevent swarms is to take a split before the bees create swarm cells (queen cells that look like peanuts built on the edge of the combs). To do this, take the old queen, along with a few frames or bars of honey and pollen, brood in different stages, and some extra worker bees and put them in a new box. The original colony will raise a new queen from a fresh egg. This most closely resembles their natural behavior.

Another vital time in the beekeeping year is at the end of the season (early fall) when you harvest. At this time it is important to judge how much honey your bees will need to make it through the winter (this varies greatly from California, where there is forage all winter long, to Iowa, with long, cold snowy winters). Be sure to talk to local beekeepers to find out the best practice in your area, since the loss of bees through winter starvation is devastating.

Beekeeping is a truly glorious undertaking, and offers a lifetime of learning. If you are interested in pursuing this as a path, there are many resources available, including books, online forums, and local beekeeping associations that offer classes for you to get hands-on learning before you start. Bees are amazing and mysterious and generous. Nothing is sweeter than honey from the hive, a cozy beeswax candle in the middle of winter, or the feeling you get when you interact with the hive. If you aren't up to beekeeping yourself, find someone who is, and get a chance to peer into the hive. It will teach you about cooperation, purpose, and love.

Chickens Are the New Black

The backyard chicken revolution is well under way, and for good reason. Chickens fit in small spaces, offer delicious, nutritious eggs and meat, make nitrogen-rich poop that is great for the compost pile, and are far more entertaining to watch than television. They can also be helpful in turning over the soil and making your garden

Making friends with the chicken.   *Photo by Daniel Miller/Spiral Garden*

A moveable chicken coop made from scraps of recycled lumber.

Coop constructed from recycled materials fits small space in urban backyard. *Photo by Dana Yares*

This narrow alleyway between two houses provides a safe home for coop and small chicken run. Covering the area with chicken wire prevents visits from predators and pigeons.

beds ready for planting. They don't take much work once you've got their home established, and most hens will lay an egg almost every day for at least three years before they make a really good soup.

Coop and Foraging Space

Figure out the size of your coop and flock by assessing how much space you have to house the birds in relation to how much space they need to live healthy, comfortable lives. Most breeds allowed to forage outside their coop require at least four square feet of space once they are full-grown. Eighteen chickens are too much for the average backyard lot, but three or four may be just right. The size of your flock will also depend on the size of your family and your appetite for eggs. Make sure your coop and chicken run will get both sun and shade during the day.

Hens prefer to lay their eggs in nesting boxes. Build one for every 4-5 chickens. A highly recommended design feature: make a hinged door outside the nesting boxes so you can reach in to get the eggs and avoid walking into the coop each day.

Chickens need room to forage during the day. You probably won't give your chickens free range in your whole garden (unless you want them to decimate any plant under 12 inches high, and you don't mind stepping on chicken poop whenever you go outside), but creating an exclusive enclosure gives them room to move and the ability to peck, scratch, perch, and roll in the dust, all of which keep them happy and healthy and satisfied.

A coop is essentially a simple box with a door on one side to let the chickens exit, and a hatch on another side where you can pick up the eggs. Make it out of recycled wood from the dump, or wood pallets, or an old packing crate. We built ours from the recycled wood from a fence we didn't want anymore. If you're lucky, you can find a used coop online, or trade your homegrown produce with someone who loves to build. Make it small and make it beautiful.

Food and Water

Chickens need lay pellets, crumble, and calcium (often in the form of oyster shells). We recommend organic feed, but it costs twice that of nonorganic. You can grow certain plants that supplement the store-bought feed and bring good nutrition to your birds. Any of the brassica family will be well loved by the chickens. Grow an extra row of broccoli each season, or a perennial tree collard will feed them for many seasons. They also enjoy purslane and comfrey, two very easy to grow herbaceous plants that provide good nutritional supplements. They love insects, slugs, snails, and your kitchen compost as well. In addition to the proper balance of nutrients in their food, chickens need water at all times. A waterer can be hung inside the coop, or outside under a tree, and is easy to source at a feed store.

Chicken Breeds

There are many breeds of chickens to choose from. Things to consider in choosing your flock include: level of aggression, good laying (even in winter), ease of handling, ease of flocking with different breeds, lack of desire to fly away, and aesthetics. Some breeds you might enjoy are Rhode Island Reds (prolific layers of good-size brown eggs), Ameraucanas (beautiful, social birds that lay blue eggs), Buff Orpingtons (for the name alone, their beautiful shape and color, and the excellent eggs), Brahman (white eggs, beautiful feathers, easy to handle), and Australorps (brown eggs, and nice-looking birds).[53]

If you're raising chickens with an eye to eating them, Cornish, Buckeyes, Rhode Island Whites,

A variety of breeds in the yard—Brahman, Ameracauna, and Rhode Island Red. *Photo by Stacey Evans*

Orpingtons, and Plymouth Rocks are excellent. If you have the space, you can raise a multi-purpose flock of chickens for meat and eggs, with some chickens getting the hatchet at about 10 weeks, and the others living for a few years as layers. When chickens get to the end of their egg-laying days, it will be up to you to decide if you are going to keep feeding them, or if it isn't a better choice to turn them into soup.

Chickens in the Garden

Just to further add to the beauty and wonder of the chicken, you can also invite their help with your garden tasks. A small chicken tractor will provide a moveable run for the chickens that can be placed in areas that need turning over and fertilization. It's like a no-work, no-noise, non-electric rototiller digging in your backyard while you are busy at some other task. A chicken tractor is a great homesteading item to share with other folks because your need for it is intermittent and can easily be balanced with the needs of other gardeners.

The chicken tractor evolved through an assessment of the needs and yields of the chicken. A chicken will happily peck in the dirt in search of insects and organic matter to eat. When we use this yield of the chicken, we cut down on our workload while making positive connections between the yields and needs of the chicken and the garden. This is solid permaculture design; it exemplifies the way connections between elements in our gardens amplify productivity and the how stacking functions meets the needs of different creatures and makes less work for all.

You can also simulate the tractor effect by creating small runs for your chickens in areas you want turned over. You can do this with a moveable cage high enough to keep the chickens enclosed, and if you create this small run with chicken wire, it will move flexibly in response to the shape of your garden beds. You can use the chickens in this way to keep your garden paths mowed, or to overturn the soil in specific beds, as needed.

Urban Rabbits for Fun, Food, and Fertilizer

Rabbits are assets to the backyard homestead for some of the same reasons as chickens—they produce the best manure of all the backyard animals, they'll help you keep your lawn trimmed, they breed like, well, rabbits and thus are a sustainable source of meat, and they are much beloved by children on the homestead. If you are willing to consider butchering your own meat, or if you just want the benefits of their rich, pre-composted manure, rabbits are an excellent choice for the urban homestead. Taking care of rabbits is easy, and this task can

The chicken tractor was designed to rest on this raised bed. Chickens live here for a number of days, turning over the soil, eating small insects, and leaving behind their high-nitrogen manure. When the chicken tractor is moved, the bed is well-prepared for planting.

Constructed out of chicken wire, a chicken run keeps the chickens where you want them in the garden, and away from small seedlings that can't survive their pecking. This kind of run can easily be moved and placed in other parts of the yard when you want to avail yourself of the Power of the Chicken to prepare and fertilize beds as needed.

Three bunny hutches in a shady spot in the yard provide plenty of space for living, romping around, and breeding. Bunnies are let out during the day and protected from predators by fencing around and overhead.

be handed on to children, giving them a sense of purpose and belonging in the work of homesteading, as well as feeling the pride and sense of accomplishment any child gets from a job well done.

Housing

Rabbits need some basic care and secure housing, but once you have gotten it set up, they require little more than a few minutes a day. If you are raising them for meat, you will want at least one male and one female or consider a trio of one male and two females. If you just want them for entertainment and fertilizer, get two of the same gender. If they grow up together from a young age, they should remain compatible.

Rabbits can live in a wood or wire hutch. Rabbits need shelter from drafts, wind, and rain, but unless your temperatures go below freezing, rabbits do fine outside in the winter. If your temps do go below freezing, consider moving them indoors or weatherizing your hutches. Most importantly make sure their water source does not freeze. Each rabbit should have six square feet or so within their hutch (except nursing mothers with their kits), plus an opportunity to range farther in an open pen or run.

Rabbits prefer shade and can overheat and die if left in direct sun. They enjoy having places to hide (like a box or a burrow), especially in the heat of the day. They are perfect for a shady part of your garden that is otherwise unused for growing. They enjoy being let out of the hutch and running around in an enclosed area; use a low moveable fence to mark off an area where they can romp around during the day. When they are very small, this is not recommended, as they can become prey to hawks and other birds, but as they get older, and especially if you provide a box or crate they can crawl under as needed, there should be no problems with predators. If you're going to leave your bunnies out of their coops during the day, remember to lock them up again at night so that they do not become someone else's dinner.

You can also contrive a moveable enclosure of chicken wire to keep them penned in where you want them. Rabbits are great at mowing down anything growing in the enclosure, so only put them where you want them to munch and chew. A moveable rabbit tractor will serve the same function as the chicken tractor.

Food and Water

Rabbits have specific nutritional needs and it is unrealistic to consider growing everything they need on a small urban lot. Purchase a good-quality organic pellet, if you can find it. They are sold in 16 percent and 18 percent protein; 16 percent is sufficient for most rabbits. If you can't get a good organic pellet specifically for rabbits, mix 14 percent organic livestock pellets with 18 percent rabbit pellet to get a half-organic 16 percent pellet. For Mama and baby rabbits, supplement with Calf Manna. Rabbits also

An animal cage on wheels can be placed in different parts of the yard that need mowing. Rabbits keep the lawn trimmed and deposit high-quality fertilizer as they go.

enjoy greens from the garden, fruit tree branches (anything but stone fruit, which is toxic), alfalfa, or timothy hay. They will also chew on stale bread. You can toss them garden scraps, especially greens, brassica leaves, carrots, radishes, and celery. Whatever they don't eat can be raked up and placed in the compost pile. Rabbits need a plentiful and constant source of fresh water. Some people like to get an auto waterer, which works well if your pen is close to a hose outlet; otherwise the larger ball-tipped kinds work well.

Breeding and Babies

When you are ready to breed your rabbits, you can either let the male and female into a run together, or bring the female to the male's cage. A female rabbit is quite territorial, so it isn't a good idea to introduce the rabbits to one another on her turf. Once they mate, count four weeks from that date and put out a nesting box for the mama-to-be. In the interim, it will be hard to know for sure if she is pregnant, but you can try putting her with the male bunny two weeks from the first mating attempt to maximize the chances of impregnation. If she rejects him, she is almost surely pregnant. If you don't want baby bunnies, keep your male and female rabbits separate *at all times*. Breeding age males and females may also not get along well, so you may need a separate cage for each adult rabbit you want to maintain. That said, they are social and may be depressed if they don't occasionally get to see, smell, or touch other rabbits. We've had rabbits living alone, and rabbits living in pairs, and the rabbits in pairs are much more contented, less likely to try to escape, and live longer.

The mama rabbit will kindle (have babies) sometime between day 28 and day 31. Place a nesting box in the cage with the pregnant doe at this time. In the winter you may want to line the box with cardboard, or put a heating pad under it. At any time of year it should be lined with straw. The mamma will pull fur from her neck and belly to line the nest. This will also expose her nipples for better nursing. Once she has her babies, it is best not to disturb them much, though you should check to see that they are all alive.

In the first couple of weeks it will seem that the mama rabbit is ignoring her young. Rabbits do this instinctually as a way to discourage predators. She will nurse them privately at night. At about day ten the baby rabbits will open their eyes. At about three weeks they will come out of the nest. At this point, stop putting greens in the cage for the mama because the babies need some time for their digestive flora to develop before they eat fresh foods. Take the nest box out at six weeks and at eight weeks, sex them and separate the males and females. Keep your mama rabbit to rebreed for the next round of bunnies, and decide if you are going to slaughter all the babies, keep others for breeding purposes, or give them away to friends.

Newborn bunnies in a warm nest.

Ducks—The Perfect Permaculture Pets

Ducks in the backyard homestead also have a number of uses and benefits. Like the chicken, they scratch and peck, and they are extremely good at picking snails, bugs, and insects off your vegetables. (The permaculture cliché goes like this: "You don't have a snail surplus. You have a duck deficiency.") Ducks are also good at eating little sprouts, so you want to set them out in the garden after your plants have reached a pretty good size, or make a contained run for them in places where you want them to go. They happily waddle through the garden in a little flock, seeking out food and nesting under any available straw or leaf pile. They have great personalities and are fun to watch, much like their avian cousins. They lay eggs, and can also be turned into fancy food products if you are so inclined.

Housing, Food, and Water

You can house a small flock of two to four ducks in a large doghouse (easy to find at a recycling center or a lucky street giveaway) because they don't need much inside space to be happy. You can use an old box crate or desk, predator-proof, and out of the hot sun for much of the day. It should have good ventilation, and be a bit off the ground. Ducks will wander through the garden and seek shelter when needed, but are mostly out and about during the day.

Unlike the chicken and the rabbit, the duck needs a water habitat to be happy. You can set up an old bathtub filled with water-loving plants for use as a multipurpose duck habitat, plant growing, and rain storage feature in your yard. Or use an animal water trough and keep it filled for them, or a bucket, or even a kiddie pool. The exception to this is the Indian Runner duck, which needs less water. The Indian Runner is also a good choice if you are keeping ducks primarily for egg production; they are prolific layers who lay throughout the entire garden and bear some watching.

Ducks eat well on non-medicated pelleted mash as a staple, supplemented with fresh vegetable trimmings, chopped hard-boiled eggs, tomatoes, cracked corn, garden snails, worms, night crawlers, and bloodworms. Most items are available at pet and grocery stores. Protein levels are very important to your duck's stage of growth. Ducklings need starter feed with 20 to 22 percent protein for three weeks. Adolescents do best on 16 percent protein. Adult ducks need 16 to 18 percent when they are laying and 14 to 16 percent if they aren't laying. Although the duck will feed happily off your snails and aphids and other bugs, they will also need to have their diet supplemented with store-bought food.

Raising Quail for Eggs and Meat

Coturnix, or Japanese quail, are a delight to raise. Their space requirements are small, they don't eat a lot, they convert feed into protein efficiently, and they are more congenial by far than the chicken.

Ducks in indoor habitat, feeder in background.
Photo by Rachel Kaplan

Feeding the friendly duck.   *Photo by Erik Bjorkquist*

Quail have been raised under domestic conditions since the Pharaoh ruled beside the Nile. The modern Coturnix has been bred to begin producing eggs when less than two months old. Once she starts laying, the hen will produce an egg daily for at least a year. The males are equally rapid growers, being ready for the table at six to eight weeks of age.

Coturnix eggs are nearly identical in taste and nutritional quality to chicken eggs. Coturnix hens, however, need less than two pounds of feed to produce a pound of eggs. Chickens need almost three pounds of feed to make that same pound of eggs. Five Coturnix eggs equal one chicken egg. Quail eggs are all different in appearance, being speckled and mottled.

Food, Water, and Housing

Because of their small size, Coturnix can be kept in small pens, such as a wire cage, rabbit hutch, or even a small dog kennel. Cages can be raised or can rest on the ground. If the cage is raised, it will be easier to clean, the birds will never be standing in manure, and their eggs will remain clean.

Quail don't need special nesting or brooding boxes; they will gladly lay their eggs in the straw or wood shaving bedding you leave for them. If your quail enclosure is small, you can simply reach into the cage to get the eggs and don't need to make a separate entrance into the cage to gather eggs. Quail are small but need some space around them, approximately one square foot per bird. For a small urban habitat, one male and three to six females is a good-sized flock that will keep the birds happy and productive.

Like most birds, Coturnix quail like to take dust baths in hot weather. You can place them in a mini quail tractor so they can run around and peck in the dirt and have a dirt bath in the sun. Always put water in the tractor with them and give them an option for shade. You can also include a small tray of soil for them inside their enclosure. Add a small amount of diatomaceous earth to further help them keep skin mites to a minimum.

Birds being raised only for meat will thrive and grow plump on a high-carbohydrate diet. Hens will need laying mash if they are to produce lots of eggs. Whether for meat or eggs, quail will thrive on a high-protein diet of 25 percent or more. Try a combination of high-protein 16 percent laying mash, some 22 percent turkey crumbles with a small amount of flaxseed, quail and dove seed mix, and if available, pure ground seed cake from seed oil extracting.

Supplement your hens' diet with chopped greens from the kitchen. Food scraps for the compost pile can be processed through the Coturnix hens first. Chop leaves and other vegetable scraps fine enough for the birds to eat easily and there will be almost no waste.

Quail can be housed safely and comfortably in a large pet carrier. Any container will work as long as it protects quail from predators, has ventilation, and is placed in a shady and protected area of the yard.

Always provide the birds with fresh, clean water. Quail will need winter protection from cold. They are territorial and will peck if you introduce a new bird into their habitat. To prevent this, move them to an unfamiliar location with a different feeder and waterer when introducing new birds to one another.

Coturnix quail in their habitat.
Photo by Susan Doering

Breeding and Raising

Quail do not incubate their own eggs, so you will have to purchase young quail to raise rather than expecting to breed them yourself. You could also invest in an incubator, gather fertile eggs, and incubate them yourself. For beginners, we recommend checking your local classified ads for live birds when you are bringing your flock together and taking on the incubation project in season two. Your quail will begin laying eggs after a few months and should continue to lay for at least a year. Frequency of laying is linked to the length of the days, so you have to artificially increase the day length with additional lighting if you want them to lay through the winter. Left to their natural rhythm, they will lay from equinox to equinox. If you are raising quail for meat as well as eggs, you can set up a rotation schedule for buying new birds for eggs and meat.

Goats . . . in the City?

We see goat herding on the outer edges of what's possible in most backyard homesteads, and are amazed to see goats gaining ground in cities around the country. Goats are curious, friendly, smart, and deeply relational creatures; they need attention, diverse forage, and one another. Goats are herd animals, and are unhappy living a solitary life. Goats will run up to you to say hello and jump up on you if you let them. They form a pecking order, like chickens, with one doe as the herd boss who butts the other goats around. Goats are territorial, especially around their children, and can be noisy and belligerent when crossed, threatened, or in unfamiliar territory.

Goats are generally kept for the milk and compost they provide, or because they are great at keeping whatever is left of the lawn in good trim. They are ideally suited for sloping areas that may be challenging for gardens, and can speedily clear a yard of bamboo or blackberry brambles. Goats are happy in marginal spaces and will quickly take over an abandoned lot. Goats demand a commitment of both time and money. While forage goats are hardy creatures that may struggle along on whatever is available, a goat that is expected to produce milk every day needs good-quality feed, housing, and medical care.

For these reasons, goat-keeping collectives are excellent ways to care for an urban herd. Members share the tasks of tending, walking, milking, and cleaning up after the goats in exchange for milk, meat, compost amendments, and good company. If you can join a goat collective or find someone who has goats who will teach you, becoming an apprentice is a good way to begin keeping goats.

City people also share herds of goats solely for lawn maintenance purposes. A shared, intermittent need for lawn mowing makes a moveable herd a quiet, fossil fuel-free, compost-producing, highly entertaining neighborhood asset.

Shelter and Range

Goats need a warm and dry place to sleep for the night, but they prefer sleeping outside and won't really use the indoor space except when it is raining or snowing. Goat houses can be constructed from used storage

Walking a goat in an urban neighborhood is an educational opportunity for all, and it gives the goats a chance to encounter other animals and humans, stand down a stray dog, and sniff and nibble in a new place.

sheds, parts of unused garages, packing crates, or other recycled materials. A 6 x 8-foot indoor area is sufficient for up to four adult goats and six kids as long as there is plenty of room outside. We know people who keep two perfectly contented city goats in a 10 x 15-foot outdoor space, but in general goats prefer more room than that.

Goats need space to range, and a diversity of forage. By far the best option for extending the range of the urban goat is simply to take the goats for a walk in the neighborhood. Walking your goats stacks functions; goats get to range and forage; neighbors get to meet the goats; friends are made. Oakland goat girl Jeannie McKenzie reports, "The goats have created so much community in our lives. Once we started walking our goats, we met everyone in the neighborhood." Wethered (neutered) goats are great pack animals. They don't like being controlled on a leash like a dog, but will stay with their leader when they walk. Once the goatherd is the established head of the herd, the goats will scamper to follow along. Walking your goats in the neighborhood every other day gives the animals the space and diversity of forage they need.

Food

Adult goats need 14 percent protein food, generally a combination of corn and kibble. But goats need more fodder than hay and store-bought feed—give them scraps from the kitchen, or they will eat any plants you don't want. (They'll also eat plants you do want, so make sure you pen them in to keep them from eating the whole garden.) They enjoy black sunflower seeds—it's good for their coats and makes their milk richer. Grow these in the garden and harvest them for the goats, but you're not likely to be able to grow them in sufficient quantity throughout the year.

Goats are messy; in a small yard, this may pose a problem and should be given consideration in your homestead design. Goat poop can go directly into the garden, but the urine-soaked straw needs to be composted first. It breaks down rapidly in two to three weeks, and provides high nitrogen compost for the garden.

ON THE MENU FOR GOATS AT GREEN FAERIE FARMS

Morning and evening, each female in milk receives a gourmet meal consisting of the following:

1½–2 cups organic goat pellet

1½–2 cups organic alfalfa pellet

1½–2 cups organic whole oats

½ cup kelp meal

½ cup black sunflower seeds

1 tbsp probiotic

Topped with a few tablespoon of molasses, and garnished with a piece of apple, cactus pad or other garden forage. These are some happy goats!

Breeds

When choosing a goat breed for the city, selecting for sound rather than size is a good mantra. Goats can be very loud, especially when uncomfortable, separate from their herd, or in unfamiliar places. A mother goat separated from her kid will scream for days. This hardly ever goes over well with the neighbors. The Oberhasli breed is quiet and highly recommended for urban use. Some urban homesteaders opt for the Nigerian dwarf goat due to its size, but this breed is quite noisy and not really well suited for city living, despite its small stature.

Got Milk?

Unlike the other animals described in this chapter, dairy goats need to be tended twice a day. Four goats can take up to an hour to be milked, and goats are notorious for being resistant to the process at times. Busy urban dwellers deal with this workload by sharing it. People take on the task of caring for the goats in trade for the high-quality raw milk. With 14 weekly milking slots, this kind of community project provides a lot of people with the chance to learn about tending and feeding goats, raking and composting manure, and the art of milking.

Once a goat has had a kid, she can usually be milked for two years. When you breed the goat, she must be allowed to dry up for two months before she kids, so expect to milk a goat for approximately 8 months out of the year. The kid commands milk for the first two months, though some goat keepers wean the baby off the mother early and do the job of feeding the kids themselves. This interrupts a certain part of the relationship between mother and kid and can make both adult and young easier to manage. Goats that are bottle-fed by humans bond with humans and this makes them easier to handle.

The intricacies of breeding goats and supporting the mother/kid relationship while getting fresh milk on

Learning to feed a baby goat at a class at Green Faerie Farm. Goat apprentices care for the goats on a schedule throughout the week.

the homestead is somewhat beyond the scope of this chapter. Suffice it to say, there is a bit of a balancing act in how we keep goats for their milk that factors in both human and animal needs. Because of their sensitive and highly interactive natures, making decisions about how to humanely manage these relationships is an important part of the art of goat keeping.

Harvesting Your Animal Friends

When we choose to raise animals for their meat, we come more into the reality of what it means to be a meat-eater. Raising the meat you eat makes good emotional sense, as well as good nutritional and environmental sense. When we feel what it means to kill what we eat, our relationship to this part of our food chain is transformed. For some, it'll be enough to turn you into a vegetarian; for others, it will reinforce the cyclical nature of life in a body, making the eating of meat a more sacred act, a living choice, not a pre–packaged, bloodless, headless dining adventure.

Learning to slaughter your backyard animals in a compassionate and timely manner is good homesteading practice. We recognize that slaughtering animals is not for everyone—some people won't even consider it, and would rather support their barnyard flocks into their dotage. And some people raise animals because of what they offer on tangible, and intangible, levels and would never consider eating these feathered or furry friends. We fully respect that, but have made another choice—for prioritizing maximum yield on our homesteads, as well as including sustainable practices for both the nurturing and humane butchering of our barnyard animals.

The careful slaughtering of farm animals is an heirloom skill that was once standard for farming folk around the country. With the mass production of meat and meat by-products and the domination of the industrial meat industry, these skills have been mostly lost, as has our connection to the cycle of living and dying, which was intrinsic to the family farm. If you're going to enter into the harvesting of your animals and do it in a responsible way, we highly recommend getting a few lessons from an expert. We print here directions on how to slaughter and clean a chicken, a rabbit, and a quail, but hope you will use these instructions in tandem with the wisdom of an experienced teacher.

Ruby says: "Before I got my rabbits, I went to my teacher's house to experience butchering and see if I could even do it. [Ruby's teacher is Jim Montgomery, see previous interview.] It was pretty intense, but Jim was amazing and modeled such compassionate care for the animals, even in the act of killing them. This surprised me. The method he uses is instant so the rabbit doesn't experience a lot of pain. You would know—a rabbit in pain screams an awful scream. He holds the animal between his knees so he is in physical contact with it while it dies. Then he thanks the rabbit for its life and for what it's giving to him. There's something really dreadful about killing an animal, but on the other hand, there's also something oddly spiritual about it. When you consciously take a life for your own sustenance, you have to deal with life and death, and what that means. If you have to kill your dinner yourself, you don't take it for granted, and in the end, you don't want to take so much."

How to Butcher a Chicken

1. Catch the chicken. Cover its wings with your hands to keep it from flapping away. Hold it snugly next to your body until it calms down. You can also cover its eyes to help it calm down.
2. Turn the chicken over and hold it by its feet. Gently hold it upside down, close to your body. You can place the chicken into a chicken cone (an ice-cream-cone-shaped tool with the bottom cut off to make room for the chicken's head), which will further help it calm down.

3.   Once the chicken is resting in the chicken cone, slit its throat and allow it to bleed out into a bucket. This will usually take less than one minute.

4.   Once the bird is done bleeding, remove its head.

5.   Plunge the bird into a large pot of water that is about 140 degrees. Submerge the bird for 10 seconds, which loosens the feathers but does not cook the bird.

6.   De-feather the bird, starting with the wing and tail feathers. The other feathers should come out pretty easily just by wiping them off. If not, resubmerge the bird for a few more seconds.

7.   Once the bird is slaughtered, scalded, and plucked, it will need to be eviscerated. (The quail is butchered much like the chicken, although standard practice is to simply lop off the head of the quail with a lopping tool.)

Eviscerating the Bird

1.   Lay the chicken on its back. Using a sharp knife, cut down the neck until you find two tubes. One will resemble a vacuum cleaner hose, which is the esophagus. Cut this off as far down as you can and discard it. The second tube will be the one leading to the crop.

2.   Follow the crop tube down until you find the attached bag, which is the crop. Pull this out as far as possible, cut, and discard.

3.   Turn the bird around, still on its back, and begin the cleaning. Start by cutting a Y, beginning with a straight line from the bottom of the breastbone, about two inches.

4.   Make two diagonal cuts down to either side of the tail. Make sure you only cut deep enough to pierce the skin. The intestines are right under your knife and you want to avoid cutting those open.

5.   With your knife angled almost straight down, cut a horizontal slit right below the anus along the tail, connecting to the two end points of your Y.

6.   You have now cut straight down a few inches, along either side of the body cavity, and then along the underside of the anus. You should be able to now gently pull the anus out, which will also remove the intestines and other organs as well.

7.   Once you have some guts pulled out, put your hand inside the bird. Using the side of your hand like a spatula, move along the internal body cavity to loosen any guts from the walls. This should help you gently pull everything out. Be very gentle with the gallbladder, which will be an almost black organ with a texture similar to the intestines. This is the absolute worst thing to puncture, as it holds all the bile.

8.   Once cleaned, the bird needs to be rinsed with a hose and plunged into ice-cold water to cool quickly.

How to Butcher a Rabbit

Commercial meat rabbits are slaughtered at around eight weeks, but you can wait until four to six months to maximize the meat production. If you plan on butchering your rabbits for meat, you will need a good heavy cleaver and, for best results, a pellet gun. While there are other ways of butchering rabbits without causing them pain, the pellet gun is quick and sure, and minimizes the animal's distress.

1.   Prepare your pellet gun and prime it.

2.   Hold the rabbit firmly between your knees and cover its eyes to calm it.

3.   Once the rabbit is calm, set the tip of the gun slightly forward of the ears and angled slightly to the front. Pull the trigger.

4.   Allow a few minutes for the bodily reactions to die down.

5.   Remove the head with a cleaver. There will be a small amount of blood. Angle the cutting board so it flows into the dirt.

6.   Remove both front paws and one rear paw at the wrist joint.

7.   Hang the rabbit from the remaining foot.

8.   Slit down the color line of the inside of the hanging leg, just cutting through the fur and skin without damaging the muscle. Work your fingers under the skin until you can touch around the leg. Cut the skin at the ankle.

9.   Then slit down to the tail, working your hands under the skin and slicing away the fur around the tail and the genitals.

10.   Snip the tail, and pull the skin from the other leg.

11.   Then pull the fur straight down. When you get to the shoulder, work your hand under the skin and pull the forelegs out of the skin.

12.   Lay the skin to the side to cool.

13.   Now, with a sharp knife, slice into the leg joint on either side of the genitals and break the pelvis with your hands.

14.   Delicately separate the digestive tract, making sure not to cut into it.

15.   Using a sharp knife or butcher scissors, cut down the center line of the belly from genitals to ribs without puncturing any of the viscera.

16.   Pull out the digestive tracts and all the offal and let it fall into your bucket.

17.   Next pull out the kidneys, liver, heart, and lungs and save in a clean dish for eating. You will need to cut and soak the kidneys before eating to remove the "pee" flavor and cut the gall bladder out of the liver before preparing.

18.   Split the body the rest of the way down the front.

19.   Rinse the meat and age in the fridge for three days before eating or freezing for later use. The organ meats should be sautéed up immediately or frozen for later use.

Wildlife on the Urban Homestead

Visits from "wild" animals in the city can be both a pleasure and a nuisance. The sight of a hummingbird enjoying the nectar from a beautiful flower is a tingly treat on any afternoon. Even if we know we've secured our animals and our produce, a raccoon's nocturnal knock on the back door or hungry foray at the chicken coop is another story. Maximizing visits from pollinators like birds, bees, and bats is easily done by choosing flowers, vegetables, shrubs, and trees that pollinators enjoy and spreading them all around your garden. Protecting against barnyard animal deaths by providing predator-proof cages also leaves us the space to enjoy visits from wild animals in our backyards without worrying about the lives of our animal friends. We feel particularly graced when snakes, hawks, and foxes pay us a visit, but only when we know that the creatures we care for are safe.

Some city predators are bold and jaded urban warriors. A raccoon won't think twice about walking into an open back door, and an opossum will come into the house to get the cat food, if it can. Keeping house doors latched and locked (standard city practice), will usually be enough to keep them out of the house. Secure any cat doors, which might provide entrances to the house. Store your homegrown edibles in predator-proof containers.

The rat stands to be the biggest beneficiary of homegrown produce, but other small predators will eat a bag of potatoes or garlic or onions pretty quickly if they aren't safely stashed.

The cat—feral or otherwise—can be one of the most disruptive animals on the homestead. A highly efficient killing machine, the cat can dispatch a mole or a mouse or a baby chick or rabbit without a qualm. On the other hand, if gophers enjoy the treasures of your backyard you can try praying at the altar of the feral cat, hoping they will do the dirty killing work for you.

Like other practices on the urban homestead, managing animals and their predators reminds us that life and death are on opposite sides of the same hand, and in the midst of our successes, we will also and always sustain some losses.

Eating Close to the Ground

Food Is a Verb

Tending to our own food needs—by buying responsibly, or better yet, growing our own food—connects us to medical, economic, political, and religious issues from a land-based point of view. This will force us to reconsider the meaning of health, and help us see that our health and the health of our communities are inextricably tied to one another.

—Wendell Berry[54]

Food has always been at the center of cultural experience, the way we express our connection to the land and to one another. Food nourishes us and gives us life. Local foods express an intimacy with and the "taste" of place. Winemakers speak of the *terroir* of a particular wine, suggesting that each sip embodies the geology, drainage, soil, plants, and weather of its harvest place. Cheese makers tend toward this kind of sensitivity to the milk they use in making their cheese; in it, the grass, the air, even the wind make themselves known. You can't fake this sense of place—it is an expression of the land that gives rise to the food we eat.

All cultures make meaning from their own historical food; what is France without its cheese, or Mexico without its corn and beans, or Korea without its kimchee? Human cultures and the processes of preserving and culturing food are inextricably linked. The Roquefort cheese we eat today, for example, was originally cultured from mold that grew in a cave in a particular region of France. The sourdough bread starter you make at home will only succeed when you invite bacteria and microorganisms from your kitchen into the process. Soup made from ingredients grown in your garden brings the taste of place into your body. When we engage in creating, preserving, and fermenting our food in these time-honored ways, we participate in growing a place-based culture that connects us to the living world around us, as well as participating in the evolving history of human habits, traditions, and ceremonies that give life meaning. The industrialization of agriculture and the widespread distribution of commercial food products may have given us more time, but it has taken away our local foods and food sources, heirloom food preservation practices, and our involvement with the food we eat. It has impoverished our human culture.

But now, everything old is new again. The resurgent interest in local foods and home-scale preservation—from canning, jamming, freezing, brewing, fermenting, and otherwise experimenting with food—is happening coast to coast. Taking up the pot and the pan, the cheesecloth and strainer, the canning jar and the wine bottle, homesteaders are beginning to reweave the web of culture lost in the toxic downdrift of the industrial food supply. Food preservation is hooked into all the values of homesteading—self-sufficiency, community resilience, DIY for fun and pleasure—a reminder that food is not something that's done for us, but something that we do with one another. Remaking our relationship to food is one of the central homesteading pleasures and practices, a radical act that can go a long way toward growing into our role as producers rather than consumers.

Persimmon in autumn.   *Photo by Datna Kory*

The Slow Food and locavore movements remind us of the cultural value of local foods, and our capacity to rebuild culture through the choices we make in the simple, daily act of eating. Both movements extol the virtues of indigenous foods and local food production, as well as the conviviality and community that come from sharing food with one another. Slow Food began in Italy in response to a McDonald's opening in a beautiful classical plaza; Martin Bové drove a tractor through the window in protest, and a movement was born. Locavores make a practice of eating close to home; some people eat only food grown within 100 miles of where they live or within a day's drive, and skip the rest.

These movements have been rightly criticized for their class politics, for advancing a laudable goal that is unattainable by many who might choose it if they could, and for consumption excesses that they justify as being local and "slow." Their essential message, however, that food is an intimate reflection of our lives and culture, is not a class-based assertion but a human one. The appropriate class critique lies in the fact that not everyone can afford a Slow Food meal or the labyrinthine lifestyle of the locavore, but the drive toward localizing our food sources and reimagining our relationship with food can be shared with everyone. Generating local food sources in order to provide food security for everyone is part of the bigger story of the urban food revival currently underway.

The politics of homegrown food have been widely recounted in recent years and the value of harvesting it yourself—the positive benefits for earth and family that come from bringing the farm closer to our tables—is becoming more widely understood. Much has also been written about the excesses of agribusiness, a degenerate system that relies on cheap oil to produce food, and in the process poisons the waterways and destroys the soil it's grown in. Industrial agriculture is one of the biggest carbon culprits in the climate change crisis (meat production being responsible for 18 percent of global warming gases produced in this country compared to the 13 percent created by our transportation choices). Low in nutritive value, industrially grown food is largely contributive to the epidemics of degenerative illness, learning disorders, alarming rates of hyperactivity among youth, diabetes, obesity, heart disease, chronic fatigue, and other immunological disorders. Industrial agriculture's reliance on fossil fuels also makes the food system dangerously unsustainable, subject to the vagaries of the market for price and availability, not to mention the challenges of bringing food to table as fossil fuel becomes more scarce.

The excesses of industrial agriculture and its shortsightedness in plowing for the bottom line rather than for highest nutritive value lead us to new choices about where our food comes from and what we want to eat. The history of traditional food has relied on a symbiosis with beneficial microbes and bacteria that can be cultivated in our own kitchens in the form of lacto-fermented vegetables, cultured dairy products, and homebrewed beverages. Industrial food has been literally processed to death and much of the food we eat is drenched in fossil fuel. Our agricultural system is broken; we can help fix it through our daily actions and choices.

Understanding our relationship to the food chain and localizing the means of production benefits us with good work, good community relations, good soil, enhanced habitat, and healthier food. Everyone growing some part of what they need and drastically reducing their purchase of industrial food would make a significant difference not only environmentally but also economically. Remaking our relationship to food—where it comes from, how it's grown, how it's processed, how we eat it, and how to make it available for everyone's benefit—is an important part of the homesteading lifestyle. Simple and banal as it may seem, changing your relationship to food is one of the most powerful ways to cut loose from an unsustainable and dangerous system and participate in making a better one. Even without growing your own food it is possible to realign your food buying and eating habits within a sustainable urban framework. Choosing with our wallets is one way each of us can nudge aside the industrial food supply.

on the ground: diy queen

K. Ruby Blume started the Institute for Urban Homesteading (IUH) in 2008 "to promote localism, sustainability, and self-reliance." An artist and activist who had run a nonprofit for almost twenty years, Ruby's recovery from burnout included deepening her connection to plants through gardening and studying botany, native plants, herbal medicine, and pollination ecology. She completed a permaculture design course and had been involved in DIY projects such as beekeeping, biofuels conversion, canning, and mead-making for several years when the idea of "urban homesteading" began to gel as a cultural meme.

"I started the school almost as a joke," Ruby recalls. "I was sitting with my girlfriends and they were saying they wanted to learn some of the heirloom skills I'd gathered. First they said, 'I want to send my kids to Ruby Camp!' Then they said, 'Wait a minute—*I* want to go to Ruby Camp.' But instead of just teaching a couple of my friends—I had to go and start a whole school! I made a fancy name that sounded serious, whipped up a good-looking postcard, and did a massive press campaign, like I would have done in my art career for a performance event.

"I knew in my gut that my timing was perfect. And I was right—my little project was overwhelmingly successful. The first year we did not sell out every class, but many were overbooked and had to repeat. And there were sizeable articles about my work or the school in a half dozen periodicals, including some national press. Our most popular classes were, and continue to be, beekeeping, gardening, canning, and cheese making. Each year I have added curriculum and teachers and in 2010 we produced sixty classes throughout Oakland, with fifteen additional teachers besides myself."

For Ruby, sustainability means "putting more into the system than you take out of it." She's running the institute on that model. "It's in my design to remain sustainable," she explains. "One of the ways to do that is to stay small. It can't be a pyramid scheme of ever-expanding profits and growth. I've been experimenting with how much I can do and still feel sustainable. I reached that limit this year—sixty classes are enough!

"I don't write any grants, I only get small contributions, and all the money comes from people paying for classes. On top of that, not only do I try to pay the teachers well, but I also have a sliding scale. It's a balance to pay the teachers and make it affordable for people on the other end. One way we do that is by having a moveable classroom. People from the public get to visit private homesteads while they learn. It's part of the intimacy of the style of teaching we do, recreating the traditional folk school."

K. RUBY BLUME ~ SECRET GARDEN & MINI-FARM
50' x 100' with cutaway ~ 4875 sq.ft ~ about 1/10th of an acre

STREET

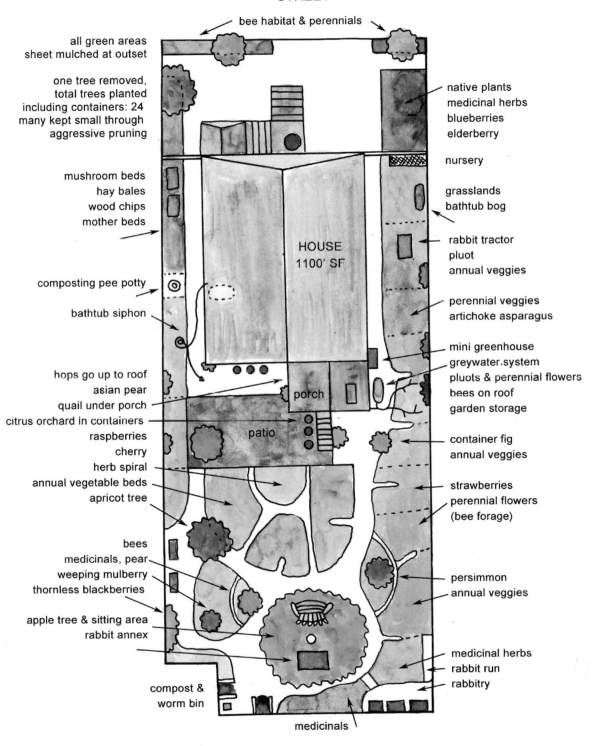

bee habitat & perennials

all green areas
sheet mulched at outset

one tree removed,
total trees planted
including containers: 24
many kept small through
aggressive pruning

native plants
medicinal herbs
blueberries
elderberry

nursery

mushroom beds
hay bales
wood chips
mother beds

grasslands
bathtub bog

rabbit tractor
pluot
annual veggies

HOUSE
1100' SF

composting pee potty

bathtub siphon

perennial veggies
artichoke asparagus

mini greenhouse
greywater system
pluots & perennial flowers
bees on roof
garden storage

hops go up to roof
asian pear
quail under porch
citrus orchard in containers
raspberries
cherry
herb spiral
annual vegetable beds
apricot tree

porch

patio

container fig
annual veggies

strawberries
perennial flowers
(bee forage)

bees
medicinals, pear
weeping mulberry
thornless blackberries

persimmon
annual veggies

apple tree & sitting area
rabbit annex

medicinal herbs
rabbit run
rabbitry

compost &
worm bin

medicinals

143

K. Ruby's Secret Garden and mini-farm, before.

K. Ruby's Secret Garden and mini-farm, after.

The mission of IUH is to reclaim the lost arts of the kitchen and the garden and to help people understand how they are part of a bigger cycle of life. Classes range from the kitchen arts to how to build a solar-panel system or install rainwater storage tanks in the backyard. The gardening and food preservation classes are the most popular, but Ruby is committed to teaching a full complement of urban homesteading skills, including natural building, medicine making, beekeeping, soil building, small animal husbandry, and more. "I teach people that everything related to food and gardening and sustainability is cyclical and seasonal," Ruby says. "Sometimes people want me to teach canning in January, and I remind them that this would mean buying tomatoes from Chile which doesn't make much sense. The semester is based on the season of the honeybee, so we start at the beginning of bee season in mid-March and end with the honey harvest in mid-October. And then we rest."

While heirloom skills and kitchen witchery are large parts of the curriculum, on a deeper level Ruby is trying to teach people how to get the answers they need. "I want to teach them how to learn. People have a question or a need, and learning how to get the information is important. This is a way of empowering and relying on ourselves, of learning we have everything inside ourselves we need to succeed."

Ruby's not called the "Queen of DIY" in her local paper for nothing. In the midst of talking, Ruby jumps up and says, "Hold on, hold on! I'm doing three things at once. I have to add rennet to my cheese!" When she sits down again, she says, "This morning I prepped the two flats of tomatoes someone brought me last night to go in the dehydrator, finished canning sauce from the class I taught yesterday, made two pints of pesto and two lacto-fermented sodas. I'm making cheese from the three gallons of goat milk my friends brought me (because, hey, we're going out of town), harvesting hops, talking to you, thinking about microbes, and wondering when I can take a ride on my motorcycle and install the windows in the downstairs apartment. Oh, and the rabbits need to be fed.

"I see my work as caretaking in whatever way, nourishing, making better, more beautiful, tending, observing. Whether you're making art or tending a garden, it's a process between your hands and your heart and your eyes, taking in and seeing what you are going to make, how you are going to craft and form something.

"If I could say anything directly to people, it would sound cliché, but it's true for me: Open your eyes and open your heart and do what you love. Like Rumi says, 'There are hundreds of ways to kneel and touch the ground.' If you're really open in your heart, you'll see what's necessary to sustain the world, for ourselves and those who come after us. There are a lot of important cultural issues out there, but if we don't have a planet or a healthy ecosystem, everything else is completely moot. That's the bottom line that motivates what I do.

"I wish I could get out there and have a global impact and change the world, but what works for me is to work small and local, to touch people who I can actually look in the eye. I don't know if it's enough to stay home and tend the garden. I don't think it is, but it's a really good start. All the skills you need to tend a garden well—like observation, and listening, and caring, and inventing, and recovering from mistakes, and repairing—are the skills we need to shift the way energy is being used on our planet."

Food As Medicine

Eating is one of life's pleasures. Good food is sensuous, comforting, nourishing, and delicious. When we eat foods that have no history or are full of processed ingredients we can't pronounce or recognize or digest, slapped together from low-quality ingredients, our bodies take the hit. Instead of thinking of food as "fuel for the engine," something we eat on the run, we can think of our food as medicine, the soil from which we grow our healthy bodies. Eating pure, fresh, nourishing food is our best health insurance and one of the surest ways to honor our bodies. Eating "close to the ground" or lower down on the food chain has the greatest impact on lowering the carbon footprint in our food, as well as having excellent health outcomes. Vegetables, fruits, and legumes are the most nutritious, easy to grow, and low-impact foods we can eat.

The higher up the food chain we eat, the more our food is implicated in poor agricultural practices, including fossil-fuel farming, greenhouse gas production, degradation of agricultural land, and the torture of animals. When we fill up on the empty calories of fast food and derivatives of corn (from high-fructose corn syrup to chips to citric acid) we're battering our earth and our bodies with the same poisonous monoculture. It doesn't work for the soil, it doesn't work for the body, and it doesn't work for the planet. The simplest rule of thumb when buying packaged products is: if you can't pronounce it or don't know what it is, don't eat it, clean with it, or put it on your body.

When we eat locally grown food that's in season—fresh vegetables in the summer and fall and soups and stews from stored squashes and root veggies as well as canned and fermented foods in the winter when the garden takes a rest—we are eating in alignment with our place. Rather than insisting on mangoes from Mexico in the middle of winter, eating the root veggies of the season lifts us out of a problematic food marketplace and into a relationship with the cycles of nature. When we are literally eating *where* we are—the *terroir* of our place,

the medicine within our food—we understand it better, and live more intrinsically with it. When we eat only what's available we really appreciate each new taste when its time comes around.

Food as medicine is a meditation on gratitude—to eat what we grow with a reverence and a connection to the process, the sacredness of the harvest and the joy of eating become an embodied experience. It's no wonder people all around the world say a blessing before eating—a way of acknowledging the work, the skill, and the love that go into putting food on the table.

Fresh runner beans from the summer garden.
Photo by Dafna Kory

Where's Your Food Coming From?

There are two main issues to note when thinking about where our food comes from: access and choice. While it's easy to agree that eating nutritious and organic food is better for both our bodies and the earth, not everyone has access to quality food. People eat factory-farmed meat or high-fructose corn syrup products and other fast foods not only because that's what's available and affordable but also because it's laden with sugar and salt and fat to entice our taste buds. Part of remaking the food system means working for affordable, available healthy food for people who don't currently have access. This involves swimming against the strong undertow of advertising, governmental food subsidies, and the pressure of the global marketplace in order to remake a local food system that works for everyone.

Numerous food security projects, whose mission is to make available healthier food to people who are "food insecure," are sprouting up in cities around the country, especially in neighborhoods where what's most available is alcohol, chips, and cigarettes. Projects like these support a wider distribution of healthy foods for more people. Gardeners with an excess of produce are spreading this wealth to people who don't have enough, in a powerful human impulse to share. Those of us with enough space can plant an extra row dedicated to providing for those who just don't have enough.

on the ground: food security—the right to good food

While some communities are bracing for a food emergency, for others, the emergency is already here. The West Oakland community—deemed "food insecure" for decades—has been served for almost twenty years by a variety of food justice organizations whose missions are to expand the availability of healthy food in chronically underserved communities. This is community homesteading in the truest sense of the word—creative and place-based strategies for providing and distributing resources to people with the highest need and the least access.

Since 1996, Mo Betta Foods has been asking the question about how to recruit the African American community to feed itself. "Fresh food was hard to find in our neighborhoods," the Mo Better website explains. "So many of the mom-and-pop markets had been converted to liquor stores, with no produce. Then we realized it was even harder to find African American farmers. We decided that if we could find African American farmers, we could connect what they grow back into our neighborhoods."

Mo Betta Foods has marketed and helped to distribute produce grown by the African American farmers of California by identifying or creating new markets for their produce for over twenty years. The organization founded a farmers' market and a cooperative grocery story in the mid-nineties. "While we agree that the tactics of teaching cooking or starting urban gardens and farms have some benefits," states Mo Betta's website, "we believe they are a Band-Aid approach to solving the health disparities we face in our communities today. The effects of the 'double blow' (banks red-lining African American neighborhoods and the years of African American farmers being discriminated against by the USDA and other agencies) will not be resolved by converting lots into a community garden, or teaching youth how to cook.... The amount of resources these programs receive and the number of people their garden or urban farms feed should not be a priority over rebuilding the food system that once existed between African American farmers and the communities where we live."

Their strategy to rebuild these relationships is a "healthy economics campaign." David Roach, founder of Mo Betta Foods, writes, "When I started teaching high school, I gained a better understanding of the role high schools play in the lives of our children. . . . In most under-served communities, there are very few actual businesses that exist. And very few, if any, would enable youth from the neighborhood to train or work there. I also realized that the majority of students attending these schools, upon graduation, sought to leave their community as soon as the opportunity would arise. This led me to view the issue from a communities perspective because education was not seen as the passport to the future of building a better community, but a way out of the community for those educated within it… A major role of our campaign is to inspire the educated ones to stay in their community to inspire others to further their education."[55]

Another community endeavor, the People's Grocery, was founded in 2002, and seeks to "build a local food system that improves the health and economy of West Oakland. Their website acknowledges that "the food system is failing to provide low-income people with the healthy foods they need. It is also failing to create good jobs and support local food businesses in urban communities. People's Grocery works toward creating a food system that prioritizes the needs of the urban poor by increasing the local supply of fresh foods; advocating for living-wage business and job opportunities; and developing strong relationships and community leadership."

Working at the People's Grocery farm.

The idea for People's Grocery was born out of a working relationship with Willow Rosenthal, a food activist and urban farmer who was developing a community garden in West Oakland. Willow went on to create City Slicker Farms while People's Grocery began its own garden in partnership with the North Oakland Land Trust. Expanding from gardening into education and outreach, People's Grocery launched the Collards & Commerce Youth Program in the summer of 2003, a program of urban gardening, community outreach, business classes, and cooking and nutrition workshops. In August of 2003, the founders and their first crew of youth launched what has become the flagship enterprise of the organization, the Mobile Market, a grocery store on wheels that sells fresh produce, packaged foods, and bulk foods at affordable prices. The People's Grocery now has a two-acre farm, various youth and adult outreach programs, and weekly cooking classes. Another successful People's Grocery program is the Grub Box, a modified CSA program designed to create access to healthy food for local residents.[56]

City Slicker Farms meets its mission to increase the healthy food supply in West Oakland by creating high-yield urban farms and backyard gardens. These gardens "provide healthy, affordable food and improve the environment, demonstrate the viability of a local food-production system, serve as community spaces, empower children and adults who want to learn about the connections between ecology, farming and the urban environment, and give West Oakland residents tools for self-reliance."

City Slickers works to capture the wealth of knowledge and information that exists in Oakland. "A lot of people we work with have been gardening and raising chickens for a long time," said Barbara Finnen, executive director of City Slickers. "Gardening was part of growing up. A lot of people in Oakland come from agricultural backgrounds. In people's lifetimes, Oakland was a place where people farmed. When you were young, that's what you did. Others need support to get started. We connect up the resources of people and place to help people grow their own food."[57]

Farmer's Markets

Farmer's markets offer a way to eat locally, access healthy food, and support good growing practices. They tend to be convivial environments where people meet and greet, sampling local food and visiting with friends. They offer a satisfying opportunity to make friends with a farmer. One of the farmers at our market runs a large orchard in California's central valley, smiles at everyone, and throws as many extra peaches or figs as he can into your basket. He grows without pesticides and sells a wide variety of stone fruit, figs, apples, and pears. His seconds are half price, which usually runs about $10 for thirty pounds of fruit. They're delicious to eat right out of the box, and great for canning and sauce-making projects that get our family happily through the winter months on the tastes of high summer.

Make friends with a farmer. You'll learn what grows well where you live, what grows when, and even what to do with it when there's a surplus.

Community Supported Agriculture

Community Supported Agriculture (CSA) projects offer another solid way to interact with local farmers, access excellent food, and bring your family into a more intimate relationship with a living, working farm. CSAs are small farms that sell "shares" to members. When you buy a share, you get just that, your "share" of the bounty. In the spring you get strawberries and in the fall you get squash and the abundance of each and every season. Some CSAs invite their members to come to the farm once a week during the harvest to pick up their food allotment. They may also offer U-pick crops, or excess produce at bulk discounts, providing affordable food preservation opportunities.

CSAs are sometimes criticized as limiting choice (e.g., "If I get another squash in my weekly box, I'm going to freak out!"), but they really do make abundantly clear what grows where and when, and provide not only an opportunity to try different kinds of fruits and vegetables, but also the chance to revel in the abundance of each season. And when the farmer suffers a crop loss—from drought, too much rain, bugs, or blight—farm members experience the loss, too. There's nothing like that kind of market reality to help us really understand the challenges of growing our own food and feel grateful for what we've got.

Community-Based Food Co-ops

Another way to limit the purchase of processed foods or unnecessary packaging is to buy in bulk. There may be a local health food store in your area that sells grains, nuts, dried fruit, flour, spices, and herbs in bulk. If not, investigate forming a food co-op to order in bulk from a distributor. You can organize your neighbors into a buying club, and on a monthly or quarterly basis place an order for the bulk foods you like to use. This is an affordable option to pricier health food stores, and by working in a collective you have more buying power and can share food that comes in bulk quantities. Our buying club purchases twenty-five-pound bags of rice and oatmeal and other grains at a time, and shares them out among members. We each get what we need in abundance, without the excess packaging or the visit to our local Whole Foods store. Our only other expenditure is large plastic bins that we use to store the grains so rodents do not come to call.

Selling the stone fruit of high summer at the farmers' market, Oakland, California.

Weekly farmstand, Oakland, California.

Foods from the Wilds of the City

When you reap the harvest of your land, you shall not reap all the way to the edges of your field, or gather the gleanings of your harvest. You shall not pick your vineyards bare, or gather the fallen fruit of your vineyard; you shall leave them for the poor and the stranger.

—Leviticus 19:9–10

In my small city, the fruit literally falls off the trees and onto the streets. Some people harvest their backyard trees, but many people let the fruit fall and rot. I'm hanging over the fences at times, longing for that last tasty fig that is destined for the ground. Fortunately, the age-old tasks of gleaning await us. All you need is a basket, a fruit picker, permission, and a little bit of time. If you happen upon a tree that is not obviously on anyone's property, it's fair game for picking. This kind of gleaning can be one of the pleasures of city living.

Foraging is an ancient, beautiful practice that traditionally supplied a portion of people's food needs over different seasons of the year. Foraging fell into disfavor in the last century through our reliance on processed foods, but was once the revered province of shamans, natural healers, and rural folks of all stripes. Herbs, medicines, tasty greens, root crops, and fruits in season are part of the forager's grocery. We reclaim the pleasure of the hunt, and pick up our forager's sticks once again. No more will we ignore this fine excuse for a productive walk in our neighborhoods. Foraging and gleaning are ways to eat local, save money, and practice our resourceful relation to place.

Gleaning

By far, the most reliable source of food in the city's gleaning web are those fruit trees standing in front of people's houses waiting for someone—you—to pick them. Don't delay. Walk up to that house and knock on the door. Ask your neighbor if you can harvest their fruit. Sometimes people who don't tend their trees are happy to share some of the excess, especially if you are willing to pick as much as you can and leave some in baskets on their front porch. Our fruit foraging practice yields hundreds of pounds of fruit for our family each year, an excellent return on the time we spend wandering the streets and picking it up.

Petaluma Bounty, a northern California non-profit dedicated to food security for all citizens, runs a program called the Bounty Hunters. After identifying trees whose fruit is going to waste, the organization makes relationships with the owners of the trees, and then heads out to pick up the extra produce. It's then given to local residents or sold at a low fee at the community farm that was established to serve food-insecure folks in the area. The Bounty Hunters picked three tons of food in their first year. In year two, four

Full-up and fantastic forage from a local orchard.
Photo by Trathen Heckman/Daily Acts

tons of food, which otherwise would have gone to waste and rotted on people's lawns, was foraged and distributed to people who need it. This program resourcefully uses the bounty around us to strengthen our community. You can start a project like this your neighborhood, too. (Check out www.PetalumaBounty.org for more inspiration.)

The Dumpster: The Last American Commons

In 1988, the Supreme Court, that great bastion of freedom for all, declared that once something is thrown away in a dumpster, it becomes common property, thus establishing the dumpster as possibly the last and largest commons in the U.S. Most people think eating or foraging out of a dumpster is nasty, but when you really get in there, you'd be amazed at how much perfectly fine food, good books, clean clothing, furniture and kitchenware the dumpster holds. "Freegans"—folks who live an anti-consumerist lifestyle of alternative strategies based on "limited participation in the conventional economy and minimal consumption of resources"—often dumpster dive for dinner.

Freegans "embrace community, generosity, social concern, freedom, cooperation, and sharing in opposition to a society based on materialism, moral apathy, competition, conformity, and greed."[58] The lifestyle involves salvaging discarded, unspoiled food from supermarket dumpsters. Freegans salvage the food for political reasons, rather than out of need. "These people don't eat out of dumpsters because they're poor and desperate. They do it to prove a political point. You wouldn't expect someone to choose a lifestyle that involved eating out of dumpsters. Kind of seems like something you do as a desperate last resort, but there's an entire society of people who willingly get their meals out of the garbage," MSNBC correspondent Tucker Carlson reported. [59]

Most of us won't go the hard-core route of the freegans, but a moderate approach to the riches abandoned in a dumpster never hurt anyone. Dumpster diving is a truly conservative approach to resource use. The ultimate reuse action, it can feed you and your farm animals quite well, while saving massive amounts of perfectly good stuff from going right into the landfill. While people may look at you askance with your feet in the air and your head in the dumpster, you are also bound to make some friends dumpster diving. And you'll definitely eat for less, and decorate your house for less, too.

Mapping the City Wilds

Using fruit as their lens, the artists' collective Fallen Fruit "investigates urban space, ideas of neighborhood and new forms of located citizenship and community. From protests to proposals for new urban green spaces, we aim to reconfigure the relation between those who have resources and those who do not, to examine the nature of and in the city, and to investigate new, shared forms of land use and property.... We consider fruit to be many things: a subject, an object, and a symbol... Everyone has a fruit story linked to place and family, and many echo a sense of connection with something very primal. One word for this thing could be sweetness."

Creating maps of public fruit—the fruit trees growing on or over public property—was this group's first

Learning to can and jam in the middle of the city—Los Angelinos slice and dice at Fallen Fruit's public fruit jam.
Photo by Fallen Fruit Collective

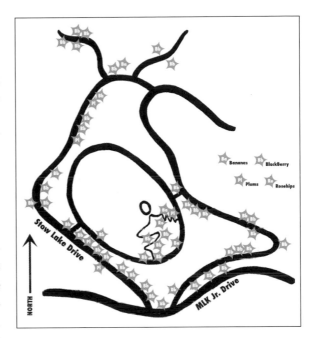

Map of Golden Gate Park's Fallen Fruit opportunities, Fallen Fruit Collective.

action. Their interests have expanded from mapping public fruit to sponsoring public jamming events to nocturnal fruit forages; community tree plantings on the margins of private property and in community gardens; and neighborhood infusions, where citizen take the fruit found on a street and infuse it in alcohol to capture the spirit of the place.[60]

Foraging for Wild Food in the City

City landscapes often provide botanical diversity that makes foraging for the city's weeds and wilds a tasty treat. Some city landscapes are more biodiverse than a forest or a meadow. Urban and suburban environments, while providing a wide variety for foragers, also provide a number of challenges. One of them is the liberal use of pesticides on many people's properties, which you definitely don't want to be eating. The other limitation is, of course, private property. It's best not to take food from someone else's tree—or garden—without permission. Beyond the problems of poison and property, foraging for wild foods has a couple of common sense rules.

- Know what part of a plant is edible, and don't experiment with the other parts.
- When you've positively identified a plant, be cautious. Take small bites at first to make sure you have no negative reaction to it.
- Plants that grow close to busy roadsides are best avoided—they're toxic.
- Eat wild foods only in season. They taste better.
- Just because something is a "weed" doesn't mean it can't hurt you. Conversely, just because it's a "weed" doesn't mean it isn't great to eat. Don't eat anything unless you know what it is and what its effects are.

When foraging, a basket, pruning shears, plastic bags, and a fruit picker may come in handy. You'll also need the more intangible skill of plant identification. An

Wild and cultivated flowers and leaves make up this beautiful summer salad.

awareness of where you walk and what you might find are also important. Having a broad palate is also helpful—if you're willing to try different things, you'll have more food for dinner. An awareness of the seasons and when certain plants are likely to be found is also useful. This is a most essential and beautiful skill to cultivate however you choose to practice it.

Here are a select few wild "weeds" common throughout North America that may be available in your city or town. A list of local available edibles—with names and identifying characteristics—will be your best source of information for foraging in your locale.

Common name	Botanical Name	Edible Part
Alfalfa	*Medicago sativa*	Leaf, flowering tops, seeds
Blackberry, raspberry	*Rubus spp.*	Fruits, leaves
Burdock	*Arctium lappa*	Root
Calendula	*Calendula officinalis*	Flowers
Chickweed	*Stellarioa media*	Leaf
Chicory	*Cichorium intybus*	Leaf, buds, roots
Chrysanthemum	*Chrysanthemum leucanthemum*	Leaf
Cleavers	*Galium trifidum*	Leaf
Dandelion	*Taraxacum officinale*	Leaf, roots, flower
Dock, sorrel	*Rumex spp.*	Leaf, stem
Lambs quarters	*Chenopodium album*	Leaf
Lemon balm	*Melisa officinalis*	Leaf, flowering tops
Lettuce, wild	*Latuca scariola*	Leaf
Mallow, cheeseweed	*Malva spp., M. crispa*	Flowers, fruit, young roots, leaf
Miner's lettuce	*Montia spp.*	Leaf
Mint	*Mentha spp.*	Leaf
Mustard, wild	*Brassica spp.*	Leaf, flower, seed
Nasturtium	*Tropaiolum majus*	Seed, flower
Pigweed, amaranth	*Amaranthus spp.*	Leaf, stems, seeds
Plaintain	*Plantago spp.*	Leaf, seed
Purslane	*Portulaca oleracea*	Leaf
Queen Anne's lace	*Daucus carota*	Leaf, flower, root
Shepherd's purse	*Capsella bursa-pastoris*	Leaf, roots, seeds
Sorrel, sheep	*Rumex acetosella*	Leaf
Stinging nettle	*Urtica diocica*	Leaf (cooked)

Urban Farm Kitchen

Right Food, Right Action, Right Awareness.

—*Masanobu Fukuoka*[61]

As our culture developed in the middle of the twentieth century toward more convenience and speed, heirloom skills of canning, fermenting, culturing, drying, and storing our extra produce were sacrificed. It's easy to see why some of the kitchen appliances and fast foods invented in the middle of the last century became popular and why the old skills were abandoned. They spared (mostly) women from hours in the kitchen, canning, pickling, baking, and fermenting the harvest. It's surely easier to open a can of soup than to cook it up at the end of the day, but it rarely tastes as good. What was lost in exchange for convenience is knowledge about our food sources, nutritious food grown in healthy ways, and the security that comes from being able to tend to life's essentials ourselves.

Homesteading returns us to these practices in the kitchen, in abundance. Some of the easy-to-learn heirloom skills are canning, drying, and freezing the abundance from the garden harvest. Food preservation processes, like fermentation and lacto-fermentation, which require a little heat and only a heart and hand for experimental kitchen chemistry, are also important heirloom practices. The living cultures of fermentation create a kind of medicinal food that has largely been replaced by over-processed and homogenized food at the expense of our health and well-being.

Some of your choices about how to preserve the harvest will come down to your available space, your personal preferences about energy use (human and electric), and the kinds of produce you have in abundance. Some things freeze well (juice, fruits, soup), and some not so well (except in soup, broccoli or zucchini, in my personal opinion, are better fed to the rabbits than the freezer). With all food preservation, and food growing for that matter, start with what you like. Don't waste your time or energy preserving something you won't cook with—unless

153

The harvest bounty of the garden in late summer.

you are thinking of giving some of these goodies away to friends and family as gifts. Food preservation is done when you have excess from the garden, foraging, gleaning or the farmer's market; don't run out to the store to buy extra produce and then preserve it. If that's your situation, save your kitchen time for fermenting projects or other kitchen chemistry experiments.

Freezing Food

Freezing food is a simple way to preserve the harvest. It uses the energy it takes to prepare the food, and the energy in the freezer. For folks with extra space in their freezer, this is the least labor-intensive way to store food, and it's so simple a child can do it (which may give you some ideas about how to get your budding homesteaders involved in the fun). You may want to invest in a small, extra freezer for preserving the harvest; if you are growing animals for meat, a freezer will be essential. Some won't want to add another appliance to the homestead, and will only freeze what fits in the freezer they already have. Freezing tends to keep the nourishment level of the food higher than if you cook it up for canning, or preserve it by drying.

When freezing food, choose unblemished, high-quality food that has been well cleaned. As you put it in the containers for freezing, make sure it's in the right size serving, remove as much air as possible from the container, and leave some room for the expansion that comes from freezing, especially if you are freezing food in glass jars. Use food-grade rigid plastic containers with airtight covers that seal securely, wide-mouth freezer jars with straight sides, or food-safe resealable plastic freezer bags. Butcher paper can be used to securely wrap solid foods. If you make tomato sauce, make more than you need and put the extra in a plastic container and freeze it. For optimum energy efficiency, keep your freezer as well stocked as possible.

If you are freezing uncooked vegetables (beans, broccoli, zucchini), blanch them first in boiling water for thirty seconds. Drain the blanched food and then "shock" it in ice water and drain well again before freezing. Beans, as well as strawberries, apricots, or plums, can be frozen on cookie sheets then packed into bags or boxes for easy retrieval and less water-logging. Tub freezing works best for things like tomato and apple sauce, and soups or stews made from your garden produce. A soup made in the summer from your extra zucchinis (always plentiful) is a great treat for a cold winter day.

Drying the Harvest

Drying the surplus harvest is a simple way to preserve it for the months when things aren't growing in the garden. A solar food dehydrator will maintain a nutritious and tasty supply of high quality, locally grown foods all year long. If you live in a part of the country where summers are hot and dry, a solar dryer is a wonderful tool. When meeting your conservation goals, the choice for a solar dryer over an electric one is clear.

Building a Solar Drying Rack

Materials needed

  1″ x 1″ wood
  Metal brackets
  Food-grade plastic mesh
  Staple gun and staples
  Drill and screws

A simple-to-build solar food dryer is a rectangular box with shelves holding a set of drying racks. The bottom is open, which allows for airflow; the top can be covered with a screen to keep off insects. This dryer took a few afternoons for an imperfect carpenter to design and build.

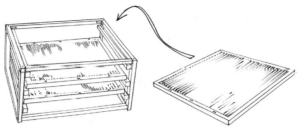

1. Construct three squares out of wood. Brace them with metal brackets.
2. Cover squares with plastic mesh by stapling mesh to the frames. Overlap mesh around wood frames to secure it. These are your drying trays.
3. Build two more squares of wood big enough to go around the drying trays. Brace them with brackets as well.
4. Turn these two squares into a cube by connecting them together with four pieces of wood in the corners, approximately six inches high.
5. Once you have made the rectangular cube, affix pieces of wood the length of the frame to the short sides of the cube. If you want to have three drying trays, affix three pieces of wood to the frame. These will support the trays as they sit in the rack.

How to Use a Solar Dryer

Slice the produce into pieces between a quarter and a half inch thick (depending on the fruit or vegetable) and place them on a rack. The general rule of thumb is that the thicker the fruit (like a nectarine, compared to a strawberry), the thinner the slices. This cuts down on drying time. Set up the solar dryer and place it in a sunny spot oriented to the south. If you're around during the day, speed up drying times by occasionally repositioning the dryer to track the sun as it moves across the sky. Many foods will dry out with one day of sunshine. Wet foods such as tomatoes or pears will require a second day. Stone fruit, apples, tomatoes, herbs, and vegetables can all be dried in the solar dryer.

Dried fruits can be stored in canning jars, as long as they are not placed too near a heat source. If you notice any sort of mold or insect life in the jars of dried fruit, toss the contents of the jar into the compost. Dried fruits also store well in resealable plastic bags, and can be frozen and thawed again with no ill effect

Slice fruits and vegetables about one-quarter-inch thick and let the sun do the work of drying them into tasty treats that will last into the next season.
Photo by Trathen Heckman/Daily Acts

(although drying, freezing, and then thawing again takes a lot of human and natural energy, something to consider before going into production). Some fruits, even when dried, store better than others: apples, apricots, and strawberries have a long shelf life, but pears are a bit more perishable (because they start out so juicy). Store them in the refrigerator or in sealed jars in the pantry.

	2.8	Plums
	3.0	Gooseberries
	3.1	Prunes
	3.3	Apricots
	3.4	Apples
	3.5	Blackberries
	3.5	Peaches
	3.6	Sour Cherries
	3.9	Sauerkraut, Vinegar Pickles
high acid	3.8	Sweet Cherries
low pH	3.9	Pears
water bath	3.5–4.6	Tomatoes
low acid	5.0	Okra
high pH	5.0	Pumpkin
pressure	5.0	Carrots
canning	5.2	Cabbage, Turnips
ONLY	5.3	Beets, String Beans
	5.4–5.7	Meat
	5.5	Spinach
	5.7	Asparagus, Cauliflower
	6.0	Lima Beans
	6.3	Corn
	6.9	Peas
	7.0	Fsh
	7.0	Milk
	8.0	Hominy

Canning

Canning is a simple way to preserve the harvest, and one of the central homesteading practices in the kitchen. It's a safe, energy efficient, and affordable way to provide food for your family throughout the winter months. There is a minimal amount of special equipment to purchase, and once you have tried it a few times you will be able to whip up and can a batch of compote or jam in an hour or so.

There are two basic canning methods—one for high-acid foods and one for low-acid foods. High-acids foods are all fruits, and tomatoes have the lowest acid content of the high-acid foods. Because of this, canning tomatoes requires a few extra considerations to make them safe. Tomatoes should either be cooked extensively before canning (for marinara sauce, or ketchup, which takes hours), or a tablespoon of lemon juice should be added to each pint jar to raise the acidity level. Low-acid varieties should not be used for canning projects. Vinegar-pickled vegetables are also considered high acid. Vegetables and meats are low-acid foods. Dairy should not be canned in a home environment, nor should you home-can any pureed vegetables, with the exception of tomato sauce.

Both high-acid and low-acid canning rely on heat to create a safe environment in the jar. High-acid foods can be safely canned using the boiling water bath method. Low-acid foods require more heat over a longer time, and the required temperature can be only be obtained at home using a pressure canner.

Pressure canning is a more advanced process for the serious homesteader as a good canner can cost several hundred dollars. The high heat and time involved with pressure canning means your food will be more than "well done" (as in canned corn or green beans). To preserve freshness and nutrition, fresh vegetables are much better preserved through freezing, root cellaring, or fresh pickling (lacto-fermenting). Pressure canning is well suited for canning prepared foods such as chicken soup, chili con carne, meat sauce, and so forth. It is also useful for cooking and canning cooked dry beans for ready-made use right out of the jar.

Whether you are canning low-acid or high-acid foods, the quality of the ingredients you use is crucial. Most canning guides call for fruits and vegetables in the highest state of freshness and ripeness. However, extra-ripe fruit that is on its way out can also be used for fruit compotes, jams, or sauces. The rule of thumb is, if you wouldn't put it in your mouth, don't put it in a jar. Canning won't make rotten fruit any better.

For jams, jellies, compotes, and sliced fruit, any type of sweetener may be used, including honey, cane sugar, agave, stevia, or maple syrup. For jam or jelly you may want to seek out a universal pectin that requires

Preparing cherries for cherry jam.

less sugar than traditional recipes. Universal pectin will use around one cup sugar to four cups fruit—this may sound like a lot, but traditional recipes call for seven cups of sugar to the same amount of fruit. Pay attention to processing times and the headspace you leave at the top of the jar. Processing time refers to the length of time the food in jars should stay in the boiling water bath. Headspace is the amount of space that should be left between the top of the food in the jar and the lid to allow for expansion and contraction during the canning process. The proper amounts for both of these will vary from project to project and also depend on the size of your jars.

Canned food is safe to eat as long as the seal holds. With each passing year, however, the quality of the food—color, texture, taste, and nutritional value—is diminished. For best results, try to consume your canned goods within one year of canning them. As with the garden, you will learn how much to can as you go along—some seasons there's just too much applesauce; in other seasons, there are not enough pickles. Once you feel you've got it figured out for your household, it will be easier to know how much work you will have to do in the summer putting the food up for winter. Consider putting up just what you will use from one season to the next, and store your canned goods in a cool, dark place for best keeping.

A note about food safety and botulism: The biggest danger in canning is botulism. Botulism is a bacterium that is present on most fresh foods. We eat it every day, but in an anaerobic environment, botulism can multiply to toxic levels. Fortunately, botulism spore cannot survive under conditions of acidity and heat, so if your food is in the acid range and you process it for the correct amount of time, it is extremely unlikely that you will poison yourself. If you have any doubt about the food, cook it for 10 minutes after taking it out of the jar to re-sterilize it.

In most cases it will be obvious if a jar of canned food has gone bad. The middle of the lid may be popped up, or the food will look and smell bad. Sometimes the metal inside the jar will rust. This is not poisonous, but certainly unsavory. If you have the slightest inkling that the food might not be good, toss it.

Tools You'll Need

Without a lot of expensive new tools, you can safely preserve your garden produce by canning. Here are the basics:

Clean mason jars (up to a one-quart size)
Jar lids and rings
Clean towels
Canning funnel
Lid magnet
Jar lifter

For water-bath canning: Two large pots (one for your food and one for your water bath). You do not need a special canning pot! If the jars rattle on the bottom during processing, just put a dish towel or round cake rack under them. Make sure that your pot is big enough to cover the tops of quart jars with water.

Canning Supplies:
A. Jar lifter
B. Mason jar with lid & ring
C. Large Pot
D. Canning funnel for filling jars

For pressure canning: Pressure canner in good condition with all rubber bits intact. A high quality metal-to-metal seal is safer than a rubber to metal seal. Older used canners should be refurbished before use.

Summer carrots pickled in salt brine.   *Photo by Dafna Kory*

Pressure canner for preserving low-acid foods.

TWELVE LIFE-SAVING CANNING RULES

Adapted from the *Encyclopedia of Country Living*

Follow these instructions to the letter to gain a wide margin of safety with your canning projects.

1. Do not use jars larger than one quart. The inside will not get hot enough.
2. Use water-bath canning only for high-acid foods: high-acid tomatoes, fruits, rhubarb, vinegar-based sauerkraut, and pickles. Chutneys and jam. *Vegetables, meats, and stews must be pressure-canned.*
3. Use only modern canning recipes from reliable sources.

4. Never reuse jar lids. Never use rusty or bent screw bands.

5. Don't use antique or French-type canning jars. They are not as safe as modern mason jars.

6. Check each jar rim carefully for chips by running your finger around it. Even a small chip can make it impossible to create a perfect seal.

7. Raw pack is not safe for low-acid vegetables such as beets, greens, potatoes, and squash.

8. Don't begin counting your processing time until after your water bath comes to a boil or until after your pressure canner has vented for at least ten minutes.

9. Always check the canning recipe for correct headspace.

10. Process for the fully recommended time, keeping the boil or pressure at the proper level during the entire time. If it drops in the middle you must start over and reprocess for the entire recommended time.

11. Lift each jar individually from the water bath with a jar lifter, keeping it upright. Do not tip jars until fully cooled. For pressure canning, do not open the canner until pressure has released and the pot is cool to touch. *Always open so that any steam vents away from your face.*

12. If the jar didn't seal (the center of lid is not popped down), discard the lid, put on a new one, and reprocess, or eat the contents immediately instead of storing.

How to Water-Bath Can

1. Prepare your jars. Make sure there are no chips in the rims and the jars are clean and dry and have new lids. The rings may be reused as long as they are in good condition, but lids should be replaced every time.

2. Prepare your food. Refer to a modern canning book for tried-and-true recipes, process times, and headspace.

3. Put your cooked food into clean jars using a ladle and your funnel. Use a clean rag to wipe off any food or juice that gets on the rim of the jar.

4. Place the lids on the jars with your lid lifter, being careful not to touch inside the ring. Tighten the rings finger tight.

5. Place the jars in the hot water bath, making sure the tops are covered with water. Count your processing time from when the water starts to boil.

6. When the processing time is up, lift the jars from the water using your jar lifter, without tipping the jars. This is important as the jars are not yet sealed and food could still get between the jar rim and the lid. Set on a counter to cool. When the lid pops down (you may hear a ping!) the canning process is complete.

Tomato sauce, fruit compote, apples in syrup.

How to Pressure Can

Follow steps 1 to 4 in "How to Water-Bath Can."

5. Place your jars in the canner and add water up to about two inches above the jar tops.

6. Close the canner tightly.
7. Turn the heat on and wait until steam starts to blow out of the pressure valve.
8. Count ten minutes from this time and put the weight on the valve.
9. Once the gauge reaches the target temperature, start counting your processing time. You may need to adjust the heat to maintain the temperature.
10. When the time is up, turn off the heat and let the canner cool down without removing the weight or the lid.
11. When the canner is cool, you can open it and remove the jars.

FRUIT CANNING PROJECTS

Compote: Any fruit, chopped, mashed, or chunked, lightly sweetened. You can use nectarines, peaches, pears, cherries, plums, apples, and apricots alone or in combination to make delicious fruit compote.

Fruit in Water, Juice, or Syrup: Whole, halved, or sliced fruit can be canned in syrup, which can be water, unsweetened fruit juice, or water with sweetener dissolved in it (see chart below). Agave, cane sugar, honey, stevia, maple syrup, or any other sweetener may be used. Hot-pack or raw-pack methods can be used.

Syrups for Canning

Type	% of sugar	Sugar	Water	Yield
Extra-Light	20	1 ¼ cups	5-½ cups	6 cups
Light	30	2 ¼ cups	5-¼ cups	6 ½ cups
Medium	40	3 ¼ cups	5 cups	7 cups
Heavy	50	4 ¼ cups	4-¼cups	7 cups

Jam: Fruit cooked with optional sweetener, and jelled with pectin. Any kind of sweetener may be used. Use universal pectin for success with low and no sugar recipes.

Jelly: Fruit juice or strained fruit with sweetener and pectin added.

Fruit Butter: Fruit, sweetener, and spices reduced to a spreadable consistency.

Juice: Juice made in an electric juicer or stovetop steamer.

RAW PACK AND HOT PACK

Raw Pack: Raw fruits or vegetables are placed in jars, covered in water, and processed in either a water bath or pressure canner (depending on the acid level). Raw pack is one of the more risky projects, so use your crafty researching skills to check on processing times for raw pack projects.

Hot Pack: Fruits or vegetables are placed in water on the stove. The water is brought to a boil and the fruits or vegetables are then packed into jars along with hot water and processed accordingly.

TOMATO PROJECTS

Tomato Sauce, Plain: Puree tomatoes and cook down on the stove with a little salt.

Tomato Paste: Like sauce, but cooked down even further.

Marinara Sauce: Pureed tomatoes cooked down on the stove. You may add up to 5 percent by weight of non-acid ingredients thinly sliced (garlic, onions, herbs, and mushrooms are typical). Do not use too much oil, as oil can trap the botulism spores and prevent them from being neutralized. "Not much" is about four tablespoons per twenty pounds of tomatoes. Official sources say to add a tablespoon or two of lemon juice. Do not add Parmesan or other cheese.

Raw or Hot Pack Whole Tomatoes: Blanch, peel, and pack into jars. Cover with water. Add salt to taste and one tablespoon lemon juice per pint.

Diced Tomatoes, Raw Pack: Blanch, dice, and pack into jars. Top off with water, salt, and lemon juice.

Tomato Juice: Peel, puree, and strain tomatoes until the juice is released. Add lemon juice.

Photo by Trathen Heckman/Daily Acts

Inversion Canning

Inversion canning is the process of putting boiling hot, cooked, high-acid foods into boiling hot, just-sterilized jars, screwing on the lid securely, and flipping it upside down to cool and seal. Given that the food and jar are hot enough, the inversion process sterilizes the small amount of air that is trapped in the jar when you close it. Use this method at your own risk. It is quick and usually effective, but it is not considered 100 percent safe by the USDA. It is widely used in Europe, as it has been for decades, and until about five years ago, was also recommended in the United States. We use it all the time and find it simple and effective. If you choose this method, use the following guidelines to be as safe as possible.

1. Sterilize your jars in boiling water for at least 2 minutes, or in water at 169°F and above for 5 minutes. This will ensure sterility. Use a pair of tongs to pull the jars and lids from the boiling water when you are ready for them.

2. Never touch the rim, inside the rim, or the inside of the lid with your fingers as you are sealing it.

3. If you wipe the rim, use a clean towel dedicated for that purpose.

4. Make sure whatever it is you are canning is at a good boil for at least 10 minutes before processing. The combination of sustained heat plus acidity is what makes this process work.

5. Work quickly and efficiently. Once you have put the fruit in the jar, screw the top down within seconds.

6. Once the jar is filled and the top is firmly on, flip the jar over. The heat from the food and the jar will seal the jar. Leave the jar turned over until the lid has sucked in and the jar is cool enough to touch.

7. When you turn the jars over, they should be sealed just as they would be if they had been water-bath canned. If you ever have *any* doubt about the seal, put your jars in a water bath to ensure the seal.

Root Cellaring

If you live in a cold winter climate and are lucky enough to have a cellar, storing root vegetables in this cool part of your home is a great way to preserve the harvest. Different winter vegetables need different things (carrots like to be stored in sawdust, and squash are happy in a bucket or barrel), but all can be stored in a cool place for many months after they are harvested. Apples, citrus fruits, grapes, pears, beets, cabbage, carrots, cauliflower, celery, corn, dried beans and peas, endive, onions, garlic, potatoes, winter squash, root crops, sweet potatoes, and mature green tomatoes can be stored in live storage. For those of us living in the temperate climates of the West, root cellaring is less necessary as we benefit from a yearlong gardening calendar that makes it possible to have fresh veggies on the table in all seasons. Even in sunny California, though, you can store onions, garlic, and potatoes over winter in a cool garage, as well as apples, winter squash, and beets. Make sure they are protected from rodents, or any kind of moisture.

You can create live storage for a variety of crops in a dry, clean area of a basement or garage. Partition off an area where you can place shelves and keep the space free of humidity. Use storage containers such as plastic containers with airtight lids, wooden crates, cardboard boxes, baskets, trash cans, mesh bags, and plastic stacking bins. Make sure the site is accessible, clean, and that you can check your produce frequently. If you store something and it starts to go bad, the decay can quickly spread to the rest of your stores.

Create a buried storage bin by using a wooden barrel, box, or galvanized tin garbage can filled with food and buried to form a mini root cellar. Make sure the container is clean and nontoxic. This method will not work well in regions where wintertime is a rainy affair, but in very cold climates, it's a good option. Once you open a storage bin that has been buried, you should empty it within one to two weeks or the produce will spoil. For this reason, and if you have the space, it may make more sense to dig a few buried storage bins.

Home-scale root cellars can be constructed from old refrigerators buried underground, on top of a bed of rocks and gravel for drainage. A storage barrel can be buried underground at a 45-degree angle, and banked with rocks and gravel for drainage. The top of the barrel is covered with soil held down with a retaining board. A vegetable basket can also be buried within a dry, lined hole. When creating buried root cellars, take into account the water retention of the soil, the presence of predators underground, and the ability to keep the produce dry to prevent rot.

Andean purple potatoes, fingerlings, and Yukon golds, as far as the eye can see. Store them in a dry spot for eating in autumn and winter.
Photo by Trathen Heckman/Daily Acts

The garlic harvest, braided and ready to hang in storage.

In-garden storage for root crops like carrots, parsnips, and turnips works in many climates. Some of these crops improve in flavor after there has been a frost. Cover the root crops with a foot or more of straw, hay, or leaves once the ground begins to freeze in the fall.

FROM CONSUMPTION TO PRODUCTION

When refining your practices around sourcing, eating, preserving, and preparing your food, think about changes within a realistic timeframe.[62]

Six-Month Transition Timeline

Wean off packaged foods

Buy from farmer's markets or join a CSA farm

Buy bulk from cooperative grocers

Begin to seek out space to grow your own

Research nutrition

Assess your diet—can you bring it closer to the ground?

Learn some gardening basics

Take basic cooking classes

Try baking or fermenting or bread making

Eat 50 percent or more of your meals at home

Twelve-Month Transition

Locate space for growing your own: Front, back, sidewalk, rooftop, vacant lot, community garden?

Plant Zone 1 bed—herbs, root crops, and greens

Plant fruit and nut trees

Basic meal planning to source food locally

Eat 75 percent of meals at home or in community

Buy your food in bulk

Attempt preservation skills—canning, drying, and freezing

Solar-oven cooking

Gain local wild foods awareness

Thirty-six to Sixty-Month Transition

Develop crop rotation plan to grow as much food as you can in the space you have

Develop additional urban gardens in your neighborhood

Practice urban forage and gleaning

Plant a perennial food bed

Plant a food forest

Become a beekeeper

Take on chickens, rabbits, or quail

Develop local exchange for surplus produce and/or preserves

Innovate vertical garden experiments to maximize growing space

Living Cultures

Culture always originates in the partnership of man and nature. When the union of human society and nature is realized, culture takes shape of itself. Culture has always been closely connected with daily life. And so has been passed on to future generations, and has been preserved up to the present time.

—*Masanobu Fukuoka*[63]

People have co-evolved and partnered with countless microbial life forms in mutually beneficial relationships for a long time. Human culture has evolved with the living cultures of bacteria and yeasts, the beneficial microorganisms responsible for the transformation of fresh into fermented food. Every society developed unique fermented delights, many of which are treasured to this day. The long list of fermented foods that many of us know and love includes olives, cheese, sourdough bread, dry meats, wine, fish sauce, ketchup, vinegar, coffee, tea, vanilla, ginger beer, tofu, soy sauce, miso, natto, tempeh, sauerkraut, and chocolate. Other simple-to-cultivate treats include yogurt, kefir, kombucha, kimchee, and lacto-fermented vegetables and sodas. Over time, fermented foods have entered the lexicon of global good taste, becoming identified with a place and its people. It is no coincidence that our living language coincides with the processes of living cultures—they have evolved and grown in tandem with one another and weave together some of the distinctive threads of social traditions.

When we are culturing and fermenting foods, we recruit the microbes and beneficial bacteria from our own kitchens in the process. Fermenting and culturing are collaborative processes we enter into with the microbial world around us. By now, you're sure to recognize this as one of our central themes—the benefits we derive, personally, economically, and environmentally—by remaking our relationships to the essential sources of life that surround us. In the case of fermented foods, not only do we save money, and positively impact our environment, we also get to eat fabulously delicious food.

These traditional foods also enhance our health and the proper functioning of our bodies. Unseen, beneficial microbes enhance the flavors and nutritional quality of our foods, transforming fresh food into something tasty, healthful, and new. We have dozens, if not hundreds, of symbiotic relationships with the communities of bacteria and yeasts that live within and around us. Making friends with these microorganisms enhances our personal health and the health of the planet. Many of the microbes and fungi involved in fermentation processes partially digest our foods, making them easier to break down in our own systems and more nutritious for us to eat. Many also help preserve our foods, extending their shelf life over months and in some cases years, while often enhancing their flavor. These foods are alive—vital, complex, and complete, a perfect example of "whole food."

Cucumbers, dill, and onions make a simple summer pickle. They ferment for a short time before they are done—a day or two, and not more than one week.

Before pasteurization and the ability to preserve food through refrigeration, fermentation was one of the principal ways people extended the harvest. The industrialization and overprocessing of food puts these pioneer species at risk, and in the last fifty years or so, many have disappeared from our tables. Fermented food, unlike pasteurized food, is alive with microorganisms that are of considerable benefit to our overall health, especially our digestive health. Some researchers and nutritionists believe that the rising prevalence of colon cancer is related to the lack of fermented foods in our diets. Fermented foods aid in the predigestion of certain foods that can be difficult to digest, including soy, grains, and milk. Fermentation also makes minerals and vitamins in our food more easily assimilated.

The enzymes created in the fermentation process catalyze countless bodily processes. Microbial organisms in fermented food create the B vitamins that support better digestive and overall health. Anticarcinogenic lactobacilli in fermented foods create omega-3 fatty acids essential for cell membrane formation and immune system function. Living foods remove toxins such as gluten, cyanide, and phytic acid.[64] When cabbage is fermented to make sauerkraut, the vitamin C content is increased by a factor of ten. Fermented foods replenish the flora of our digestive tracts. There is also research suggesting that the over-pasteurization of our food has created conditions in our bodies that are literally too "clean," lulling our immune systems into disuse and giving rise to a whole host of autoimmune illnesses. For the health of our bodies, we need to relearn our partnership with beneficial microbes in our environment.

The Process—What Makes It Work

Before cultures were freeze-dried and prepackaged, the cheese, wine, or sourdough starter was a cherished family heirloom. Even today, living cultures continue to be collected, fed, and passed on, and you can participate in this age-old tradition in your own kitchen. The process of fermentation happens naturally as bacteria in the environment seek out their own favorite foods to grow in. In the process, bacteria eat available sugars in food and transform them into acids. When we intentionally ferment our food, we create the conditions for these natural bacterial reactions. These conditions are created with salt, which pulls the water out of the food, makes it crunchier, and gives desirable bacteria a competitive advantage; temperature, which speeds up or slows down the process; and the creation of an anaerobic environment, which is necessary for the fermentation process and limits the interaction of air and thus airborne mold spores with the food.

Lacto-Fermentation

When lacto-fermenting vegetables, we are inviting lactic acid forming bacteria to transform our fresh veggies into products like sauerkraut, kimchee, and a variety of relishes and pickled preserves. Lactic acid is a natural preservative that inhibits putrefying bacteria. Starches and sugars in vegetables and fruits are converted into lactic acid by the many species of lactic acid–producing bacteria. These lactobacilli are ubiquitous, present on the surface of all living things and especially numerous on leaves and roots of plants growing in or near the ground. The culturing process is safe; the bacteria that acidify food eliminate unhealthy bacteria that cannot thrive in an acidic environment.

The process of lacto-fermentation is very simple. Vegetables are washed, cut up, mixed with salt and herbs and spices, and then pounded or massaged briefly to release their juices. They are then pressed into an airtight container and placed in cool spot out of direct sunlight. Salt inhibits the putrefying bacteria for several days until enough lactic acid is produced to preserve the vegetables for many months. Some fermentation processes take a few days (sauerkraut), others a few weeks (kimchee), and some take years (miso). There are many recipes for fermenting, and many different fermenting projects; here are a few simple ones we enjoy at home.

You'll need a few simple tools to begin fermenting fruits and vegetables at home. The most important will be a fermenting vessel—you can use either a crock or a jar. Look for a ceramic crock at the hardware store, or if you're lucky, score one at the thrift store, though they are getting harder and harder to find. If you use a glass jar for fermenting, one with a wide mouth will work best. Jars one quart or larger are recommended. Other than the fermenting crock, you'll need clean towels, salt, and produce for fermenting,

Lacto-Fermenting Vegetables and Fruit

This is the basic process for making sauerkrauts, kimchee, and fermented chutneys. The many different recipes for these delicious foods will be at your fingertips once you master the process of lacto-fermenting.

Ingredients

Fruit or vegetables to be fermented. Try a combination of kale, cabbage, and kohlrabi. Or ginger and carrots. Or plums and ginger. One pound of vegetables will make approximately one pint; two pounds will make approximately one quart.

Salt

Whey (optional)

1. Use 1 to 1½ tbsp of salt per quart, half that if using whey.
2. Grate or slice veggies into a large bowl or pot.
3. Add salt to taste. A little more if the weather is warm, a little less if it is cool. Add spices or herbs if desired. (See our suggestions below.)
4. With clean hands, mix, mash, and massage the veggies until they start to release their juices.
5. Pack veggies into the fermentation vessel, pressing down to close up any air pockets.

Mixing up cabbage for sauerkraut.
Photo by Diane Dew Photography

6. At this point the veggie juices will now mostly cover the vegetables. If they do not, add a bit of salt brine or whey.

7. Use weight or other means to keep the vegetables submerged in the juice. If your fermentation vessel is a jar, another jar filled with water that fits inside the opening of the first will work well.

8. Do not cap tightly.

9. Let sit at room temperature, no more than about 80 degrees F, until the mixture starts to bubble and has a tangy fermented smell.

10. Put in the refrigerator or other cool place, such as a basement (ideally no more than 55°F) and eat when you like the way it tastes.

11. For fruits, refrigerate and start to eat after two weeks.

Carrot, ginger, and beet kimchee.

Some people like their ferment quite young while other people don't think they're "done" unless they've been sitting at least a year. The longer you let it sit, the tangier it will become. Vegetables stored this way can be preserved for several years. If a little scum or discoloration forms on top, don't worry—just skim it off. If it's a really smelly mess, compost it.

You can do this fermenting project with a wide variety of vegetables, including cabbage, carrots, garlic, cucumbers, celery, turnips, beets, onions, greens, radishes, rutabaga, or kohlrabi. Traditional picking spices include juniper berries, cinnamon sticks, coriander, fennel, mustard seed, peppercorns, whole cloves, fresh dill, whole allspice, bay leaf, ginger, and coriander. Adding these spices will change the flavor, but not the process.

Traditional sauerkraut and kimchee

Traditional Sauerkraut

> 1 medium cabbage
> 1 tbsp caraway, mustard seed, and/or juniper berries
> 1 tbsp salt
> 4 tbsp whey (add additional tablespoon of salt if whey is not available)

Kimchee

> 3 parts Napa cabbage (coarsely chopped)
> 1 part carrots, daikon, scallions grated, chopped, or sliced
> Garlic, ginger, spicy peppers (to taste)
> Salt
> Optional additions: chili paste, sweetener, fish sauce

These recipes for traditional sauerkraut and kimchee use the process outlined above.

Making Whey and Cream Cheese

The amount of salt can be reduced or even eliminated in lacto-fermenting if whey is added to the pickling solution. Rich in lactic acid and lactic acid producing bacteria, whey acts as an inoculant, reducing the time needed to ensure preservation. The use of whey will result in consistently successful pickling; it is essential for pickling fruits. Keep the vegetables at room temperature for the first few days; afterwards, they must be placed in a cool, dark place for long-term preservation.

Use two quarts whole milk buttermilk, yogurt, or raw milk. If you are using yogurt, no advance preparation is required. If you are using raw milk or buttermilk, place the milk in a clean glass container and allow it to stand at room temperature one to four days until it separates into white curds and yellowish whey.

Line a large strainer set over a bowl with a clean dishtowel. Pour the yogurt or separated milk into the strainer, cover, and let stand at room temperature for several hours. Yogurt can take up to 12 hours to drain through the towel. The whey will run into the bowl and the milk solids will stay in the strainer. Tie up the towel with the milk solids inside, being careful not to squeeze. Tie this little sack to a wooden spoon placed across the top of a container so that more whey can drip out. When the bag stops dripping, the cream cheese in the bag is ready.

Store whey in a Mason jar, and cream cheese in a covered glass container. Refrigerated, the cream cheese keeps for about 2 weeks and the whey for about one month, after which time the strength of the culture will begin to degrade. Use the whey in the lacto-fermented recipes described here. Whey can also be used in lacto-fermented sodas (below), or to soak grains overnight to make them more digestible.[65]

Lacto-Fermented Sodas

Naturally lacto-fermented sodas are both tasty and healthy. They are great to drink with food as they assist digestion. And kids love them too! Because they are fermented by lactobacilli and not by yeast, they are non-alcoholic and packed with beneficial bacteria. Infusions for homemade sodas can be made from the wide range of what's available, including sassafras, sarsaparilla, ginger, elderflower, elderberry, valerian, cinnamon, spruce, juniper, spearmint, wintergreen, hops, star anise, yarrow, rosehips, hibiscus, most fruits (mashed), lemon, orange, grapefruit, raspberry, strawberry, peach, and pear. Don't forget to ferment whatever's good and growing in your neighborhood.

For all the lacto-fermented sodas described below, you'll need the following equipment to get started.

Equipment

Pot
Funnel or sieve
Fermentation vessels—jars, crocks, carboys, buckets, etc.
Measuring cups and spoons
Bottles and caps

The ingredients for all lacto-fermented sodas are the same, and vary only in the kinds of herbs or fruits with which you infuse your beverage, and your choice of sweetener.

Ingredients

Non-chlorinated water
Fruit or herbs

Sweeteners—agave, cane, maple, honey

Starter cultures—this can be whey (described above), a commercial culture, your last batch of soda, or make your own

Ginger (optional)

The basic beverage-making process is as follows:

1. Wash your fermenting vessels in hot soapy water.

2a. Make an herbal infusion by placing herbs into hot water, as you would for tea. Let the herbs steep to taste, then strain them out of the water OR

2b. Place fresh fruit or fresh fruit juice, up to two cups per half gallon, into your fermenting vessel.

3. Add approximately one cup sweetener per gallon.

4. Add two cups starter culture or one cup whey per gallon to fermentation vessel.

5. Add infusion to jar until close to the rim of the jar or, if making a fruit soda, top the jar off with pure non-chlorinated water.

6. Add two or three slices of fresh ginger and/or a pinch of salt. (This is optional, but it helps with the fermenting process.)

7. Cover loosely (you can use a canning lid, just don't screw it on all the way).

8. Let sit for two to six days, until the mixture in the jar starts to get fizzy. Timelines will vary with temperature and cultures. The only way to tell how the culturing process is going is to check your ferments every day by tasting them until they've reached the taste you like.

9. Bottle, refrigerate, and drink. If you seal in airtight containers, you can let it sit out for a few hours so the carbonation builds up, but be careful—a potent lacto-fermented soda can be quite explosive! If your container is airtight, it is best to release extra carbonation, even if refrigerated, if you don't drink the soda within the first few days.

As you can see from this recipe, proportions are approximate and vary according to taste. The lactobacilli do need sugar to get started, however, so if you like a tarter drink, just let the bacteria ferment longer. Once you master the basic process, you'll feel more comfortable making some of these other choices about taste and fermentation level.

Making Your Own Starter Culture

Lacto-fermented sodas need a starter culture to begin the process of fermentation. You can make one at home for sodas or ginger beer. Every beverage will need either a starter culture or whey to begin the fermentation process.

1. Grate or finely dice fresh ginger root and put a tablespoon of it into a mason jar three-quarters full of water, along with two teaspoons white sugar.

2. Add another two teaspoons each sugar and ginger every day for a week, at which time it should become bubbly with a pleasant odor.

3. If it gets moldy, dump it and start over.

Once you have your starter, use it to make a small quart-sized batch. Then use two cups of this for a gallon batch. (You can drink the rest.) Save a pint from each batch you make to inoculate the next batch. Use two cups of this starter culture or one cup whey per gallon of soda. Both whey and starter can be stored in the refrigerator for a few weeks before the cultures use up all available food and start to die off. You can stave off having to "catch" a new one by adding a tablespoon each of sugar and ginger.

Lacto-Fermented Lemon Grapefruit Soda

On a hot summer day after working in the garden or the beehive, our family loves this lemon-grapefruit soda. It's refreshing and simple to make with fruit that grows in our neighborhood. If you have access to lemons, grapefruit, or other citrus, we recommend this recipe.

Ingredients for one gallon

5 lemons

1 to 2 grapefruits

½ tsp sea salt

1 cup whey

1 cup agave

Filtered water to taste

1.   Squeeze juice of lemons and grapefruit into container.
2.   Add body of two of the lemons and half of one grapefruit.
3.   Add rest of ingredients, and fill with filtered water.
4.   Stir or shake to dissolve sugar.
5.   Put in a warm spot for two days or until fizzy.
6.   On the third day, strain into glass bottles with tight-fitting lids and top off if necessary.
7.   Put in warm spot for another one or two days. Test a bottle to see if carbonation has developed. If so, refrigerate. If not, put in warm spot for longer.
8.   Experiment with many variations of juice, flavors, and spices. Enjoy daily.

Kombucha

Kombucha is another delicious fermented drink you can make at home. The Kombucha culture is a SCOBY, a "symbiotic culture of bacteria and yeast," that make up a thick glutinous pad sometimes called a mushroom (although, to make things more confusing, it contains no fungi). Kombucha is made by feeding the culture pad its preferred food, a mixture of white sugar and black tea, which it then ferments and transforms into a slightly effervescent, tangy beverage. Kombucha has been a staple in numerous cultures around the world for a long time, and has numerous claimed health benefits, primarily the production of glucuronic acid that is said to be a powerful aid to the body's natural cleansing process, a boost to the immune system, and a proven prophylactic against cancer and other degenerative diseases.[66]

To start making kombucha at home, you have to source a kombucha SCOBY (sometimes called the kombucha "mother"). Try craigslist.org, mail order, or asking around in your community for someone who already has one. The kombucha mother proliferates freely—anyone making kombucha at home will be able to spare some to help you get started. The original culture pad separates easily into numerous SCOBYs that can be passed on to others.

Ingredients

> 1 gallon water
>
> 1 cup white sugar. Other types of sweetener will not work. Organic evaporated cane juice is great if you can get it.
>
> 5 teabags of black or green tea, or 5 tbsp of loose tea. You can use any combo of black, green or white tea, but this must be from the actual tea plant. Other herbal infusions will not work.
>
> ½ cup vinegar or already-made kombucha
>
> Kombucha mother

1. Boil water and remove from heat.
2. Add sugar to water and stir until dissolved.
3. Add the tea. Allow the tea to steep until the water has cooled to 90 degrees or below.
4. Remove tea bags or strain tea.
5. Pour liquid into fermenting vessel.
6. Add vinegar or already-made kombucha once the tea is at room temperature.
7. Place the kombucha mother on top of the liquid.
8. Cover with a cloth. Hold cloth in place with a rubber band around the mouth of the fermenting vessel.
9. Place in a cool spot out of the sunlight and allow the beverage to ferment. Don't let the kombucha sit too close to compost or fresh foods. This can sometimes add unwanted bacteria to the process.
10. It will take between two and eight days for the kombucha to ferment. As with other fermented beverages, some like it with some residual sweetness and some like it quite tart. Taste it along the way until it tastes "done." If you let it go too far, it will ferment into a vinegar-like drink that is a bit strong for drinking but works fine in salad dressings.

You can infuse kombucha with other teas—hibiscus, chamomile, or peppermint—as well as fruit juices, as long as you start with the basic recipe outlined above. Save half a cup from each batch to use as a starter for the next batch.

If you want to take a break from kombucha production, store the kombucha mother with some kombucha liquid in a container in a cool, dry place until you are ready to make more. If the kombucha mother begins to turn black or if the resulting kombucha doesn't sour properly, it's a sign that the culture has become contaminated. Compost it, and get another mother to begin again.

Farmstead Cheese

If you have a goat, or access to raw milk, making cheese will be not only natural, but also necessary, as you won't want to waste any of the precious milk. Cheese making is an elegant culturing process that takes many years to master and perfect. Some of the conditions for making cheese in a noncommercial kitchen are less than ideal; for this reason, we want to share some simple techniques with you for making cheese you can succeed with right away, and that save money and taste better than the cheese you can buy in the store. Included in this section are recipes for yogurt, cultured butter, and feta cheese (a personal favorite).

Cheese making is the process of separating the milk solids (curd) from the milk liquids (whey). There are two ways to do this—a strong acid plus heat or an acidifying bacteria (lactobacilli) plus a coagulating agent (rennet). Acid cheeses are not made with living cultures so they lack that added benefit and don't keep as long. However, they are easy and satisfying to make, a good first step for beginners.

Cheese made with cultures may be fresh or aged. Fresh cheese keeps about as long as milk in the fridge. Aged cheeses were developed as a way to preserve the abundance of milk during calving and kidding times in the ages before refrigeration. The cheese recipes we share here are farmstead cheeses, which farm wives have been making since cows have given milk. They are easy and forgiving.

There are two basic types of culture: mesophilic (which thrive in mid-range temps of 72ºF to 105ºF) and thermophilic (heat-loving, from 90ºF to 120ºF). Not surprisingly, cultured dairy from Mediterranean climates like Greece and Italy are often made with thermophilic cultures, while northern traditional cheeses are made with mesophilic cultures. People traditionally cultured at "room temperature." You can purchase these cultures in freeze-dried powdered form from a cheese-making supply house or you can simply use buttermilk (mesophilic culture) or yogurt (thermophilic culture) in any recipe that calls for these cultures.

Besides milk, buttermilk, and/or yogurt, you will need rennet. Rennet is a natural complex of enzymes produced in any mammalian stomach to digest the mother's milk, and is often used in the production of cheese. Rennet contains many enzymes, including a proteolytic enzyme (protease) that coagulates the milk, causing it to separate into curds and whey. The active enzyme in rennet is called *chymosin* or *rennin*, but there are also other important enzymes in it, such as pepsin or lipase. There are nonanimal sources for rennet that are suitable for vegetarian consumption but are often made from genetically modified organism (GMO) sources. Rennets vary in their strength (ability to coagulate milk). For best results, we recommend freeze-dried or liquid calf rennet. Keep liquid rennet in the refrigerator. Keep powder and tablet rennet in the freezer.

Basic Equipment for Cheese Making

Stainless steel pot

Strainer or colander

Bowls

Cheesecloth (The real kind, which must be ordered from a cheese store. Loose weave muslin will work.)

Dairy thermometer (with a temperature range from freezing to boiling)

Measuring cups and spoons

Forms for holding the cheese (bowls, mousse cup, moulds)

Jars (for yogurt)

Aging container

Draining board

Whisk

Curd knife

Slotted spoon

For yogurt: small camping cooler OR heating pad OR foam cooler incubator

Recipes

We like the following recipes for simple-to-make yogurt, butter, and cheese.

Yogurt

Ingredients

> 1 gallon milk
>
> 1 cup fresh plain cultured yogurt

This recipe will make nine pints of homemade yogurt, or four and half quarts. We recommend mason jars for storing the yogurt once it is prepared. Lids and rims do not have to be new, but everything has to be extremely clean.

1. Thoroughly wash all equipment with soap and hot water before you begin.
2. Heat milk on low to medium heat to between 185° and 195°F. Do not burn!
3. While the milk is heating, boil water and sterilize mason jars by setting them in the boiling water (anything over 180°F is fine) for one to two minutes each. Pull the jars out with tongs and set them upright on a clean cloth. Sterilize lids and rings as well.
4. Cool the milk to between 122°F and 130°F.
5. Gently mix one cup of the yogurt with a cup of the cooled milk. Add this mixture to the rest of the milk and stir to mix.
6. Pour into mason jars and seal.
7. Incubate for three to four hours until gelled. This can be done in several ways. If you have a gas oven with a pilot light, it will work to set them directly in the oven. You could also pour the water with which you sterilized your jars into a camping cooler and add cold water until the temperature is about 130°F.
8. Place jars in cooler and close.
9. Remove jars from incubator and place in the fridge.
10. Save one pint to start your next batch. You can do this three or four times before the culture becomes contaminated or too weak to reuse; then buy fresh yogurt from the store again. Yogurt will keep four to six weeks in its sealed jar.

Another method is to take a foam cooler or cardboard box, cut a hole in it for a 25-watt light bulb, and turn it upside-down over your yogurt. You can also turn your oven on to 100 degrees and leave the yogurt in there for four to five hours. For all methods, monitor the temperature with a thermometer. Be sure temperature range hovers around 100 degrees.

Yogurt Cheese with Herbs and Garlic

Ingredients

> Quart of yogurt
>
> Herbs: choose from rosemary, dill, basil and salt and pepper to taste
>
> Garlic

1. Set a large strainer over a deep pot.
2. Line with cheesecloth or muslin.

3.   Pour in yogurt.

4.   Tie ends of cloth and hang, using a wooden spoon set on the rim of a large pot, or on an S-hook hanging from a cabinet handle.

5.   Set in a cool place and let drip for twelve hours.

6.   After twelve hours, scrape the cheese from the cloth and season to taste with salt, crushed garlic, and fresh or dried basil, thyme, tarragon, oregano, or herb blend.

7.   This cheese will keep for seven to ten days in the fridge.

Homemade yogurt.   *Photo by Dafna Kory*

Lemon Cheese (Whole Milk Ricotta)

Ingredients

1 gallon of milk

¼ to ½ cup lemon juice (or buttermilk, citric acid, or vinegar)

Herbs to taste

1.   Heat one gallon of milk to 175°F.

2.   Add one fourth to half a cup lemon juice and stir.

3.   Within five minutes the cheese will curdle.

4.   Drain in cheesecloth until it is the consistency you like.

5.   Break it up with your fingers. This is known as "milling" the cheese.

6.   Flavor as you wish.

For extra creamy ricotta, add a fourth cup of whole whipping cream and mix thoroughly. You can also make this cheese with a half cup vinegar, one tablespoon citric acid, or one quart cultured buttermilk in place of the lemon juice.

Ricotta Salata (Salted Ricotta)

Ingredients

1 gallon of milk

¼ to ½ cup lemon juice (or buttermilk, citric acid or vinegar)

Herbs to taste

1.   For a salted cheese that will last a bit longer, start by following the Lemon Cheese recipe above, up to Step 4.

2.   After draining, mill the curd and mix in one tablespoon salt.

3.   Press into a mousse cup, a bowl, or a cheese mould and set a weight on top. A mason jar filled with water will work.

4. Press on each side for one hour, then flip it one more time and let it sit in the mould, refrigerated, for twelve hours.

5. Pull out and lightly rub the outside with salt every day for one week, returning it to the fridge after each rubbing. Wrap in plastic or keep in an airtight plastic container and let it age for two to four weeks.

This will be a very salty, crumbly cheese that can be added to a salad like feta cheese.

Chèvre

Chèvre is traditionally made with goat milk, but the recipe will also work with cow milk.

Ingredients

1 gallon milk

¼ cup fresh cultured buttermilk

2 tbsp liquid rennet mixture (put 3 drops liquid rennet into 1/4 cup clean cool water and mix. Use two tbsp of this mixture).

Salt, herbs, or other flavorings

1. Warm milk to 72°F.

2. Add buttermilk and stir thoroughly.

3. Add rennet.

4. Let sit at room temperature for twelve to twenty-four hours, until the curd has gelled. You can test this by sticking your finger into the cheese and pulling it through the curd. If it is ready you will get "a clean break" meaning that your finger will slide through cleanly, without leaving any bits of curd on your finger, much like when you test a cake with a fork to see if it is done.

5. Once you get a clean break, line a colander with cheesecloth and put the curd into it, much as in the yogurt cheese recipe above.

6. Tie ends together and let drain eight to twelve hours, depending on whether you want a dry or moist cheese.

7. Remove cheese from the cloth, flavor with herbs, salt or other flavorings, and refrigerate.

The cheese lasts one to two weeks in the fridge and may be frozen for up to six months.

Feta Cheese

Ingredients

1 gallon whole milk

¼ cup buttermilk or ⅛ tsp freeze-dried mesophilic culture

¼ to ½ tsp lipase powder

¹⁄₁₆ tsp dry rennet OR ⅛ tsp liquid rennet in ¼ cup water

Fine iodine-free salt

Brine (1 pint water plus 1 to 1½ tbsp salt depending how salty you like it)

1. Heat one gallon milk to 86° to 90°F.
2. Add mesophilic starter and lipase powder.
3. Let ripen one hour, maintaining the 90-degree temperature. This may be done in an oven with a pilot light or in a water bath in the sink.
4. Add ½ teaspoon liquid rennet diluted in ¼ cup of water.
5. Continuing to maintain the temperature, let ripen one hour, or until you get a clean break.
6. Cut curd into half-inch pieces and let sit ten minutes.
7. Stir gently for twenty minutes.
8. Drain and hang for four hours.
9. Flip in cloth and hang another four to twenty-four hours to let flavor develop.
10. Cut into slices.
11. Sprinkle with salt.
12. Set on racks at room temperature until pieces dry and ripen (two to three days).
13. Pack cheese into a quart jar and cover with brine. Brine mixture is 1 tablespoon salt per 1 quart of water.

Cultured Butter

Ingredients

1 quart heavy whipping cream
¼ cup buttermilk OR ¹⁄₁₆ tsp mesophilic culture OR 4 tbsp yogurt

1. Mix all ingredients well.
2. For mesophilic version, let sit in a warm spot (72°F to 80°F) for six to twelve hours until gelled. For yogurt butter, incubate at 100° to 115°F for three to six hours just like you would for yogurt. Chill to 55 to 60°F.
3. Agitate until butterfat separates from the whey. You can do this in a food processor, a blender, or a butter churn, or simply shake it in a jar. When the butter separates, drain it in a strainer lined with butter muslin. Use a spatula to scrape and mix butter to let all the buttermilk drain.
4. Press butter into a small bowl or a butter form, and chill. The leftover buttermilk is "real" cultured buttermilk. Use it to culture cheese, make buttermilk pancakes, or make delicious buttermilk dressing.

Feta cheese, cut and drying.

What is satisfaction? When we try to get satisfaction by getting and buying and spending, we're constantly hungry. If you only eat junk food, you're eating empty calories and you're always hungry for more because your body isn't getting what it needs to sustain itself. It takes much less of the homesteading DIY kind of pleasure to really fill up.

People's Medicine Chest

The word health belongs to a family of words including: heal, whole, wholesome, hale, hallow, holy.

—*Wendell Berry*[67]

For millennia, gardeners have tended herbs to heal themselves. Today, these powerful plants with centuries of embodied human awareness are easily cultivated in our yards. We can learn how to use herbs simply and effectively to keep our families well. People around the world have relied on herbs to relieve pain, heal wounds, increase strength and stamina, stop bleeding, stimulate digestion, expel parasites, resist infection, stop bleeding, reduce fever, aid in pregnancy and childbirth, monitor menstrual and menopausal cycles, balance hormones, and cope with all the other accidental and incidental conditions of living. Herbs heal our bodies. Let's become the keepers of our own medicine again. Even in a small amount of space, you can grow plants that will help keep you and your family well.

The herb garden is a sacred medicine place, and growing herbs for medicinal use is a beautiful DIY practice. Herbs are often simple to grow, and many of them naturalize and spread easily through the garden. Herbs are also grown for their culinary uses, as well as for use in salves, balms, tisanes, and tea. The practice of growing herbs has numerous stacking functions—health of body, health of habitat, health of earth. An integrated approach to health care, including care about the food we eat, the exercise we get, and the medicine we grow is a wise response to the limitations and toxicity of the medical industrial complex and the lack of genuine health care in this country. If you find your interests lie in learning how to use herbs for healing, there are many great teachers who can escort you into the holy sanctuary of identifying, growing, harvesting, and using herbs for healing. Even without intensive training, you can learn to grow and use herbs to keep you healthy and well.

Saint-John's-wort and feverfew in the garden. Saint John's wort is useful for depression, viral infections, and is a muscle relaxant. Feverfew is useful for migraines and arthritis, and increases a sense of well-being.

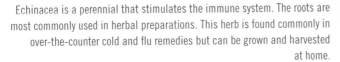

Echinacea is a perennial that stimulates the immune system. The roots are most commonly used in herbal preparations. This herb is found commonly in over-the-counter cold and flu remedies but can be grown and harvested at home.

How to Use the Herbs You've Grown

Over thousands of years, herbalists have been learning and refining their knowledge of how to derive the most benefits from plants. Herbal practices are similar the world over, and are simple and easy to learn.

Water-Based Herbal Preparations

Water is almost universally the medium for herbs, as the medicinal properties of plants are highly soluble in water. The water-based herbal preparations are teas, infusions, and decoctions.

Tea: Tea is made by pouring a cup of boiling water over a teaspoonful of plant material and steeping it from five to thirty minutes, preferably covered.

Infusion: An infusion is a stronger medicinal brew, sometimes referred to as a "standard brew." Boil two cups (one pint) of water, pour it over an ounce of plant material, cover, and let steep for six to eight hours. Use a canning jar as a steeping vessel because it has a tight-fitting lid, is heat resistant, and will not crack. If you are using a recycled glass jar, put a long-handled wooden spoon in it before you pour the hot water to avoid cracking the glass. You can steep a standard brew in a vessel of stainless steel, enamel, or ceramic—anything that does not interact with the strong medicinal compounds in the herbal brews.

Cover both teas and infusions while you brew them, to keep the volatile medicinal compounds within the remedy. The difference in strength of preparation for teas and infusions is reflected in the different ways we use these substances. Teas are everyday beverages and can be used to supplement your diet (as in the case of comfrey

GROW YOUR OWN HERBAL MEDICINE CHEST

Here's a short list of herbs that are simple to grow in a wide variety of climates and have a variety of medicinal properties. As an herbal medicine chest, they cover some basic, non-emergency health care needs. Some of these you will be familiar with, some you may not recognize. With all herbs, make sure you understand their effects before using them. Know the dosage, learn the application, understand which parts of the plant to use, and be humble. Just because they don't come in a plastic jar with a sealed lid and an expiration date doesn't mean they aren't potent.[68]

Common Name, Latin Name	Medicinal Actions
Arnica (*Arnica Montana*)	Calms and reduces bruises, shock, injury
Blackberry (*Rubus villosus*)	Treats sore throat
Black cohosh (*Cimicifuga racemosa*)	Calms nervous conditions associated with menopause
Calendula (*Calendula officinalis*)	Heals wounds and skin conditions
Catnip (*Nepeta cataria*)	Encourages good sleep
Chamomile, German (*Matricaria recutita*)	Encourages digestion and relaxation
Cleavers (*Galium aparine*)	Reduces inflammation
Comfrey (*Symphytum officinale*)	Treats bruises, sprains
Crampbark (*Viburnum opulus*)	Relaxes muscles
Dandelion (*Taraxacum officinale*)	Diuretic, blood purifier
Echinacea (*Echinacea purpurea*)	Stimulates immune system
Elder (*Sambucus nigra*)	Treats cold symptoms
Fennel (*Foeniculum vulgare*)	Encourages digestion
Feverfew (*Tanacetum parthenium*)	Treats migraines, arthritis; increases sense of well-being
Ginger (*Zingiber officinalis*)	Treats motion sickness and digestive disturbance
Hawthorn (*Crataegus oxyacanthus*)	Promotes heart health
Hops (*Humulus lupulus*)	Remedy for insomnia and nervous tension
Lemon balm (*Melissa officinalis*)	Controls high blood pressure, migraines, depression Externally—helps herpes.
Marshmallow (*Althaea officinalis*)	Treats sore throat, respiratory problems, common cold
Mugwort (*Artemisia vulgaris*)	Stimulates digestion, quiets nervous system, anxiety, and stress
Mullein (*Verbascum spp.*)	Treats sore throat
Nettle (*Urtica spp.*)	Diuretic, liver tonic
Oregon grape root	Treats poor appetite, indigestion, bronchitis
Peppermint (*Mentha piperita*)	Stimulates digestion, calms nerves
Pennyroyal (*Mentha pulegium*)	Insect repellant, and for delayed menstruation
Plantain (*Plantago lanceolata* or *P. major*)	Heal wounds, stings, burns, minor cuts, and infections

Common Name, Latin Name	Medicinal Actions
Raspberry leaves (*Rubus idaeus*)	Useful in treating menstrual issues
Red clover (*Trifolium pratense*)	Bronchitis, whooping cough, blood cleanser
Rosemary (*Rosmarinus officinalis*)	Headaches, indigestion, stimulates nervous system, and increases circulation
Rue (*Ruta graveolens*)	Reduces hypertension, diabetes, allergic reactions
Sage, Clary (*Salvia sclarea*)	Improves eyesight and clairvoyance
St. John's wort (*Hypericum perforatum*)	Relieves depression, antiviral, muscle injuries
Skullcap (*Scutellaria lateriflora*)	Eases muscle tension, sedative, headaches
Valerian (*Valeriana officinalis*)	Sedative and antispasmodic
Yarrow (*Achillea millefolium*)	Reduces inflammation

or alfalfa teas). Infusions are for herbal *treatments*, especially for acute problems. They will be stronger to the taste, and can be bitter.

Decoction: Concentrating an infusion, usually by slowly simmering the liquid until the volume is down to half the starting amount, makes a decoction. Decoctions are often taken in half-cup to full-cup doses, whereas infusions are commonly taken at the rate of two cups a day to deal with acute problems.

Cooking down the herbs into tinctures and decoctions. Spiral Garden's "Summer Sundays on the Farm."
Photo by Daniel Miller/Spiral Garden

Cutting and stripping herbs for herbal preparations. Spiral Garden's "Summer Sundays on the Farm."
Photo by Daniel Miller/Spiral Garden

Non–Water Based Herbal Preparations

Tinctures: These are non–water based herbal preparations in which the medicinal properties of the plants are extracted by alcohol or an acid such as apple cider vinegar. Place the herbal material in a clean, wide-mouthed jar; cover it with alcohol, cider vinegar, or any other non–water based vehicle. Six to eight ounces of plant matter to one quart of liquid is an approximate proportion that yields a strong concentrated tincture. This will produce much more than is needed for an individual. Use the one to four ratio, plant to liquid, and adjust it to an appropriate amount to satisfy the needs of your family or household. Or you can make a large batch and share it with your friends and neighbors.

Tinctures can also be made using 80 to 100 proof alcohol (vodka, gin, brandy, or grain or grape alcohol). Chop herbs finely, using fresh when possible. Place the herbs in a clean jar and pour the alcohol two to three

Passionflower and vodka combine in the first step of tincture making. Add leaves and flowers to the alcohol and steep for four to six weeks in a dark place. Strain herbs from liquid when done and store tincture in an amber-colored bottle or jar for use throughout the year.   *Photo by Dori Midnight*

inches over the herbs. Place in a warm spot and let the batch stand for three to six weeks to produce a good, concentrated tincture. Shake it daily. When the formula is done, strain off the liquid that contains the plant's medicinal properties and compost the plant matter. This process is called decanting. Store the tincture in dark bottles with an eyedropper, as the dosage for tinctures is much smaller than that for teas, infusions, or decoctions. Do not store these tinctures in plastic bottles, as they will negatively interact with the herbal material. Tinctures have a shelf life of about one year, so this is a long-lasting herbal preparation. If you are alcohol sensitive, use a drop in hot water when you take this kind of remedy.

Herbs and plants vary from locale to locale and season to season, and are also affected by soil quality and water quantity. Your herbal preparations should respond to that. If you roughly follow the four to one ratio in creating these tinctures, the outcome will yield potent, long-lasting medicines that are beneficial in small quantities. Tinctures travel well, need no refrigeration, and take up no space. They can be used for emergency situations (as long as they have been prepared in advance). They make economic use of plant materials, use the plants at their medicinal peaks, and are a palatable means of getting the herbal properties of some hard-to-swallow plants into people who need them. Some herbs are just too bitter and unpleasant to be taken as a tea. Tinctures are an answer to this problem.

Saint-John's-wort harvested from the summer garden and made into tincture for the dark days of winter.

Other Methods

Syrups: Syrups are prepared by placing two ounces of an herb into one quart of water and simmering the liquid down to one pint to make a concentrated tea. For each pint of liquid, add ½ to 1½ cups of honey or maple syrup. You can warm this sweetened mixture for about twenty to thirty minutes to thicken the syrup. You can also add fruit concentrate to flavor, or peppermint essential oils. A little brandy will preserve the syrup. Store in the refrigerator where it will last for months.

Steam Bath: This is another good way to get the benefits of herbs into your body. Simmer leaves of an herb, or its essential oil, in hot water. Place the hot water in a bowl, put your head over the bowl, drape a towel over your head, and inhale the steam. Eucalyptus, peppermint, cedar, mullein, lavender, thyme, and chamomile work well in this formulation for relieving sinus congestion and headaches.

Oils, Salves, and Ointments: These herbal preparations can be used externally to treat skin conditions and wounds. An **infused oil** is simple to make. Take half to two pounds of plant material, stuff it into a quart jar, and fill the jar with olive oil. Make sure the oil covers the plant material to prevent spoilage. Cap tightly and let stand in a cool, dark place for three to six weeks. Decant the oil and store in an amber glass jar. This is excellent for massage oils and other healing oils. Many different herbs, including Saint-John's-wort, calendula, and mullein can be prepared in this way.

Charmed honey. Infuse local honey with herbs and spices, add plant, gem, and environmental essences and cast your spell. Dori says, "I have made love spell honey, abundance honey, fertility honey, get me a lover honey, honey for health, honey for creativity, honey for love. The love spell honey has essences of apple tree, hibiscus pollen, pink yarrow, wild rose, rose quartz, and ruby." *Photo by Dori Midnight*

Salves and ointments are also made with herbal-infused oils. Once you decant the infused oil, mix it with a small amount of beeswax. Using local beeswax confers additional healing properties. Add one third cup of grated beeswax to one cup of infused oil. Heat this mixture in a double boiler until the beeswax has melted, but don't let it boil. Decant this mixture into small, wide-mouthed jars with tightly fitting lids. As the beeswax cools, you will have a salve for use on bruises, chapped

Herbal salve in base of olive oil and beeswax. Salves are effective for skin conditions, bruises, and sore muscles, and are easily made and stored. Combinations of herbs like calendula, yarrow, mint, wild rose, and lavender make a sweet-smelling and soothing salve. Add a drop or two of essential oil to deepen the brew. *Photo by Dori Midnight*

lips, dry skin, and sore muscles. Some herbs well adapted to salves include lavender, yarrow, Saint-John's-wort, calendula, goldenseal, comfrey, mugwort, chickweed, and plantain. These products maintain their potency for about one year, but usually you'll finish it before that, especially if you put them in small-sized jars. They make welcome gifts to friends and family.

Poultice: Poultices are an old-fashioned way of using herbs. They are simply an external application of herbs, or cloths soaked in herbal brews and applied to the body. They are useful for warming, relaxing, soothing, and drawing out inflammation or foreign matter from the body. These are simple and truly local medicines you can turn to in time of need: a mud or clay pack when bitten by a wasp, or jewelweed compresses for poison oak and ivy will pull the poisons out of your body quickly and painlessly. You can also poultice with raw vegetables if you are in need of pulling a foreign object like a splinter from an inflamed part of the body. Raw potato, pumpkin, onion, or winter squash nicely do the trick. These kitchen-based remedies are the essence of homesteading; they make do with what's on hand, the earth-based remedies of our mother's mothers.

Drying Plants

Drying is a useful way to preserve herbs, especially if you live in a climate where you do not always have access to fresh herbs. If you can collect fresh herbs, do so after the dew has dried off the leaves. With some plants you want the flowers (lavender, rose geranium, yarrow), and with others, you want the leaves (peppermint, lemon balm, mugwort). Bundle your herbs in small bunches, tie them with a string or rubber band, and hang them upside down in a dry, dark place. They will dry fairly quickly. Don't leave them hanging around too long; if you pick them in June and don't store them until August, the medicinal value of the plant will be gone.

Once dry, crumble the leaves off the stems and place them in a brown paper bag or a jar. The brown paper bag method allows the plants to breathe, does not collect moisture, and takes up less space on the shelf. In general, the dryness of the plants is more critical than their container.

These are the basic preparation techniques of herbalists worldwide. You will find cultural and climatic variations and personal preferences, but essentially, this is how herbs are prepared on our green planet. The methods are simple, require few rudimentary tools, use local resources, and can be learned by anyone who wants to try.[69] These basic preparations yield effective low-cost medicines that can be used by children, adults, and elders.

Simple Medicine from the Kitchen[70]

We can keep ourselves well and take care of each other by using a good deal of what we have in our kitchens and making simple medicines out of everyday foods. Making our own remedies and keeping ourselves well is seriously affordable health care. We highly recommend using organic or wild crafted sources whenever possible. If you balk at the prices of organics, remind yourself that food and herbs are homegrown health insurance and can keep you from frequent, expensive doctor's visits.

Treating Colds and Flu

The best way to prevent the onslaught of colds, flus, and wintertime ills is to strengthen your immune system *before* you get sick.

1. Get enough sleep and rest and reduce stress as much as possible. Winter and fall are the right time to go inward and do less.

2. Stay warm (head, throat, feet, and kidneys).

3. Eat warming food and eat with the seasons. Try garlic, onions, ginger, cayenne, cinnamon, turmeric, cardamom, soups, stews, pumpkins and squash, and fresh greens.

4. Drink plenty of water and hot liquids.

5. Strengthen your immune system with immune-boosting herbs, foods, and tonics (such as mushrooms and astragalus). In the late summer and fall, add teas high in vitamin C.

6. Exercise! Dance! Walk! Sweating is good to clear toxins.

7. Eliminate foods that are stressful for you. You can tell when food isn't working for you when you feel tired after eating, bloated, or heavy.

8. Laugh, sing, eat with friends, play inside, and get happy.

9. Begin taking herbs at the onset of feeling sick. Know your signs, such as tickly throat, exhaustion, crankiness, headache, or sneezing.

When you feel like you are getting sick:

- Stop! Figure out what is being affected. Know your body.
- Sleep! Take herbs to sleep (chamomile, valerian, kava kava, vervain).
- Water! Take a hot bath (not too hot, and not longer than twenty minutes), hot fluids, steam, gargle, and saline flush.
- Support! Ask friends for help. Take herbs to stimulate the immune system.
- Eat light and avoid dairy, sugar, and other mucus-producing foods.

Here's a list of affordable and accessible items that are great for preventing and treating colds and flu:

Garlic

Ginger

Onions

Cayenne

Salt

Honey

Soups in the freezer: chicken (made with bone broth) or mushroom broth

Basic herbs to have on hand: peppermint, thyme, sage, rosemary, echinacea, elder flowers and berries, yarrow, licorice root, chamomile. (Many of these are easily grown in the garden.)

Essential oils: lavender, thyme, tea tree, eucalyptus, and peppermint

Water

Apple cider vinegar

Vinegar infused with herbs is good for cooking and can be used in some home-based remedies as well. Dried kitchen herbs from the garden, like rosemary, dill, basil, oregano, thyme, marjoram, and mustard seed can be placed into the apple cider vinegar and left to steep. *Photo by Dori Midnight*

on the ground: tracking the wild in the heart of the city

Dori Midnight is a city witch, living in the heart of the hustle and bustle of San Francisco's Mission District. "My street stretches across this part of the city like a river of bicycles, cars and trucks, parades of drums and flags—an unconscious homage to the (now) underground Mission Creek that once provided drinking water and bathing for the Ohlone people. Living between what used to be ceremonial burial ground and the site of colonial occupation, I feel and acknowledge the painful history that is always part of living on this American soil every time I sink my invisible roots down into the earth."

Dori's city witchery is grounded in personal morning practices, as well as daily practices in her work with people. "When I start my morning practice, I imagine I have this magic tool bag, like Mary Poppins, and I pull out different pieces each morning. I always start by grounding into the earth (a practice that was given to me by my teacher Cybele, from the Reclaiming tradition). I imagine/feel/see a tail or an anchor or a root moving through the three stories of my building, past the foundation and cement, and sinking into the rich soil beneath. I let my root unfurl and widen, relishing its descent through the earth. When my root finds the right layer I let it rest there and exhale all the stuff that doesn't serve me or isn't mine, to be composted into new life under the earth.

"When I'm ready, I breathe in and pull up rich nourishing energy from the living earth and let it fill me up, re-mineralizing my blood and bones. I breathe this dirt all the way to my crown and let myself grow a magnificent trunk that branches and leafs out into the sky. I stretch out further into the dark starry mystery beyond our sky and begin to take sips of light—moonlight, starlight, and sunlight—into my branches. This light rivers down my body, inside and out, whooshing through the dirt through my feet and down my roots. I stand there, filled up with these two energies, letting my body be a place of exchange.

"Often, I do a movement practice that expresses gratitude and to connect to the five directions. From there, I am guided to sing, dance, stretch, meditate, pray, or lie on the floor with my cat. Often I will take this time to ask for help and support for the people I work with. Then I make my morning porridge, walk the dog, and start my day!

"Witches (by some definitions) are folks that ground their spirituality in the living earth. So what do city witches ground into and connect with—concrete? Wi-Fi? Trees blackened with exhaust and roots popping up pavement? Aren't we supposed to be living in little magical huts in the woods, gathering our herbs in the meadow and dancing in circles around bonfires? This sounds dreamy, but the reality is that most of us live in cities and are desperate to find ways to reconnect with magic in a place that is inherently disconnecting.

"I am a working witch. As an herbalist and intuitive counselor, I see clients four days a week out of the enchanted parlor room of my drafty Victorian flat. I also teach workshops on handcrafted herbalism and medicine making, folk healing, and magic for kids and adults. My work feels akin to the work of village healers in the old days (or what I imagine to be the old days). My kitchen is my office—I make tea and give cookies to the people who come to see me, like an old-fashioned healer, and I often find baskets of offerings at my doorstep. The folks I work with are artists, activists, body workers, sex workers, farmers, kids, and parents. Everyone comes to me to heal from their disconnection from the earth and one another.

"I consider my work as a witch to be a form of homesteading because I am drawing on ancient skills, like making my own medicine at home, but also because I think that part of the shift that needs to happen is not just about 'greening' our lives and making practical changes, but shifting our minds and hearts. If we don't acknowledge that we are part of a web of life, what motivation do we really have to make such immense changes that will most likely be inconvenient at first? So I work with folks to help them reconnect with their own magic/spirit/self, with other people, and with life.

"From a very young age, I was in connection with the invisible. Many children are, but not only is it not encouraged, it is often pathologized or ridiculed. Kids are expected to grow out of this and mature into realistic, productive grown-ups. For ten years I have taught Magical Arts to kids in San Francisco and I can't tell you how many adults whine about how they wish they could be in my classes! We *crave* it and we need it—I believe it is our birthright to be in communion and communication with everything. I can't believe no one is more freaked out about how we walk around with little boxes that send messages and images to space and back, which is totally weird magic, and don't believe we can talk to trees. This just illuminates how deep and pervasive the un-learning of these skills goes. This is the secret to it—we can communicate with it all when we begin to remember that we are not separate.

"Herbal medicine is powerful stuff—we are talking about **plants** that bring down fevers, boost immunity, strengthen bones, heal wounds, support heart function—it's a kind of a miracle when you think about it. While I am supremely excited and grateful that there seems to be a shift from Western and allopathic medicines to the more natural, I also believe we need to shift our relationship to these materials, out of capitalism and taking something to fix us, toward seeing our imbalances as something to work with and plants as allies in healing. When we run to the health food store and then gulp down echinacea to get rid of a cold we are afraid of catching from our co-workers, we are still engaging in the same framework: this thing outside of me will fix the bad thing inside of me. This is the legacy of Western medicine, the history of which is rooted in the undermining and suppression of community healers and being in right relationship with the natural world."

Homemade Recipes from Dori's Kitchen

Fire Cider

> 4 to 5 mashed cloves garlic
> 1 horseradish root, grated or chopped finely
> 1 ginger root, grated or chopped finely
> 1 onion, chopped
> ¼ tsp cayenne
> 2 cups apple cider vinegar
> 2 cups honey

Put all the ingredients in a jar. Cover lid with plastic wrap or wax paper, as the vinegar eats away at the metal. One traditional way to make fire cider is to bury the jar in a double plastic bag under the earth for a season, or steep it in a warm place for two to four weeks. Take one tablespoon every day as a preventive, and more if you are feeling sick. Don't take if you have a heat condition like hives or fever. It can also be used as a marinade or salad dressing!

Pickled Garlic

This is ideal for people who can't eat raw garlic. The vinegar mellows out the flavor, but you still get the benefits of the raw garlic.

1. Cover fresh, peeled garlic cloves with apple cider vinegar or tamari.
2. Let steep two to four weeks.

You can also preserve garlic in olive oil or honey. Keep it in the fridge when using these mediums, as they will spoil otherwise.

Learning about herbs and kitchen witchery in Dori's Magical Arts class gives children permission to do what they do best—taste, smell, experiment, combine, and believe. *Photo by Dori Midnight*

Garlic Drops

1. Mash two heads of fresh garlic.
2. Steep overnight in one part apple cider vinegar, one part water.
3. Squeeze out the liquid and add molasses or honey to sweeten the mixture.
4. Take a few drops every few hours or just slather it on toast. Kids love it!

Honey Drippin' Onions

1–2 onions
Honey

1. Thinly slice one onion and place in a bowl.
2. Cover the onion with honey, and put a plate over the bowl.
3. Leave overnight (eight hours or so).
4. Take one spoonful of the honey. This is especially good for sore throats.

Hot Ginger Lemonade

Lemon
Ginger
Honey
Cayenne

1. Grate fresh ginger.
2. Simmer in water for fifteen minutes.
3. Squeeze lemons in.
4. Add honey or cayenne as you please.

Grandma's Magic Cold and Flu Tea

Equal parts elderberry, elderflower, yarrow, and peppermint
Echinacea if desired

1. Steep ten minutes and drink 1 cup every hour (while in bed!). This recipe is especially good for a fever.

Elderberry Cordial

1 cup elderberries (fresh are best, dry is totally fine, just use less)
1 cup brandy
Honey to taste

1. Combine elderberries and brandy and let steep four weeks.
2. Strain and add honey to taste.

You might have to warm the liquid a little so the honey blends well, but don't cook it. Honey is best raw. Sip throughout the winter or if you feel a cold coming on. Warming and protective!

Elderberry Syrup

> 1 cup elderberries
> 2 cups water
> 1 cup honey or other sweetener

1. Simmer on low for twenty minutes, strain the seeds, and press out with cheesecloth.
2. Add honey (or sugar, agave, or maple syrup) to sweeten, half a cup of sweetener for every cup of liquid.
3. Bottle and pour on pancakes, or eat by the spoonful. It's totally delicious.

A good way to get people in your life who won't take medicine to take it—on their pancakes.

Cough Syrup

> 1 cup herbs (elderberries, wild cherry bark, cinnamon, fennel seeds, mullein, ginger and licorice)
> 2 cups water
> ½ cup raw honey
> ¼-cup brandy

1. Simmer roots and bark for fifteen minutes, or steep flowers and leaves for twenty minutes.
2. Strain.
3. Add raw honey and brandy as a preservative.
4. Keep in the refrigerator.

High-C Tea

> Dried hibiscus, lemongrass, orange peel, peppermint, and rose hips
> Boiling water

1. Steep herbs for ten minutes
2. Drink as a vitamin C boost.

These simple recipes are effective ways to keep our immune systems strong and responsive. When coupled with the herbs you grow in the garden, their simple transformations into earth medicine form a good beginning to a homegrown medicine chest you can use throughout the year. Don't rely on "health care" that just lines the pockets of the medical industrial complex. For simple illnesses, coughs, colds and flus, it's safer and more affordable to rely on the health care given to us by the earth.

The House

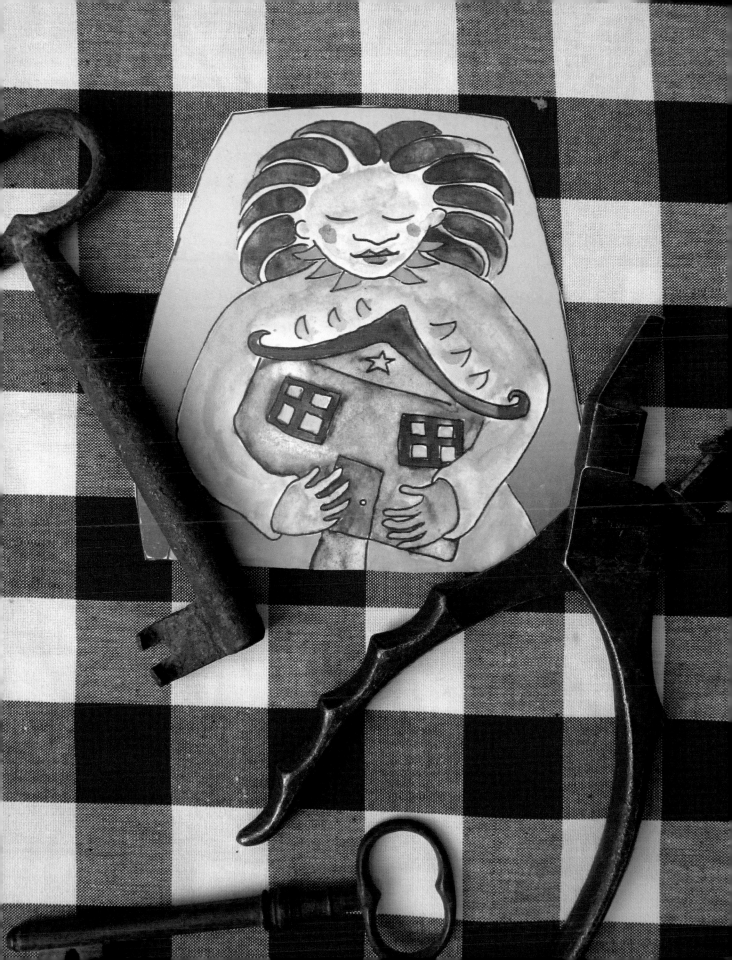

From the Ground Up

Small is beautiful.

—*E. F. Schumacher*[71]

Natural building has expressed itself across cultures throughout human history and is currently evolving to fit our contemporary needs in gorgeous and creative ways. Earthen building, rather than wood or stick construction, is the predominant type of building in the world; 75 percent of non-Western cultures currently build with natural materials as they have for millennia. Most of the building materials used by natural builders are as old as the hills—earth, clay, lime, sand, stone, thatch, straw, bamboo, twice-used wood—and many of the techniques for turning these earthen materials into shelter for humans are ancient.

Casting our eyes back over thousands of years of culture, we can participate in evolving these traditions and use what's available to create beautiful and usable structures in the city. As with all earth-based homesteading practices, natural building is entwined with the social justice issues of our time, particularly issues of affordability, access, and the right use of materials. The more we can do for ourselves, the less we need to depend on the power of the extractive economy to control how we live. We can build homes that fit our needs without sacrificing our values of living simply and well with the earth.

The vernacular practices of natural building return us to one of our favorite themes: Listen to the earth. Natural builder Tracy Theriot says, "Building with earth, shaping our structures, and feeling the dirt with our hands and our feet, is a fantastic opportunity to listen. If you're aware of the earth you're building with, it's going to make a shape that makes sense. If you go at it with your ideas, without listening, it's mostly not going to work. Natural building is a practice of presence and place—the soil in your backyard will feel different than the soil in your neighbor's backyard. One size does not fit all. Every soil has its own spirit, its own nature, and its own voice. If you listen with your body, your soil will talk to you, telling you the story of the ground under your feet. Once you get to know that, it tells you what it wants to do."[72]

Natural building in the city is highly adaptable to urban needs: issues of scale, recycling of materials, and adaptation to small or unusual spaces are but a few of its appealing qualities. Natural building is largely DIY, and can be happily shared with friends and neighbors. Most urban homesteaders won't have the opportunity to build entire houses from alternative building materials, though small structures are increasingly more common. In the city, people are using natural building to make chicken coops, studio space, office space, saunas, or to renovate and renew existing buildings with earthen plasters, clay paints, and earthen and recycled building materials. Some people build small sleeping rooms in the backyard to make it possible for more people to share their space; others use recycled and earthen materials to make ovens, or benches for the backyard, front yard or community neighborhood meeting place. These structures are beautiful, livable, and affordable. With a little knowledge, some time and some good friends, you can make safe, homegrown structures to suit your needs.

First, let's note the difference between "natural" building and "green" building: Natural building is done with resources from the earth: dirt, clay, straw, bamboo, sand, stone, lime. It also accesses recycled materials of all sorts, often from the waste stream of conventional building projects. Green building is a standard within the conventional building industry that co-opts some aspects of sustainability, while remaining quite mainstream in other ways. A "green" house may use passive solar energy, or natural paint or bamboo flooring. Maybe a water wise landscape will be installed (but only if code demands, and it usually doesn't) while still using an abundance of new and highly manufactured materials. Many of these houses are overly large, negatively impacting the carbon footprint of the structure.

Natural building made from adobe, recycled wood and glass, bamboo grown in the backyard, and roofing materials.

A cob oven built for the Petaluma Bounty Farm. The chicken is the icon of Petaluma, once a major chicken and egg production site for the entire West Coast.
Photo by Miguel Micah Elliott

Small backyard structure, in process. Materials include cob, light straw, recycled wood, and urbanite.

Natural building provides an affordable solution to the problem of resource waste and pollution in constructions. "Green" buildings tend to be overly large and continue the trend of excess waste and resource use in building practice. Permitting code in most parts of the country leans heavily toward conventional building practices using nonrenewable resources, mostly wood, brick, cement, metal, stone, paper, and plastics. This kind of construction is highly specialized, resource intensive work. It's expensive, and rightly belongs in the hands of people with know-how and skill, but this specialization makes housing unaffordable and inaccessible to many.

While skill is needed in situating natural buildings and making sure that foundations are properly established and roofs will protect the structure, natural building is something you and your friends can do in your spare time. It won't cost you much, and it will let you express your creativity in a way that more standardized structures rarely do. Using local and renewable resources, you can design and build appropriately scaled structures that fit into the contours of your space.

Many natural builders merge different materials in their structures. Design decisions are based on the location, and are dependent on what's available, cheap, structurally sound, and beautiful. If your curiosity leads you to natural building, source what's local and make choices based on your values of conservation, affordability, and need. Learn about the local building codes, and if you're going to defy them, take it on as a political project as well as a building project. A well-made natural structure can become a great model for changing local codes and can have a big impact on the viability of natural building in your area. Natural building retrofits are a good fit for the city because they minimize resource waste and cost by using recycled building materials. Since a city has an excess of buildings, retrofitting what we have is a wise approach to resource conservation.

on the ground: urban retrofit

Villa Sobrante is a collective household in El Sobrante, California. The resident owners are four young women—Sasha, Massey, Trilby, and Lindsey—joined in a shared love of natural building and homesteading. They decided to stake their claim "outside the borders of Berkeley" and bring the good news of natural building to the outlying suburbs. "When we decided to do this we wanted to find out how much we can do for ourselves as urban homesteaders," Sasha explains. "We want to learn how to put things into practice, how to live with them, observe them, and learn to do it with other people.

"We wanted to do this in a place where it isn't being done much, and to see what kind of influence that has. Partly we chose this site because it was a good place to regenerate. It was such a problematic property when we arrived, so we were welcomed here. If this place had had a perfect lawn in the front of the house when we bought it, the neighbors would not have welcomed us as much. Since it was in such disrepair, even a sheet mulched lawn was better than a foreclosed property that was a target for homeless people. We were seen as regenerating the neighborhood just by being here."

Villa Sobrante, which houses the first permitted natural building retrofit in Contra Costa County, is the most substantial natural building urban retrofit we found in our area. "We're trying to experiment with alternative housing in an urban setting in a way that's legitimate," Massey says. "We're taking this 800-square-foot house and making it livable for four or more people, within code. When people who have

Roof detail in one of the outbuildings at Villa Sobrante. This building utilizes a combination of bamboo, cob, and light straw.

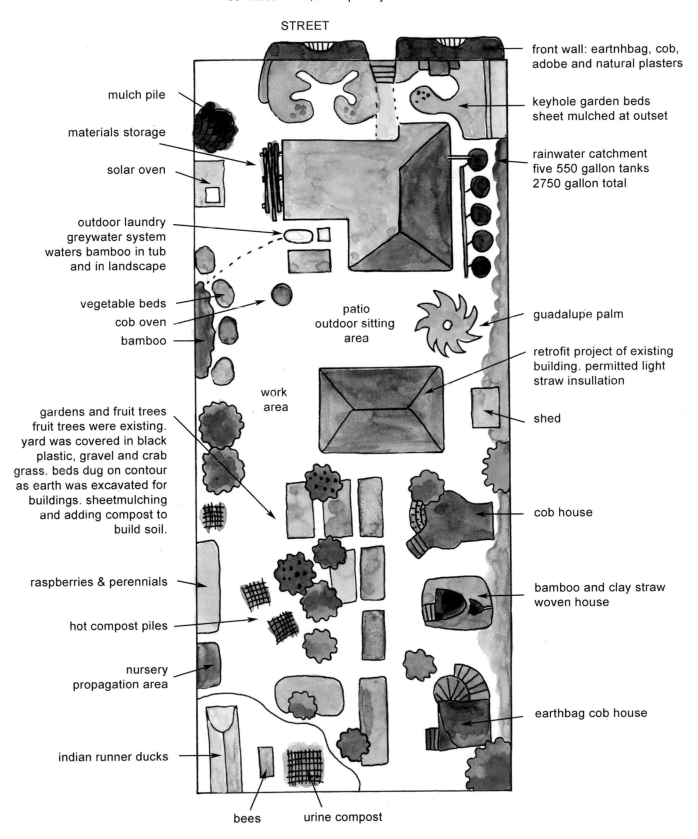

VILLA SOBRANTE
60' x 200' ~ 12,000 sq.ft. ~ just under 1/3 acre

STREET

front wall: eartnhbag, cob, adobe and natural plasters

mulch pile

materials storage

solar oven

keyhole garden beds sheet mulched at outset

rainwater catchment five 550 gallon tanks 2750 gallon total

outdoor laundry greywater system waters bamboo in tub and in landscape

vegetable beds

cob oven

bamboo

patio outdoor sitting area

guadalupe palm

retrofit project of existing building. permitted light straw insulation

shed

work area

gardens and fruit trees fruit trees were existing. yard was covered in black plastic, gravel and crab grass. beds dug on contour as earth was excavated for buildings. sheetmulching and adding compost to build soil.

cob house

raspberries & perennials

hot compost piles

bamboo and clay straw woven house

nursery propagation area

earthbag cob house

indian runner ducks

bees

urine compost

Cob benches can incorporate ovens and fireplaces and make a beautiful center to a garden or yard. This structure is designed in such a way that some of the excess heat from the oven warms the bench.
Photo by Trathen Heckman/Daily Acts

power over the permitting process get to see that these kinds of structures are sound and safe and a viable alternative to the waste and expense of building, and the problems of creating affordable housing especially in very dense urban areas like this one, the laws will change. We care about the legality issue because we're trying to do this in a way that can be a model. Working with the county and the existing laws is the only way to get them to change."

Other small buildings on the property are made from natural materials like cob, straw, earth bags, and locally sourced bamboo. Materials are reused and recycled from the dump, Craigslist, other building projects, and the overflow of their retrofit. Alongside the building projects, residents are creating gardens, digging swales for water catchment, and learning to live together. "The swales and earthworks were a natural thing as we started working with the dirt—when you dig a hole, there's a lot of extra dirt, so what we didn't use in the buildings, we used in shaping the land," Massey explains. "And since growing food was one of our goals and there are so many issues with water in our area, managing the dirt so that it helps sink the water into the ground was a smart design. It helped us really see our property in a different way—how does the water flow? Where does it go? What's the best way to use the earth we have to the most effect? How do you go about designing what you have in a way that's efficient and doesn't waste a lot of resources or energy?"

Villa Sobrante residents share living space, garden space, financial commitment, and workload for the project. "Working with the county is a learning experience and learning to make a great garden bed is a challenge too, but probably the biggest learning edge is in living together. In the beginning, we were living pretty much on top of one another," Lindsey recalls. "That helped us decide pretty quickly to build structures where we could have some privacy while living together so intensively. We hope the natural building we do here serves as an inspiration for people, but sometimes we think the way we've learned to live together is just as important as a model for another way of living in our culture."

Homesteading has become meaningful for Sasha, Massey, Trilby, and Lindsey in ways that are personal and political. Sasha summed it up when she said, "When people are more involved in the systems that sustain their own lives, something more freeing happens than when we live in a more constricted way. When we are involved in how our food grows, where it comes from, how our structures are built, where we live, some deeper part of our humanity comes through. There is so much to learn from being connected to your environment in this way, spending a lot of time outside with different systems and being more connected to the root of everything that provides for our basic human needs. If you are more connected to them, you feel more alive."

Building with Dirt and Water and Straw

Cob is the most user-friendly and DIY natural building material because it's just soil dug from the backyard and mixed with water, sand, clay, and straw. This humble formula often prompts jokes about mud huts or makes you wonder if cob will dissolve in the first rainstorm. But cob is very durable and requires little upkeep. A fire won't burn it, termites leave it alone, and it's as cheap as, well, dirt. It's nontoxic, creates no waste, and requires minimal tools to construct. Thousands of cob houses have weathered rainy England for hundreds of years, and a recent renaissance of cob building centered in the West Coast is joyfully exploring the modern artistic and architectural possibilities of the material.

In addition to its affordability and beauty, cob is easy to learn to do and easy to repair, add on, change, or replaster. The clay in cob absorbs toxins, and the walls "breathe," which creates a more environmentally healthy place. Cob has high thermal mass, and absorbs heat during the day and radiates it out at night. Its design can be round, smooth, soft, and sensual. The round shape helps with earthquake resistance, and cob structures, when properly protected, are water resistant. It's simple, fun, efficient, and affordable to build with cob.[73]

Creating cob structures is a great community effort because it really does take many hands and feet to make a cob structure. Beautiful benches, ovens, altars, and small structures for animals and people are easily made and shaped. The sense of belonging that is conferred when a group of people work on a project like this is powerful, much like an old-fashioned barn raising. Anybody can help mix and sculpt cob—children, the elderly, and everyone in between. Cob is an amazing material that will respond to your creativity and imagination. Bury magic things in the wall. Make up a ritual while you cob. Use the experience to invite sacredness and beauty into your life.

How to Make Cob[74]

Mix the cob as close to the building site as you can to minimize dragging the building materials any further than necessary.

1. Throw some soil on a tarp.
2. Break up the clods with a tamper or your feet.
3. If you need to add sand or clay to your soil, measure it out by shovel or bucket until you can do it by eye and by feel. When adding sand, shovel it onto the tarp and press it into the crushed soil until the two are well mixed. You can pull up the edges of the tarp and roll the ingredients around on the tarp.
4. If you need to add clay, add clay mixed with water (clay soup) or dry powdered clay to the dry sand. Wear a mask when adding and stirring dry clay to your mix. It's not healthy to inhale clay dust.
5. Stand on one edge of the tarp and lift the tarp on the opposite site of the mix. Pull it towards you, turning the mix over onto itself. Don't lift the edge of the tarp at your feet or push the mix to turn it. It's too hard on your back. The more you roll the tarp, the faster the cob mixes.
6. Add water until you've got a cookie-dough consistency. Tread on the mix with your feet. It's easier to feel the mix with bare feet, and bare feet mix the cob faster than shod feet. Stir the mix often by pulling up the edges of the tarp while you're treading. It's impossible to tread the mix too much, so have fun playing in the mud. The mixture's plasticity as you tread on it.
7. When your mix begins holding together in a giant loaf when you turn it with the tarp, it's time to add straw. This will dry out the mixture quite a bit. If the mix is too dry to add lots of straw, either add less straw or more water.
8. Flatten the cob loaf down by treading on it. Grab an armful of straw and sprinkle some onto the mix. Step on it so it sticks into the cob a little and pull the tarp up, making a cinnamon roll of dirt-dough and straw-sugar. Tread on it until it's flattened out into a pancake again, and add more straw.
9. Make another cinnamon roll. Do this three or four times until you have enough straw in the mix. Too much straw will cause the mix to fall apart. Try adding a lot to a small mix so you can see what too much straw feels like.

If your cob mix is too dry it won't stick together very well or integrate the straw. Add water and stir until the mix is like cookie dough. If your cob is too wet, it won't keep its shape because it can't hold up its own weight. If it's too wet, add a little dry sand,

Mixing the cob by foot.
Photo by Miguel Micah Elliott

dirt, or straw and tread it in. If you add a large amount of dry stuff, use roughly the same ratio of dirt and sand as your original recipe. You can also dry a mix by treading it into a pancake so the air can get to it, and letting it sit for a few hours or overnight. It's easier to mix the cob if it's wet, so you may want to work that way and let it sit for a few days after mixing it before you use it on your building.

Curing the cob is not necessary. It's fine to make a mix and put it directly onto the wall, but you might want to experiment with letting it cure overnight and seeing what works best for your project. Avoid leaving the mix covered for too long because the straw will start to rot and stink, and lose some of its strength. If you do this accidentally, put it on the wall anyway. Once it's dried, it won't have a smell.

All natural buildings need a strong foundation. A cob structure is no different. This is the part of the building practice that needs a trained eye and a bit of skill. You can source foundation stones for a cob structure locally, using bricks, natural stone, or urbanite. You can also make cob and adobe bricks that will provide foundational support for the building. Make sure that your building area is level, that foundation stones are placed carefully and sunk as needed to provide the best base for your structure. Cob is strong and will last for many years if done properly, but, like the rest of us, it won't stand for long without a good foundation of support.

Building the Structure by Foot

1. Drag the mix on the tarp to the spot where you want to build. If the foundation is low enough, two people can roll the mix off the tarp and onto the foundation. If the foundation or walls are too high to roll the cob up onto, make a ramp to drag it up or lift it by the handful, tubful, or bucket onto the foundation.

2. You can cob by foot by simply squishing the cob mix down onto the foundation of the building and stepping on it. This pushes the cob into the spaces between foundation stones.

3. Keep adding cob and compressing it together with your weight. When the sides start bulging out, get a friend to use his or her feet to create a form. Let the cob harden for a day or two once this starts happening. You can always cut the extra cob off the wall later with a big knife or spade before it's completely hardened.

Building by Hand

1. Roll the mix in the tarp like you would for stirring. Lift handfuls of cob onto the walls and massage it with your hands and a stick. Thumbs are great for this kind of work, though they do get tired. You can use your palms and knuckles too. One hand can hold the cob as the other compresses it and incorporates it into the cob below it.

2. It's important to make your structure one solid hunk of cob, avoiding weak connections between the layers. You may need to re-wet the last application of cob so you can massage the new cob into it.

3. Do not smooth out the surface of the cob too much or it will dry as a hard skin, which will slow the drying in the wall beneath it.

4. Get in the habit of making a flat surface on the top edges of the wall so the next layer of cob has enough surface area to sit on comfortably instead of sliding off.

Putting the Cob to Bed

When you're finished for the day, poke finger-sized holes with a stick in the top of the wall. Make them three or four inches apart, especially along the edges. The holes provide a key for the new cob to hold onto. When you return to cobbing, push the fresh cob down into the holes.

Building the cob structure by hand.
Photo by Miguel Micah Elliott

When you're ready to cob again, fill the holes with water if the old cob needs re-wetting. A hose with a spray nozzle is perfect for the re-wetting job, or keep some squirt bottles full of water around the building site for re-wetting smaller areas. You'll soon get the feel for how much to wet it so that you can work the new mix into the old.

To control the drying of the wall, cover the tops of the walls with dampened straw, burlap sacks, and/or tarps and kiss it good night. If you suspect heavy rain is coming, cover the walls with tarps or sacks. To speed up drying, place lots of dry straw on top of the wall under the tarps or sacks. Rain won't hurt your wall but it will slow down the drying time.

Keep your fresh walls out of strong direct sun and harsh winds that will dry the outside surface too quickly, and cause cracks on the dried surface as the inner cob dries and shrinks. Protect the walls with shade or by covering them. If you're cobbing where the temperature drops below freezing at night, pile straw bales or loose straw on the sides and on top of the walls to prevent them from cracking. Use old sleeping bags or anything that insulates. Take the covers off during the day so the walls can dry.

If the cob dries out completely before you get back to it, fill the holes in the top of the wall with water until the cob is rehydrated. Incorporate the first new layer of cob well by using a stick and pushing the new cob into the old. You can also sew the layers of cob together with straw, pushing the straw through the new mix into the holes in the old cob.

Other Natural Building Possibilities

While there are differences between them, natural building materials such as adobe, cob, plaster, light straw clay, and wattle and daub are all made with the same basic ingredients: clay (also called site soil or binder); sand (aggregate); and straw (fiber). The binder may include site soil, bagged clay, and lime. The aggregate adds body and limits the shrinkage and therefore the cracking of the binder and may include rocks, small sand particles, mica, rice hulls, pumice, or crushed glass. Fibers are used mostly for tensile strength and to limit shrinkage. These may include plant fibers of many lengths (straw, coconut, jute, or hemp); animal hair or dung (which also has some binder qualities and beneficial enzymes); glass or poly fiber; herbs; human hair; or paper pulp.

For each technique, ingredients will have differing proportions and differing sizes of materials, particularly of sand and straw. Many variations depend on the soil composition and on the builder/creator. For example, adobe options depending on site soil include 1) site soil only; 2) site soil and sand; 3) site soil and straw; and 4) site soil, sand, and straw. With cob, options include 1) site soil and long straw; 2) site soil, sand, and long straw; and 3) site soil, sand, and chopped straw.

Student mixing earthen plaster for adobe building.
Photo by Sasha Rabin/Vertical Clay

Adobes drying in parking lot, at the end of a community adobe-making festival.
Photo by Sasha Rabin/Vertical Clay

Front gate made from repurposed bamboo grown on site.

These site-specific variations point to the need to listen to the dirt and what it wants. How much sand and fiber do you need, and at what length? Natural building encourages our creativity. You will learn what the material can do. This reinforces the listening part of the process. And finally, remember that natural building materials are very forgiving and can be applied, repaired, and changed as needed.

Adobe, bamboo, earth bags, light straw, straw bale, and recycled materials provide additional natural resources for building. Adobe bricks are made with a completely saturated mixture of clay and sand poured or pressed into forms, then laid on appropriate foundations and affixed with mud-based mortar. The thick earthen walls of adobe structures provide thermal mass to modulate interior temperatures during the heat of the day and the cool of the evening. This building material has been used for many centuries in North Africa, the Middle East, and South America, and has a rich history in California and the American Southwest. All of California's missions are adobe, and are restored with traditional adobe building techniques, including brickwork, floor, and wall restoration. While there are many centuries-old adobe structures throughout the world, in most parts of our country this building material is still illegal.

Bamboo is a quick-growing grass that provides a renewable material for building, tools, and utensils, as well as food. Bamboo is sturdy, strong, and beautiful, and is becoming more popular with builders in this country. It is often used for flooring, and can be used in place of rebar to create trusses and other structural parts of the building. You can grow bamboo in your backyard and harvest it regularly to be used for structural members, molding, framing, window and door bucks, trellis poles, and scaffolding. Timber bamboo is strong enough to be used as a building pole. If you're growing it yourself, make sure you plant the clumping kind and not the invasive runner kind.

Earth bags are soil-filled fabric sacks or tubs used to create walls and domes. Like cob, they are affordable and easy to use, a quick and forgiving technique based on minimally processed soil and few tools. Moistened soil is placed into a bag set in place on the wall; the bag is lowered into place, and then compressed using a hand tamper. Earth bags are useful for foundations of straw bale or cob structures, and can be used to create a structure built into a hillside.

The use of baled straw to create heavily insulated walls has become extremely popular in recent years. The straw used in these building is mostly agricultural waste that is often burned at the end of a growing season,

First natural building urban retrofit in Contra Costa County. Light straw—a mixture of straw and clay—is most often used in urban retrofit situations. Other materials in this structure include recycled and repurposed lumber and bamboo. Note building permit hanging in the middle of the wall.
Photo by Sasha Rabin/Vertical Clay

Adobe bench built in a public park, part of a larger community building project by Massey Burke. Adobe forms the base; recycled wood forms the seat. *Photo by Sasha Rabin/Vertical Clay*

so the reuse of this building material has excellent environmental as well as structural features. Baled straw is currently included in the code in many places in the U.S. and is in the final stages of clear building standards being adopted by international code. None of these materials are illegal, but in some counties, permitting officials may not be familiar with them and may need to be educated about their applications in a building project.

There are many approved source sites throughout the natural building community to lend credibility to anyone having a hard time getting a project approved due to the building department's lack of knowledge. In many places, homes are built using straw bales as infill in a post-and-beam structure. When covered with natural plasters, they make an aesthetically pleasing, environmentally friendly building material.

Working with a wooden structure, light straw is used to infill the walls of the retrofitted building and then covered with cob to make durable walls. This material is up to code in many locations, utilizes renewable materials, has high durability and energy efficiency, and is easily used by expert and amateur builders alike.

Many structures, especially in urban areas, can be effectively rehabilitated, saving immense amounts of new construction costs and maintaining important cultural links, while avoiding massive amounts of landfill waste by using recycled building materials. The dismantling and reuse of many older buildings funnels materials into this resource stream. Wood, windows, doors, and other fixtures can be found at recycled building centers in urban areas around the country. There are obvious environmental, aesthetic, and cost benefits to this way of building.

Natural building holds a lot of possibility for the city homestead, especially when you want to shape a small, site-specific structure where you live. Whatever materials you choose to work with, building with natural materials is one solution to the need for healthy, creative, and affordable housing in the city, while offering another chance to get into rhythm with the earth, and live well, in place.

on the ground: everything here was once something else

Tina and Troy's homestead is on a large lot where they have lived for the past fifteen years. Troy is an avid builder; Tina holds down the gardening part of the homestead. They are raising their two boys among the chickens, blackberries, and natural buildings on their homestead. Their shared love of the natural world, their passion for recycling, and their commitment to doing as much as they can on their own have yielded a beautiful, fertile site filled with fruit trees, abundant squash harvests, and a lot of love.

"Our greatest investment, while living here, has been in the growing and perfecting of skills," Tina says. "We had to learn how to do things for ourselves with inexpensive materials. We love to salvage quality wood, stone, old house parts—whatever someone decided was trash. For our first project, the entire tool shed was built from an old redwood fence and deck. We were even bending out nails. Then we went on to the other structures, all of them built with the logistical and artistic challenge of finding strength and beauty in pre-used stuff. At first, we were overwhelmed to see what was being wasted out there and brought too many things home. Now we have less material and we're better organized. We learned that you can usually go out and find exactly what you need with a short search. You don't need to keep it around your house all the time, as if you're going to run out any time soon. The community of builders and businesses who appreciate quality used things is growing."

When they first bought their house, Tina and Troy didn't have the skills to do the much-needed repairs on their house. Troy did what he could on his own, but when the needs of the house were beyond his skills and he had to hire someone to help, he'd apprentice himself to them and learn how to do it from a master. In this way, he grew into his career as a contractor and stonemason. "I grew to a deep love of the craft of building, and making things that were strong and beautiful using the

Retaining wall constructed mostly of recycled materials brings new shape to the garden.
Photo by Troy Silviera

old methods. You can see in this old house that all the things that were done in the seventies were falling apart, and the stuff done in the twenties was still standing. Old time craftsmanship understands the strengths and beauty of the essential material. Once you start to appreciate the old things, it changes everything.

"If you're a builder, you can practice on your own house. I wanted to learn how to make a wall out of Sonoma fieldstone, so I built one here. Or I built all these levels of urbanite walls in the backyard. There is the great joy in doing things yourself. It's quite a motivator, and the antidote for feeling like we don't know how to take care of ourselves."

Looking around his house with pride and delight in his eyes, he says, "If you stay here long enough, I'll tell you a story about every object here. Now, that stone came from a job I did, but originally it came from a cobbled street in France . . ." In this way, their homestead is layered object by object, by the power of story.

In the backyard, while looking at the little garden Tina and her sons planted on the roof of the chicken coop, Aaron, her oldest son, said, "We got 123 potatoes off that roof last year!" Tina waters that little garden by hand, and covers the potatoes with straw to keep the water in. Her chickens live a happy life pecking the comfrey and purslane, and she butchers them when their egg-laying days are done. "I feel pretty comfortable harvesting the chickens. It seems like the right thing to do. But it's a balance—you should never get accustomed to taking a life. It's always something to take seriously." Tina's garden is full to overflowing with squash, tomatoes, and blackberries, all tucked into little nooks created by the homemade stonewalls, half-finished buildings, and found objects throughout the yard.

Troy is wandering through the yard, moving some stones for another project and pushing the chickens along the path. At the suggestion that he'd filled up every inch on his busy, creative homestead, he looked shocked and said, "Oh no! There's always room for more!"

Natural Earthen Paints and Plasters

Natural paint and plaster comprise the biggest growth area of natural building. Build your house with dirt and cover it in clay, all from your backyard—urban dirt farming at its finest! These plasters are made from dirt sourced from the backyard, clay (available at ceramics stores), and water. Natural plastering can be completely DIY, and because the plasters are layered on an existing surface, no structural or engineering skills are needed. Earthen paints and plasters have many environmental benefits: they are nontoxic and contain no VOCs, they don't off-gas, they are odorless, and they absorb odors in the house. They are minimally processed, have low embodied energy and environmental impact, and wick environmental moisture from the walls. They create a beautiful, matte finish on outer or inner walls which, when properly applied and/or sealed, can last for years with minimal upkeep. Like building with dirt, painting with dirt links you to your place, helping you understand and connect to the nature of the soil beneath you.

Making Earthen Plaster for the Exterior of Cob Structures

Underneath the beautiful topsoil you've been growing in your garden is soil you can use to plaster your house. You can make an earthen plaster for the exterior of a cob structure or natural building with dirt from your backyard. It's easiest to manipulate loose dirt that's dried out for this project. Once you've dug the dirt, use a window screen attached to a wooden frame so you can sift out the rocks and big particles of dirt from the mix. When you're done sifting, you'll have finely sifted clay.

1. Mix sifted clay in mixing box.
2. Add sand to the clay in an approximate ratio of 30 percent clay to 70 percent sand. Use a medium-fine sand that has some sharp edges but small enough particles to make a nice smooth surface. If you don't use enough sand, the plaster will crack.
3. Add flour paste to help make the plaster waterproof and help the elements bind together better.
4. Add manure that has short straw already cut up and live enzymes in it. This helps strengthen the plaster.
5. Add water to the mix.
6. Mix it up with a hoe or shovel, or your bare feet. You're looking for a smooth consistency in your plaster.
7. When applying the plaster, wet the cob structure with a sponge because it needs to be moist for the plaster to stick to it.
8. Take plaster and spread it out along the cob about an eighth of an inch thick. Keep applying the plaster, adding water as necessary. Use your hands to feel what's "right."
9. Let it sit for about five minutes as it hardens up a little bit.[75]

Natural plasters for the outside of non-cob buildings are generally thicker bodied than the clay and earthen veneers or paints that are for interior use. The first step in making natural plasters is to do a soil test for clay content, because you'll need to use clay-rich soil for your plaster. Clay is the binder in an earthen plaster and if your soil is not clay-rich you'll need to source soil from somewhere else so you can be sure to have a plaster than will stick to the walls of whatever you are painting. You will want your plaster to be made from site soil that passes the following tests.

Shaker Test

1. Place about a cup of subsoil in a mason jar; add clean water to two or three inches above the soil.
2. Seal jar and shake vigorously to dissolve all particles. Let rest. It may take a day, but most soils settle within forty-five minutes.
3. Look at the layers of settled materials. Rocks and sand will be at bottom, then silt, followed by clay, and organics will be on top. Measure and record the layers or save the jar to determine the percentages you have of silt, clay, sand, and stones.

When you first start doing this it may be hard to determine what is clay and what is silt. Ask around your natural building community or consult with a pro. Many professional earthen plasterers are happy to look at samples of your soil and within minutes can tell you a lot about it and save you time.

Elasticity Test

1. Take a handful of the clay soil in your hand, and moisten it until you can shape it into a pencil-thick worm. If it holds its shape when you wrap it around your finger and does not crack, you have a great clay soil. Soil with a little bit of cracking will also work.

Earthen wall after an earthen plasters workshop, the last of a series of workshops in which the wall was constructed.
Photo by Sasha Rabin/Vertical Clay

Adhesion Test

1. Take a handful of clay soil and moisten until you can roll it into a ball without it being too moist or crumbly.
2. Flatten ball into a pancake and slap it against the palm of one hand. Hold your hand out palm down and count how many seconds it takes for the pancake to fall off your hand. Anything over ten to fifteen seconds has relatively good adhesion. Many clay soils will hang onto the hand for thirty to sixty seconds.

There are many varieties of silica sands, both white and beige, which are easily found at landscape or masonry supply stores and which can be used in recipes for earthen plasters. We don't reprint any specific recipes here because there are many, many recipes, most of which fit only one type of soil. We recommend instead experimenting with your soil, and testing the different ratios as follows for your plaster, rather than trying to fit your soil into a specific recipe. Working with and knowing your soil cannot be overrated. Every soil has its own properties and will perform differently from every other.

Testing Your Plaster

Mix enough for a test wall at least four feet square at ⅜ inch to ½ inch thick

1:1 test

    1 part clay soil

    1 part plaster sand

    ½ to 1 part chopped fiber

    add water to consistency

1:2 test

    1 part clay soil

    2 parts plaster sand

    ½ to 1 part chopped fiber

    add water to consistency

1:3 test

    1 part clay soil

    3 parts plaster sand

    ½ to 1 part chopped fiber

    add water to consistency

Apply test plaster to a board or onto a frame. Measure against a ruler and score one-inch scratches along one edge. Once dry observe how it performed. Did it crack? How much? Re-measure. How much did it shrink? Is it dusty? Does it crumble easily? Can you scratch or break it without much force? Use all of the information to modify your mix. If plaster is too sticky and cracks while drying, it needs more sand. If plaster is soft and crumbly,

Massey Burke applying a style of mud work from the Gujarat region of India.   *Photo by Sasha Rabin/Vertical Clay*

Figuring out the right earthen plaster mix takes some experimentation. *Photo by Sasha Rabin/Vertical Clay*

it needs more clay. If plaster can break easily, it needs more straw or fiber.

Run multiple tests when you are starting out. The more you work with the materials, the more you will gain understanding and knowledge of the materials and how you can work with them. Once you have gotten the correct consistency, test out the plaster on a small part of your building and let it dry. Make sure it performs the way you want it to before applying to the rest of the building.

Interior Plaster Recipes

Nontoxic clay plasters are great for the inside of the house as well. This *clay alis* can be made from local clay-rich dirt or from bagged, powdered clay available in pottery stores. Using local clay-dirt to make alis is more time intensive than using bagged clay, but it's cheaper and requires less energy. Nontoxic clay-dirt is easily sourced near where you live and brings no adverse chemicals into your home. These plasters are easy to make and use, and yield beautiful aesthetic results.

Simple Plaster 1

Here's a simple recipe from Carole Crews, earthen plaster expert from the Southwest, where the indigenous knowledge of natural building is still alive and well, evolving into the twenty-first century.

1 part kaolin clay

1 part 325 mica

1 part flour paste

Purchase the kaolin clay and mica from a well-stocked ceramics supply store. If you don't have one in your local area, you can find these supplies online.

Simple Plaster 2

Here's another recipe from Carole Crews.[76]

1 gallon water

5 quarts powdered kaolin clay

2 quarts fine sand and/or (preferably) whiting

2 quarts mica (fine flakes or powder)

1 handful of chopped straw (optional)

1 handful of mica chips (optional)

Slaked pigment as needed for color (optional)

1 quart cooked starch paste (wheat, rice, or cornstarch) OR 1 cup casein or buttermilk OR a combination of binders

Measure water into a bucket, and whisk in remaining ingredients. This six-quart batch will cover approximately 150 square feet.

Mixing pigments for interior plaster.
Photo by Tracy Theriot/Tactile Interiors

Range of pigments for interior plasters.
Photo by Tracy Theriot/Tactile Interiors

Flour Paste Recipe

Flour paste is used as a binder for earthen plasters and clay alis. It is easy to make and use; it's also inexpensive, and best of all, sticky.

6 quarts water

1 quart white flour (not whole wheat)

1.   Place four quarts of water in a large pot on the stove and bring to a boil.

2.   While waiting, place two quarts of cold water into a smaller bowl or container and slowly whisk in one quart of white flour until it is lump-free.

3.   When the water on the stove has come to a rolling boil, slowly add the cold flour and water mixture into the pot with the hot water. Mix this together, stirring or whisking constantly. This mixture should become translucent and thicken.

4.   Turn the heat off. If the mixture does not feel thick enough, let it remain on the heat, but be careful not to burn the mixture. If it burns, it may turn brown and smell.

5.   Remove pot from the heat and let it stand to cool. If the mix ended up lumpy, simply run it through a screen before adding it to your plaster or paint.

Flour paste will last for a few days in moderate temperatures, and longer if refrigerated. Make sure to clean all your tools directly after making this, as it will dry hard like glue. Once the flour paste is combined with the clay alis, you can apply the natural interior plaster much the way you apply paint.

There are literally thousands of clay paint recipes. As with natural plaster, experiment with the materials until you find a consistency you like. Don't forget to test the paint recipes before you apply them. You can add pigments usually used in painting projects to any of these plasters to change their colors, although you may want to stick with the earth tone that comes from your own soil.

Natural building and plastering projects are fun and creative ways to lay claim to your homestead. Cob benches, saunas, and chicken coops; natural buildings for sleeping, working, or art-making, natural plasters for the inside and outside of your home, all leave a heartfelt mark of the earth on the places we call home.

Powering Down

So when Brian Williams of NBC News is asking me about what's a personal thing that you've done [that's green] and I say, "Well, I planted a bunch of trees." And he says, "I'm talking about personal," what I'm thinking in my head is, "Well, the truth is, Brian, we can't solve global warming because I f---ing changed light bulbs in my house. It's because of something collective."

—Barack Obama, quoted on Newsweek.com in 2008[77]

Transforming our cultural addiction to the fossil fuel that powers our entire economy—our homes, our transport, our food and manufacturing systems—is one of the pivotal challenges of the twenty-first century. We have a lot of work ahead of us if we are going to rescue our culture from the depths of this addiction. As I sit writing, millions of gallons of oil are pumping into the Gulf, and the resistance at the federal level to obvious solutions to this tragedy indicates the challenges we face in turning around the ship before it's too late. While big answers for new power sources do not rest entirely in your hands because the development of large-scale energy infrastructure is so highly politicized, there are still many actions we can each take in our daily lives that make a difference and begin to forge a more ecological and ethical way of living. Ultimately, we need large-scale participation to curb the destruction stemming from our addiction, but even without collaboration at the highest levels of government and industry, we can begin the powering-down process at home.

Numerous visionary citizens are pursuing high-quality alternative energy scenarios, working toward harnessing viable renewable energy sources and building a reliable renewable energy grid. Some of these options are getting more affordable and available, though many still remain out of reach for most people and are nearly impossible to implement if you live in a rental situation. But you can change your personal energy output today by changing some of your daily actions.

Some of the following behavioral fixes are things you've undoubtedly heard before. Committing to them takes a willingness to get over your cynicism and imagine that changing your habits can make a difference. That's it's not enough, or that corporations waste more in one day than we can ever save, can't be the reason for doing nothing. Too depressing. Most of the people interviewed for this book reported that getting more conscientious about energy use, rather than being a hardship, became a practice of intention, a daily wake-up call for purposeful action.

Energy Use at Home

Fossil fuel use is highly implicated in how we live in our houses, how we transport ourselves, and how we eat. Homes account for about 20 percent of the United States' total annual energy demand. Heating and cooling the interior of our homes consumes the largest portion of residential energy—about 44 percent. Lighting, cooking, and appliances consume one-third of our energy. Water heating consumes 14 percent, and the refrigerator about 9 percent. Although each home is different, this data alerts us to the big energy consumers at home and around town, and can help us target the greatest potential energy savings.[78] The following are some ideas to start chipping away at our energy use at home. They are some of the simplest things we can do, wherever we are, today, a way to participate in the inevitable powering down process that is the not-so-distant future.

- *Turn it off.* It's so simple it barely needs repeating. When you walk out of the room, turn off the light switch. There's a common misperception that turning the lights on and off takes more energy than just leaving them on. If you're going to leave them on for more than five minutes when you don't need them, you'll save energy if you turn them off instead. Better yet, only turn it on when it's really dark.

- *Shut down the phantom loads.* The energy used by appliances, stereos, televisions, and the Internet even when they are turned "off" drains 5 to 10 percent of your household energy. This energy is called the "phantom load." Put your appliances on a power strip, and turn it off when you don't need it.

- *Turn it down.* Keeping the thermostat low and wearing socks, slippers, and sweaters saves energy. Lowering your thermostat by only 2 degrees in the winter saves 353 pounds of carbon dioxide a year (approximately 170 pounds per degree).

- *Change the lights.* Changing those light bulbs really does make a difference. Compact fluorescent bulbs save 75 percent of the energy of an incandescent bulb, and LED bulbs save another 10 percent. Both of these conservative bulbs last far longer, which diminishes waste to the landfill as well as saving energy.

- *Use the sun.* Cooking a warm pot of soup in a solar oven uses no energy at all, and warms you up when you eat it. You can also use the sun to heat your shower, grow plants in a greenhouse, or warm the water you use at home.

- *Retire the dishwasher.* Using the dishwasher produces approximately two pounds of CO_2 every time you use it. Hand wash your dishes in one tub of hot soapy water, and rinse them in a second tub. When you're done with the rinse water, you can put it in your garden, or use it to water the houseplants. If you do use the dishwasher, make sure it's full, and dry the dishes by air, rather than by using electricity to dry them. (We use our dishwasher as a dish rack, which not only saves energy, but also makes more counter space.)

- *Take a shorter shower.* If you can stand it, turn the water off when you're shampooing your hair. You can also save the water at the bottom of the shower for siphoning into your garden, or using it to flush your toilets. Both these fixes stack functions, saving energy and water. More on this energy and water saving trick in Chapter 17, Sourcing the River.

- *Change your laundry habits.* Hang your clothes on the line and retire your dryer. In summer, use the clothesline, and in winter, put the laundry on a rack in front of the heater. Not everything needs to be

Accessing the sun wherever we can to do our daily tasks is one of the easiest ways to begin powering down. *Photo by Rachel Kaplan*

washed after each use—only wash what's really dirty. If it's only dirty in a small spot, wash the spot, not the whole shirt. Only run the washer when it's totally full, and when you do, use cold water. If you're getting a new washer, get a front loader. It uses much less water and energy. Some cities offer rebates when you replace your old washer with a new one. Use them.

- *Drive less*. Driving represents a huge carbon input for most people. Changing our transportation habits needs to happen. Some of these require big fixes, like better public transportation within and around cities, but there are some things we can do day-to-day. Bike, walk, or use public transport when you can. Even changing your habits by reducing one or two car trips a week makes a difference.

One of the best things about city living is the proximity of services close to home. Enjoy the walk to the grocery store and the exercise when you bring your purchases back home. When you do have to drive, carpool or stack up your errands so you use the car less. Plan your trips through town to take the most efficient route, the one that minimizes idling and stopping and starting as much as possible.

- *Make your own fuel*. Home-brewing biodiesel saves money and reuses waste products to run your car. And buying biodiesel at the pump is getting easier all the time. Biodiesel is made by a simple, two-step esterification process using lye and methanol to break the fatty acid chains in vegetable oil. This renders the oil less viscous, so it can run in any diesel engine. Older engines will also need some of their hoses changed out, as biodiesel is a great solvent and can dissolve rubber and some other hose materials. Newer diesel vehicles (1994 and younger) should be fine to run on biodiesel without any modifications to the engine. While biodiesel is thinner than vegetable oil, it can thicken in cold winters, so be sure to inform yourself more fully before trying it out.

Another option for fueling a diesel engine is to use filtered but unprocessed waste vegetable oil (WVO) that can be sourced at local restaurants that otherwise pay to have it removed. There is a famous tale in the biofuels world about some young folks who drove across the country fueling themselves on waste oil from McDonald's French fries (surely easier to find than sustainable agriculture in the "heartland"). Using WVO does require modifications to your vehicle, since cold vegetable oil is too thick to flow through the fuel lines. There are a number of kits and conversions on the market. Most of these "grease cars," as they are called, utilize a back flush system, so that the car starts on regular diesel and once it is warm enough, runs on vegetable oil. When the system shuts down, it flushes the veggie oil out of the lines, so that petroleum diesel will be in the lines for the next startup. Some systems use two tanks—one for each fuel; others run on the Elsbett system from Germany and use one tank and a coil to heat the veggie oil.

While biofuels are gaining popularity, there are still relatively few public pumps and many of them use commercially produced biodiesel made from corn or soy oil. Biodiesel from commercial sources does have a lower carbon footprint, but corn and soy crops are firmly embedded in the nasty practices of industrial agribusiness with its high environmental costs. And while the fuel itself is created from materials

If you can find one, an electric car saves gas, money, and precious parking space. You can't take them out on the freeway, but they'll lower your impact around town.

BioFuels Oasis provides biodiesel at the pump for biofuel users.

more renewable than fossil fuels, there are still many debates about whether it really burns that much cleaner than conventional fuels. Biofuels may be a good short-term solution along the road to giving up our car habits and our reliance on fossil fuels, but they aren't the final solution. Bicycling, public transport, and walking still rank higher on the list.

WVO and biofuels in general are still pretty much a DIY proposition. For the mechanically inclined, the grease car is superior to buying biodiesel at the pump. If you are serious about biofuels and looking for a new venture, you might think of starting your own biofuels co-op using WVO from local sources to make biodiesel for your community. The San Francisco Bay Area hosts at least one. Check out www.biofueloasis .com for an example of an urban biofuel station selling only WVO biodiesel (and urban farming supplies).

- *Fly less.* When my family evaluated its carbon input, our airplane trips to visit our East Coast family from California was our biggest carbon sin. This stumped me a bit, as I am personally unready to stop visiting my cross-country family. Is there any true mitigation to this problem? The best way to go is to simply fly less. If your lifestyle includes a lot of flying, cut carbon in other places to compensate, and bring your work life and vacation pleasure closer to home.

Powering Down the House

Reducing energy output is the first step in powering down—changing your house or car makes no sense if we continue to use energy at an unsustainable level. Once we've changed what we're doing and gotten our use down as low as possible, then we can change our houses by insulating them, making them more efficient, upgrading appliances and water heaters, and installing solar panels. Not all of these fixes are available to everyone—especially when money is an issue, or if you are a renter. There are more and more rebates, however, for insulating and weatherizing your home to get some of this done, as well as incentives for solar installation. You can do some of the smaller efficiency measures in an afternoon; others are the quick work of a contractor and are generally worth the financial investment over time.

on the ground: zero net energy house

George and Ellen live on a small corner lot across the street from the local high school, in close walking distance to town. When they purchased their house in 1998, it was in need of an overhaul, inside and out. They undertook the work with a commitment to learning to live sustainably. A green architect and building visionary, George was admirably suited to the task at hand, and Ellen's spiritual connection to the earth as a living organism and her intention to make every daily action count express this couple's deep practice of learning to live in harmony with creation. "Do we know how to do that yet?" asks George. "No. But we're working on the edge of sustainable society."

In 2010, their house is terracotta red with a lichen green roof artfully designed with solar electric panels and a solar thermal energy array. The garden around their house is dense and wild and filled with edible plants and trees. This house is a rarity: a zero net energy home for electrical energy. This means that the Beelers are able to make all the electrical energy they need on-site, so their house has no carbon footprint. They subscribe to a strict definition of zero net energy that excludes conventionally sourced electricity as well as fossil fuels like natural gas. They are replacing their natural gas use with solar water heating to be backed up in the future with a "district heating" system with a carbon neutral fueled boiler.

George and Ellen love their house, but they love more what it represents: right use of resources, right use of action, and right use of relationship. For Ellen, "It's more the things I can't see that I believe in. It's a spiritual, intuitive kind of thing." Learning to live an energy-efficient life style is about living intentionally, about being aware of every single thing she does. So while energy efficiency is about how often she turns on the heat, it's also about how she talks to people about how they live, and how she conducts herself in the world. "The Earth is a living being. I am part of the Earth. Every one of my actions affects the Earth, so I am working on conscious awareness in my actions at all times," she said.

Some of George and Ellen's initial investments in the energy signature of their house included replacing windows, insulating walls and attic, replacing the roof and adding solar panels and a solar thermal air heating array. They insulated and added a separate street entrance for the basement and created a suite of rooms (meeting, design studio, and library) for a home office (often referred to as "live/work"). They also installed a super-efficient heat furnace in the basement to back up the solar air heater for cloudy days.

While this remodel is neither DIY, nor particularly affordable, none of the infrastructural changes we can currently make to existing buildings fall into either category. This house is shown as a model of what's possible given the current state of energy use, efficiency, and natural and green building. While some of the fixes in this home are beyond the means of many, others (like energy efficient lighting, windows, and window coverings) are more accessible.

Another dream the Beelers share with their neighbors is a neighborhood-based energy sharing collective to lower their use of natural gas. This vision of "district heating" is a group collective of close neighbors, where each household would buy shares in an efficient system designed to conduct hot water to different buildings. Neighbors would agree about the placement of the energy efficient carbon-neutral boiler, the infrastructure for piping the energy from house to house, and a cost-sharing arrangement. Each home would get energy from the boiler, and all expenses would be shared. District heating is a good way to install energy-efficient systems at lower cost, generate interdependent and resilient structures between neighbors, and begin to significantly lower the carbon footprint at home. Costs always go down when people share them, fewer resources are used, and the workload and cost for maintaining the system is spread among many. Similar neighborhood arrangements take advantage of shared solar arrays for generating power.

Not surprising, large utility companies have already passed laws restricting such neighborhood co-ops for energy sharing. One would have to become an independent utility company to make this work, and that's pretty rough going. In California in 2010, the local industrial energy corporation placed on the ballot a measure which would have strengthened the utility's monopoly on energy provision and make it nearly impossible for counties to figure out ways to provide their own alternative energy. Fortunately, this ballot measure was overturned by a majority of Californians, but political initiatives like this must be carefully scrutinized and vigorously opposed.

Before embarking on this kind of collective enterprise, make sure you know the laws in your municipality and within your utility district. This is an obvious place for legislative agitation: either utility companies need to provide people with alternative energy, or they need to allow people to provide it for themselves. We advocate such community-based solutions over personal and private property solutions at all times. Working with others in these kinds of small-scale collectives is a direct way to take action on the issue of significantly reducing nonrenewable energy use at home.

Simple Solar Solutions

Until the beginning of the twentieth century, human beings lived in a solar-based world. They rose with the sun and slept with the dark. They used the sun to dry their clothes and preserve their food, and to order their

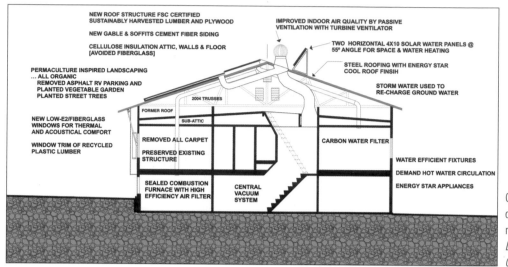

NEW ROOF STRUCTURE FSC CERTIFIED
SUSTAINABLY HARVESTED LUMBER AND PLYWOOD

NEW GABLE & SOFFITS CEMENT FIBER SIDING

CELLULOSE INSULATION ATTIC, WALLS & FLOOR
[AVOIDED FIBERGLASS]

IMPROVED INDOOR AIR QUALITY BY PASSIVE
VENTILATION WITH TURBINE VENTILATOR

TWO HORIZONTAL 4X10 SOLAR WATER PANELS @
55° ANGLE FOR SPACE & WATER HEATING

STEEL ROOFING WITH ENERGY STAR
COOL ROOF FINSIH

PERMACULTURE INSPIRED LANDSCAPING
... ALL ORGANIC
REMOVED ASPHALT RV PARKING AND
PLANTED VEGETABLE GARDEN
PLANTED STREET TREES

STORM WATER USED TO
RE-CHARGE GROUND WATER

2004 TRUSSES

FORMER ROOF

SUB-ATTIC

NEW LOW-E2/FIBERGLASS
WINDOWS FOR THERMAL
AND ACOUSTICAL COMFORT

WINDOW TRIM OF RECYCLED
PLASTIC LUMBER

REMOVED ALL CARPET

PRESERVED EXISTING
STRUCTURE

CARBON WATER FILTER

WATER EFFICIENT FIXTURES

DEMAND HOT WATER CIRCULATION

ENERGY STAR APPLIANCES

SEALED COMBUSTION
FURNACE WITH HIGH
EFFICIENCY AIR FILTER

CENTRAL
VACUUM
SYSTEM

Green features
of Fair Street
redesign.
*Drawing by
George Beeler*

daily living. We are now thoroughly unaccustomed to living a solar lifestyle, with electric lights, twenty-four-hour Internet access, and late-night movies. Getting reacquainted with the sun is good homesteading practice.

Try this experiment: commit to living with the sun for one week with just your rising and sleeping patterns. Feel how different life is when the sun is the herald of the day, and dark is the herald of sleep. If it suits you, add some more challenge to the experiment: turn off some of the appliances in the house. Start with the simplest ones, like the stereo, or the washing machine. How is it to tune your life to the sun? What do you miss? What do you find you can simply live without? See how your habits change. See how what you call *need* changes as well. If you're digging the experiment, go hard core. Turn off the refrigerator. Turn off the Internet. Practice using the sun as your sole source of energy. See how it feels. Perhaps this experiment will bring you closer to understanding which "indispensable" parts of your lifestyle are actually easy to live without.

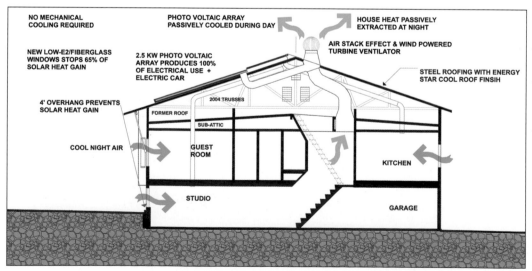

NO MECHANICAL
COOLING REQUIRED

PHOTO VOLTAIC ARRAY
PASSIVELY COOLED DURING DAY

HOUSE HEAT PASSIVELY
EXTRACTED AT NIGHT

AIR STACK EFFECT & WIND POWERED
TURBINE VENTILATOR

NEW LOW-E2/FIBERGLASS
WINDOWS STOPS 65% OF
SOLAR HEAT GAIN

2.5 KW PHOTO VOLTAIC
ARRAY PRODUCES 100%
OF ELECTRICAL USE +
ELECTRIC CAR

STEEL ROOFING WITH ENERGY
STAR COOL ROOF FINSIH

4' OVERHANG PREVENTS
SOLAR HEAT GAIN

2004 TRUSSES

FORMER ROOF

SUB-ATTIC

COOL NIGHT AIR

GUEST
ROOM

KITCHEN

STUDIO

GARAGE

Features
of the Fair
Street
house in
summer.
*Created
by George
Beeler*

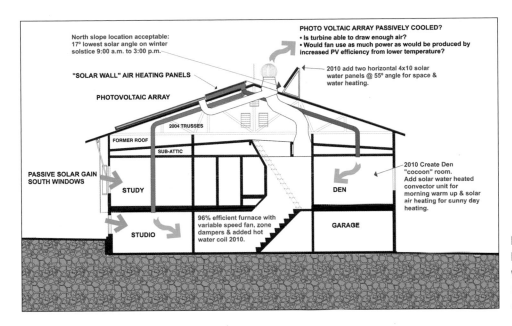

North slope location acceptable: 17° lowest solar angle on winter solstice 9:00 a.m. to 3:00 p.m.

PHOTO VOLTAIC ARRAY PASSIVELY COOLED?
• Is turbine able to draw enough air?
• Would fan use as much power as would be produced by increased PV efficiency from lower temperature?

2010 add two horizontal 4x10 solar water panels @ 55° angle for space & water heating.

"SOLAR WALL" AIR HEATING PANELS

PHOTOVOLTAIC ARRAY

2004 TRUSSES

FORMER ROOF

SUB-ATTIC

PASSIVE SOLAR GAIN SOUTH WINDOWS

STUDY

DEN

2010 Create Den "cocoon" room. Add solar water heated convector unit for morning warm up & solar air heating for sunny day heating.

96% efficient furnace with variable speed fan, zone dampers & added hot water coil 2010.

STUDIO

GARAGE

Features of the Fair Street house in winter.
Created by George Beeler

The point of this exercise is not to become a Luddite or even to suggest that some of the ways we use our energy are not appropriate. There is a common delusion, however, that we are going to be able to replace our energy needs with wind and solar power as the century marches on; that we'll just be able to plug into the solar grid and continue to use as much energy as we currently use. There is no energy source on our planet as dense as oil. This means that no matter how much infrastructure we generate to replace our diminishing resource base, we will not have the same amount of energy at hand as our fossil fuel supplies decline. This means that life is going to change, big time. Each of us uses far more than our share of the planetary juice, so reforming our habits of use is the first step toward a sustainable lifestyle.

Home-Scale Solar Energy

Most of us heat our houses or cook our food with electricity or gas, and most of us don't have much choice about where the energy comes from. While there is some motion to change to more alternative heat sources, any big scale energy changes are going to be slow in coming and pretty expensive for a long time. Some homesteaders make their stand on creating the most maximally efficient houses they can. Those taking this route, like the Beelers, are usually owners, and they usually go in the direction of solar energy, as the technology is most readily available and becoming more affordable.

There are many solar energy installers popping up around the country, but this is still not

Recycled solar panels power outdoor garden lights and backyard stage at Mariposa Grove, Oakland, California.

Small solar array powers entire house.
Photo by Ben Macri

much of a DIY project. The Solar Living Institute (www.sli.org) is a resource for all things solar. They train people to install solar energy, and sell materials, books, and other resources regarding solar energy. Some major home improvement stores, like Lowe's, are jumping on the "green revolution" bandwagon, and offering solar panels for DIY installation. If you go this route, be aware that you need a permit to install solar panels and that there are numerous bureaucratic hurdles to overcome. It's not impossible, just a bit complex.

on the ground: sun and wind

Ben Macri and his partner live in a suburb of San Francisco and have been moving toward sustainable living for the last twelve years. A self-proclaimed "geek" (he works at the local city college teaching industrial arts, carbon-free living, home efficiency, and green technology), Ben designed and installed solar panels for the roof of his house, as well as a small wind turbine, "just because I could." Does the solar array provide enough energy to power the house? It does. Do the neighbors complain about the noise of the wind turbine? They did, but it turned out the noise they thought they were hearing when they got hyper-attuned to the turbine was the sound of their own refrigerators.

Passionate about energy management, Ben "had a dream and went after it." Designing and building a solar array requires knowledge of electrical wiring and local codes, but it's not out of range for someone without specific electrical skills. It takes a little calculation, a little research, and a bit of inspiration. The first thing Ben did when designing the system was to assess how much energy his household was using. Once he calculated the energy load, he looked at where that could be trimmed and reduced. The older refrigerator was the biggest energy drain in the house. He replaced it and then calculated

the power the new refrigerator used. (Purchase a device called Kilowatt in most hardware stores that hooks up to your appliances. It will let you know how much energy the appliances are using.) He got rid of an old electric dryer and washer and changed the light bulbs to further reduce the energy load on-site.

Ben matched his energy needs for the new refrigerator to the potential output of a set of solar panels he installed on the roof. He made a simple frame for the panels, aimed it at the sun, got the approximate angle on the roof, and draped the wires down the roof and into the crawl space of the house. He hooked up the wires to an inverter and a battery, and it ran the refrigerator. Once he'd handled the biggest energy hog, designing the rest of the system was a breeze. He further calculated how much energy he'd need, and designed the solar array for that level of use, with a little extra thrown in, "just in case." Ben's solar array cost him about $11,000 when he built it in 1998. Rebates currently offered on solar panels and installation makes this more affordable today.

The micro wind turbine adds additional energy backup to Ben's house. It's small, only about forty-eight inches across, and stands about six feet over the peak of the roof. It's on a pole attached to the roof, and cost less than $500. It's

Small wind turbine powers computer and other small appliances.   *Photo by Ben Macri*

not a legal energy fix, but it generates energy year-round with no negative impact to the human or natural environment. The wind turbine doesn't need sun or even good weather to work, but when the wind is blowing at night, it's making enough power to run a laptop and a few light bulbs. At that size, the turbine makes more of a point than lots of power, but it's the shape of energy overhauls to come.

"Going solar really re-attuned us to our energy use. We become aware of using energy, and turning things off. This helped us become more socially conscious in our home, and the changes in how we lived were exponential. Tracking your input changes your consciousness for the better. It makes you more connected to the world around you." Ben sees a future where energy systems are more interconnected with one another. "Our smart homes will be linked to our transport system—you'll plug in your car and it will also function as a power source. When the grid needs power, it will draw it from your car.

"Everything is starting to become interconnected, which is what we need more than anything else—the lived reality of interconnectedness. We can generate energy from the sun and the wind and share it with one another, rather than continuing to do everything on our own. We don't have a problem with technology; we have a problem with consciousness. What we need is all around us. We have to change how we look, and then we'll change what we see." Ben sees a future of closed energy systems, powered by sun, wind, and water. "It's amazing how when you start one thing it leads to another. We make the path by walking. Our consciousness changes, our actions change, our consciousness changes again. And so it goes."

Cooking with the Sun

A solar cooker is an easy way to access the power of the sun. You can buy a solar cooker affordably, or you can easily build one yourself that will last for years. A solar cooker is a black painted box with one side resting on an angle and facing the sun. The top of the angled side is covered with plate glass, or plastic sheeting, and inside the box is a shelf for the food. A solar cooker bakes a chicken or potatoes, makes rice and baked beans, a fruit cobbler, and brownies. Results are often stew-like rather than crispy, but if you plan for that, you can prepare a meal in the morning, and come home at the end of the day and find something hot for dinner. It's similar to a crock pot you don't have to plug in and it works about eight months out of the year, when the sun is high in the sky.

How to Build a Solar Oven[79]

Materials

Large and small cardboard boxes
Newspaper
Aluminum foil
Nontoxic invisible tape
Cardboard
Nontoxic glue
Scissors
Pencils
Black construction paper
Staples
Black paint

1. Find two boxes (rectangles are better than squares). One should fit inside the other with a two- to three-inch space on each side. (The space in between the boxes will be insulated, and this will raise the heat in the boxes.)

Cooking with the sun in the front yard.
Photo by Trathen Heckman/Daily Acts

2.   Line the bottom of the large box with crumpled newspaper.

3.   Place the smaller box inside the large box.

4.   Fill the space between the sides of the two boxes with crumpled newspaper.

5.   Line the sides of the inside of the smaller box with aluminum foil. You can use a nontoxic tape or fold the edges of foil over the top of the box to hold it in place.

6.   Line the bottom of the inside of the smaller box with black construction paper to absorb heat.

7.   Lay a piece of cardboard on top of the large box and trace the shape of the box onto the cardboard.

8.   Add two inches around the trace line and cut out to make a reflector.

9.   Cover the cardboard with aluminum foil. Smooth out any wrinkles and secure the foil to the cardboard with nontoxic glue or tape.

10.   Staple the reflector to the outside back of the large box.

11.   Paint the outside of the box black to attract and absorb more heat.

For maximum heat, situate the oven with the box opened up and the reflector facing the sun. Place food to be cooked in the solar oven. Stretch clear plastic wrap across the top of the large box. Secure the plastic with tape around the entire box. You can also use a sheet of glass (which is more permanent and works better than plastic wrap.) Make sure you tape the edges of the glass so you don't cut yourself when using the oven.

To access the most heat, keep moving the solar oven to match the sun's angle for the required cooking time. Cooking time with a solar oven is about twice as long as in a conventional oven. Preheating takes about thirty minutes. Do not use duct tape or Styrofoam or anything that will give off toxic fumes inside the oven when heated. Do not use a solar oven for foods that must reach a high temperature or cook rapidly. The solar oven works well on a sunny summer day, but it's also effective on a sunny winter's day; you just have to track the angle of the sun to make sure you get enough of it to cook your meal.

Other Solar Projects

There are many DIY solar energy projects, including a solar shower (a thick black plastic bag that heats water in an afternoon and hangs overhead in the garden); a solar heat grabber, a simple box that pulls passive solar heat into the house; a solar water heater that can provide up to 70 percent of your hot water needs; or a passive solar greenhouse. A great website for building all things solar is www.builditsolar.com. Some of these projects are not for the faint-hearted, but even if your engineering skills are slight, you can build some of these projects in an afternoon.

What's the best choice for your homestead? When imagining all the different projects you can build or changes you can make, start with a goal in mind. Maybe you want to challenge yourself to reduce your energy use by one-quarter (an awesome start). How can you reduce in each of these areas? For transportation, look at driving less, biking more, using public transport, walking, or carpooling. For home use, look at your electric bill, home insulation, thermal windows and drapes, and heating bills. Imagine how you might increase your use of solar energy by building a solar space heater, water heater, or installing solar panels at your home. Or access the energy of the sun by cooking in a solar oven, drying clothes on the line, or preserving food in a solar food dryer. Put that in your energy reduction plan. Think of it like a game—how low can you go? When you're feeling pretty confident, lower your threshold, and invite your neighbors to play along.

CITY-WIDE LOW CARBON DIET

The City of Davis, California, is taking on the Low-Carbon Diet on a citywide scale, organizing groups of citizens toward reducing the community's carbon footprint, and creating a carbon-neutral city. The target behavior is to reduce the carbon emissions of each household by 5,000 pounds in thirty days. Residents, community organizations, and local business have been invited to participate.[80] In October 2008, the City of Davis gathered 100 households into small teams, ranging from average Davis households to city council members, city staff, UC Davis campus administrators, scientists, students, and business owners. The short duration of the program made it easier for households to commit their time. Upon conclusion of the program, households reported the amount of carbon emissions they were able to lose through an anonymous online reporting tool. Of the forty-seven survey responses received, 253,723 annual pounds of carbon had reportedly been saved as a result of this program. This calculates to an average of 5,398 pounds saved per household.

From their press materials: "Most people are aware of the consequences associated with global warming. However, the fight against global warming requires more than one behavioral change and can be an overwhelming process. Also, in order to prevent further climate change it takes more than one individual to make the change." The project was intended to help bite-size the changes people need to make, keep them simple, and create a model for city-wide energy reduction that can be replicated around the country in different communities. The City of Davis hopes to extend this program from this initial pilot project of 100 households to 75 percent of Davis households, over 18,000 homes.

The question of energy is central to the unfolding of this century. We must make concerted personal and collective effort in this area if we do not want to see our world spiral deeper into chaos, war, and environmental degradation. We have many different technologies at our disposal, some very old and available to all, and some very new and not accessible to most. While we continue to lack cooperation at the highest levels of commerce and government, we aren't going backward, we're only moving forward. Cutting back, balancing needs, and generating new renewable sources of energy and sharing them around will be the most efficient and secure path forward in a powered-down energy future.

ENERGY CONSERVATION TIMELINE

When planning to change energy conservation and use habits, plan within a realistic timeframe.[81]

Six-Month Transition

Monitor and measure current kWh and/or therms used

Practice basic conservation measures

Switch to LED or CFL bulbs

Put appliances on a power strip—turn it off when not in use

Launder in cold water

Use solar clothes drying

Utilize rechargeable batteries

Examine insulation/leaks

Move from car to bike or public transit for 75 percent of trips

Research car sharing

Research moving closer to work place

Eat more raw, fresh food

Twelve-Month Transition

Research passive solar retrofit

Research options for PV panels where appropriate

Advocate for community choice aggregation in energy supply

Limit car trips to three to four times per month

Walk, bike, bus, or train as appropriate

Aim towards one or fewer plane trips per year

Eat more local, raw, fresh food

Move to location near where you work

Thirty-six to Sixty-Month Transition

Advocate for 100 percent local municipal renewal energy profile

Reduce kWh by 90 percent from baseline use

Use appropriate technology for energy use:

    Bike powered clothes washing?

    Rocket stove?

    Solar oven?

    Mass heater?

Use evaporative cooler for limited refrigeration needs

Work from home if possible

Sourcing the River

We never know the worth of water till the well runs dry.

—Thomas Fuller[82]

Water once flowed in an interconnected, serpentine wave across our globe from the mountains down through rivers, ponds, lakes, and streams to the sea. Water is the lifeblood of our planet, swimming also in the veins of our bodies. We have dammed and channeled and blocked and piped and transformed the waterways of our planet until we can no longer trust either the source or the sufficiency of our water. In their 1985 book *Rivergods*, Richard Bangs and Christian Kallen wrote, "High quality water is more than the dream of the conservationists, more than a political slogan; high quality water, in the right quantity at the right place at the right time, is essential to health, recreation, and economic growth. Of all our planet's activities geological movements, the reproduction and decay of biota, and even the disruptive propensities of certain species (elephants and humans come to mind)—no force is greater than the hydrologic cycle."[83]

Access to clean water and healthy watersheds is a human right, yet global access is the exception, not the norm. One billion people worldwide do not have safe drinking water within a fifteen minute walk from their homes. In California, one million people drink from contaminated wells, mostly in poor communities in the Central Valley that provide food to the majority of Americans. "The sheer scale of dirty water means more people now die from contaminated and polluted water than from all forms of violence including wars," the United Nations Environment Programme (UNEP) said.[84] The fight for water is a social justice struggle fought by people around the world; the denial of this basic human right has given rise to the concept of water justice—the right of all people to collectively control local water sources and the watersheds that sustain them.[85]

Water is a strategic political and environmental issue not only in the Third World but also in the developed world as drought, shortages, and problems with municipal water infrastructure increase. "Water is the most critical resource issue of our lifetime and our children's lifetime," said water management expert, Luna Leopold in his book *Waters, Rivers and Creeks*. "The health of our waters is the principal measure of how we live on the land."[86] We can assume that climate change and the deteriorating global economy will continue to impact access to clean water for millions of people worldwide.

For nearly two centuries it was a given that public utilities were the best vehicles for providing water resources to people, but this is no longer the case. Beginning in the 1990s, water utility privatization began to be imposed upon nations as a condition of structural adjustment and debt-relief packages. Corporate double-speak promoted the argument that public services for essentials like water were inefficient, corrupt, and subject to political manipulation. The road from that ideological cant to water privatization was short and direct, and a handful of multinational water utility companies became aggressive players in the global market for water and sanitation service provision, further degrading ecosystems and quality of life for many people around the world.

While water privatization has not been as lucrative as these corporations had hoped, the shortage of money for the public sector and the ongoing funding shortages of many cities keep the possibility of water privatization alive. As long as multinational corporations seek to control and own our waters, a resilient strategy for water management will need to be localized, and tied to home and community infrastructure. Because of this, managing water use and reuse is an important homesteading skill.

Water's imminence as a global concern is an opportunity to change how we live right now. Household management of simple greywater and rainwater storage systems will give people choice about water sources, as well as positively affecting municipal water management and supply strategies. Becoming kin to our water also helps us learn to live within the limits of the system, linking up our water with the waters of the world. There are numerous low-cost and small-scale water storage and recycling systems already in use in our cities. The rest of this chapter is full of different ways you can begin reusing and storing water at home.

The Glory of Greywater

The simplest place to begin is with the water we are already using that is clean enough to reuse. Greywater is the lightly used water that comes from the washing machine after the wash and rinse cycle, and the water that goes down the drain from your bathroom sink, or after a bath or shower. It is abundant, and can be reused with some low-budget adjustments, as well as some more permanent changes to the basic plumbing in your house. Greywater can be efficiently reused for irrigation purposes, while simultaneously reducing fresh water use. Especially in climates where water is scarce, and municipal water resources are stretched tight, there's no reason not to reuse greywater and direct it to the landscape to feed your plants and recharge the groundwater.

Safe and Legal?

A note on safety: One of the good reasons we evolved the water system we currently use was our effort to avoid breeding dangerous pathogens (from malaria to dysentery to *E. coli*). These diseases are spread from fecal contamination of water. The public health concern about the spread of disease is real and important to keep in mind, but it is also important to recognize that we have lumped together toxic and nontoxic water streams and are wasting a massive amount of water that could be more effectively utilized. Greywater was understood as sewage for so long that there is a common misperception

Photo by Rachel Kaplan

100% of the water for this garden comes from our bathtub.

Cost of system = $20

ASK ME HOW.

that it is responsible for the spread of contaminants. In truth, you would have to drink an enormous amount of water tainted with fecal matter in order to spread these pathogens. Rather than continuing to dump this slightly used water down the drain, we can redirect it safely towards our landscapes, which positively lessens fresh water use; limits use of over-taxed city infrastructure; and supports a healthier hydrolic cycle overall.

While there are no documented cases of illness transmitted from a greywater system in the United States, caution still prevails.

1. Never drink greywater. Greywater systems need to be designed so that the water stays in the pipes and soaks into the ground to prevent accidental ingestion.

2. You can use greywater to irrigate edible plants, just not the edible portion. Do not irrigate root crops, or food like strawberries that lie on the ground, with greywater.

3. Don't let greywater sit in a storage basin for more than twenty-four hours, although this is more an odor issue than a pathogen issue.

4. Make sure that the products you use in your bath or laundry are safe for plants. Look for biodegradable products that are free of salt (sodium compounds), boron (borax or borate), and chlorine bleach. These can damage your plants. Some products say "greywater safe" on the label, but may not be. Read the ingredient labels before you buy products for use with a greywater system.

5. If you're washing diapers, don't use the greywater coming from the laundry. This is the likeliest source of contamination through greywater. Other laundering tasks, and showers and baths, are safe sources of greywater.

When thinking about potential health risks associated with greywater, consider as well the risks associated with climate change, an insecure water future, and sewer overflows, all of which can be mitigated by safe greywater use. You will undoubtedly balance out the personal pros with the global cons and come up with your own set of practices.

A note on the law: In many municipalities, it is not legal to divert greywater to the landscape, but this is beginning to change as our challenges with water are exacerbated by population growth, agricultural excess, and dwindling water resources, especially in the arid western states. Arizona, New Mexico, Texas, and Wyoming have very simple guidelines to follow for legal greywater reuse that do not require a permit. California recently passed a new code allowing washing machine greywater systems to be installed, following state guidelines.

There are no laws (yet!) against using buckets and hoses to divert water to the landscape, but the diversion of laundry water and bathwater to the landscape remain controversial in many cities. Kitchen sink water is considered "blackwater" in some, but not all states, which designates it as unsafe to be reused, and toxic for land and humans. It is therefore illegal to pipe this water into the landscape, even into a place where the water would be filtered and cleaned (through an easily constructed biofilter). In some states, it is illegal to harvest the rainwater falling on your roof.

You will have to decide for yourself if your conservation ethic trumps your law-and-order ethic. Educate yourself about the laws in your state. If you feel uneasy about the legal aspects and that keeps you from being able to conserve water in a way you'd like, work to help local agencies understand the values, benefits, and low health risks associated with greywater reuse and rainwater harvesting. Support them in safely changing laws to allow the intelligent reuse of these available and plentiful waters.

The greywater and rainwater experiments presented in this chapter are applicable in a wide variety of environments, although every home will have its own challenges and opportunities. Get creative as you reinvent your water use, reuse, and conservation ethic. You'll undoubtedly find solutions in your place that will suit it best; you need only learn a few simple facts about how water flows through the systems of your house and on your land to begin to understand how to harvest, sequester, and use it more efficiently.

on the ground: haut house

Laura Allen is one of the prime movers of Greywater Action, a nonprofit organization working for water justice, changing state legislation around water use, and offering creative alternatives to the current water infrastructure in our cities. She has been evolving greywater solutions for the urban landscape for the last decade, and teaches what she's learned to urbanites and suburbanites throughout Northern California. Laura lives in the collective Haut House in Oakland, her "research and development site," where she and her collaborators have implemented many greywater solutions, beautiful food growing gardens, and creative reuse projects.

Like many urban sites, Haut House required members to deal with an enormous amount of city repair when they established their home. The site was entirely paved when they arrived, with a giant double driveway and only one patch of green lawn. They de-paved the concrete during two "jack hammer parties," where all their friends were invited to come over for food, drink, and jack hammering up the yard. They sheet mulched the lawn and brought in tons of wood chips to add fertility to the compacted clay soil that had been driven on for decades. Chickens were next, and they helped add fertility to the soil, as did all the composted kitchen scraps and the human wastes (or "humanure", a phrase coined by waste management expert Joseph Jenkins).

The water systems and perennial gardens are the most evolved parts of their homestead. "We built a composting toilet early on because it was the easiest way to cut back on water use," Laura explains. "In terms of legality, it's not illegal to have one in the city, as long as you have one flush toilet per house. It is illegal to transport human waste anywhere, so the way we compost it on-site is fine. We compost it in a 55-gallon drum, and let it sit for a year so we can feel confident that it's fully composted. We don't need to allocate too much space for this compost area, since the material amazingly breaks down to about one third of its original volume. Most of the fertilizing benefit from having the composting toilet is from the urine that we separate from the poop. Urine is an excellent fertilizer and we dilute it, then use it directly to fertilize the landscape.

"When we moved here, we made cookies for all our neighbors and got to know them. We knew we were going to be doing things that might not be accepted, so we wanted good relationships with our neighbors. It was an intention on our part—if we have bees and they swarm, we want our neighbors to come to us, and not report us to any city agencies. And our neighbors got to see who we are and what we're doing, and they like it."

Haut House residents are teachers, post-docs, and solar installers, and Laura now works as an educator and independent contractor teaching and installing greywater systems. "I've learned a lot by living here, and watching the site progress over time. I can see what happens to a system after it's been working over the years. It's been invaluable to have a space where I can experiment over time and see how things evolve. With greywater, you can't always know how it's going to work in the long run, so it's good to learn how to tinker and fix the systems you create and make future systems that work well without much maintenance."

Laura really focuses on providing workshops and education to people because she wants people to have the skills to do these things themselves. "Everyone can have a greywater system that recycles this source of water. It's simple to do, and important. It really matters for me to have my lifestyle in line with my politics. I want to live in a way that's a contribution, and do something that's beneficial. Just by being an American, there are a lot of parts of my life that are damaging to other people and places that I can't see, but this is a way of doing what I can with the things I can immediately change."

The good news about sustainable water practices is that public perceptions, and the laws, are slowly changing. Legislation in many states is moving toward legalizing rainwater harvesting and greywater reuse in residential applications. Other practices fall outside the guidelines of codes, like urine reuse and composting toilets. "But on the larger scale, the ways communities and cities acquire water, use it, and treat it mostly ignore decentralized, lower-impact options," Laura says. "In many parts of the country these practices are still rather uncommon and typically illegal. We are working to show that it can be simple, affordable, and beneficial in a lot of ways, because we think this will help people start getting used to the idea of viewing 'waste water' as a resource.

"I think people get joy from their greywater systems and feel good about shifting the way they use water in the home. People tend to use healthier products when they have greywater systems—what's better for the planet is also better for us, and people know that. I've seen many people start with a small greywater or rainwater system and then continue to change their homes to conserve and reuse resources, growing food and beautiful landscapes, while consuming less.

"As for greywater, it's moving toward legality and that's making it easier for people to do it. There are still some sticking points—cities are nervous about things they don't understand. So that's where it's our job to really offer advice and education about how to do this in a good way. I am hopeful about these changes, but of course, our entire culture needs to shift thinking about water in a big way in order to make these practices work on a large scale."

Greywater for the Renter

Renting or living in an apartment doesn't mean giving up water sustainability. Greywater harvesting from sinks, showers, tubs, and laundry can be done at low cost, and the systems are low- to no-tech, or fully reversible when you move. In an apartment where you have access to a backyard and garden space, you can set up a simple composting toilet and use the urine as fertilizer in the garden, divert the water from the washing machine, or siphon water from the bathtub into the garden without spending much more than $20. You can also harvest the water from the bathroom sink for reuse.

Here's how to unhook the P-trap underneath the bathroom sink to catch the greywater.

1. Get a pipe wrench and loosen the pipes from the sink drain to the P-trap (the U-shaped part of the pipe).
2. Turn the P-trap to the side (a little water may spill out; have a towel handy). Make sure you keep water in the P-trap. This prevents sewer gases from getting into the bathroom. Rubber band a small plastic bag over the opening and let your sewer pipes rest.
3. Insert a bucket under the sink drain to catch your greywater.
4. Use this water to flush the toilet, or to irrigate plants in the yard or the deck.

Without significantly changing your (landlord's) house you can also install a shower shutoff valve, which is a $5 piece of hardware that will reduce shower water use by 50 percent. You can wash dishes using a dishpan, and use the water on your garden. Another simple way to save water is to add bulk (a rock or a brick) to displace water in the toilet tank. This will lower the number of gallons that get flushed each time you do flush the toilet.

The following simple water-saving techniques are also useful for both renters and owners who want simple water conservation strategies that won't be too costly or take too much effort to install or use.

Simple Water Saving: Buckets, Hoses, and Siphons

The simplest way to reuse our greywater is terribly old-fashioned: the bucket. Available to all, the bucket is a slow and small solution that makes it possible to recycle our "gently used" water for houseplants and garden.

The bucket—low tech and affordable. It works.

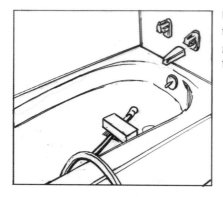

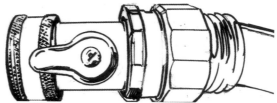

Use a garden hose to create a siphon so the water in your bathtub can be diverted into the landscape. Female to female adapters allow you to attach the siphon hose to a garden hose to prime the siphon. Once the siphon hose is filled with water, detach it from the spigot or hose and water away.

Female to female adapters extend the reach of your hose so you can divert water where it's needed.

Beyond the bucket, another low-budget system is also extremely simple—siphoning the water from your bathtub into your garden. To do this, you need a water source in the garden, and a garden hose that reaches from the water in the garden into your bathtub. You need your bathtub to be above the garden, so that gravity can do the work of moving the water once you've created the siphon.

Siphoning the Water from the Tub

1. When you are done bathing, place your hose at the bottom of the tub. Hold it down with a brick or a stone.
2. Once the hose is settled under its brick and the other end is dangling out the window, go outside to the garden.
3. Attach the other end of the hose to the water spigot. Turn it on until the water goes from the garden into the bathtub. (You will learn how long this takes after doing it a few times.) This creates a seal from the water source in the garden to the water in the tub.
4. Turn off the water, unscrew the hose at the spigot, and the water will siphon from the tub into your garden. You can direct it through the garden at will. If the end of your hose doesn't reach the spigot, you can also use a female-to-female adapter to connect two hoses to one another and extend your reach.

A family of three that take between one and three baths or showers per day can expect to recycle over 250 gallons of water per week with this system, diverting a total of 13,000 gallons of water per year from the water system and into the landscape. A family of three has an average total yearly water use of around 60,000 gallons; this simple bathtub diversion represents a sizable portion of water used. This number does not include the amount of water you would save if you also recycle from laundry and bathroom sink sources. While it's unlikely that you'll be able to irrigate your entire landscape with greywater (especially if you living in a water-conserving household or the arid West), you can use greywater to augment your water needs and cut down on outdoor water irrigation.

If you choose not to siphon your bathwater (or if it's been raining or snowing for days and you don't need to siphon the water outside), leave the water in the tub and use it to flush the toilet or water the plants. Each flush of the toilet uses somewhere between 1.6 and 3.8 gallons, so an amazing amount of water is saved this way as well. Take on the California mantra when using the toilet: "If it's yellow, let it mellow. If it's brown, flush it down." This saves a lot of water every day.

The Laundry Drum

Unless you use an efficient washing machine, you generate between thirty and fifty gallons of water per washing machine load. That's a lot of water that can be easily directed to the landscape. Water from a washing machine is diverted generally from a hose at the back of the washing machine into a drain behind the machine. To alter the system, remove the pipe at the back of the washing machine that diverts the water to the drain and connect it to another one-inch hose that runs into a fifty-five-gallon drum that has been tricked out with a spigot and a hose. When the laundry cycle runs, the water will be caught in the barrel. Once you've gathered the water in this way, you can direct it to the garden at will.

This is a low-cost, low-intervention method of catching and diverting water to the landscape. It works well for people who have the time and energy to divert the water by hand to their landscape. With this system, the storage barrel needs to be higher than the garden so gravity can do the work of dispersing the water into the landscape. In order for the water to drain, the drop needs to be at least one-quarter inch every four feet, or else the water won't flow. It is also a good choice for renters who are not willing or able to invest in long-term infrastructure at the place where they live, but still want to access greywater for the garden.

Branched Drain

If you own your house or are willing to make the investment, diverting the water from the laundry or bathtub directly into the ground with a branched drain system is a simple and efficient way to go. Rather than catching the water in a surge tank to drain later, you can install a series of pipes that direct the water to the garden each time you use water. The branched drain system also relies on gravity to disperse the water through the landscape, and only works in a setting where the water source is higher than the garden.

A branched drain system will have a diverter at the water source to direct the water to different sections of the garden. Diverters on the ground are also installed so that you can have some choice about where the water flows each time you use the system.

Direction of greywater can also be controlled from a branched drain in the garden. Each time the system is used, water can be diverted to a different part of the garden. Branches can be one-inch, two-inch, or half-inch.

Valves at the washing machine can be changed each time you do a load of laundry. Yellow valve directs water to landscape or drain. Red valves determine which part of the landscape gets the greywater.
Photo by Diane Dew Photography

Laundry to Landscape

Laundry to landscape systems use the washing machine pump to distribute the water to the landscape and do not rely on gravity the way the branched drain or laundry drum systems do. Laundry to landscape is the evolving best practices greywater system that conforms to code in a number of Western states. Without stressing your washing machine pump, laundry to landscape systems irrigate any distance downhill, or pump to an elevation two inches below the top of the washer up to 100 feet away. As with other greywater systems, you can direct the water throughout the landscape with a series of distribution and irrigation pipes.

When diverting water from the laundry either through a laundry to landscape or branched drain system, sending it to a mulch basin is an efficient way to make use of the water. A mulch basin is a trench or ring around a plant, filled with organic materials, that filters greywater through the mulch to irrigate the roots of plants in the ground. You can plant trees in a mulch basin and set up the distribution system to bring water to the roots of the tree. You can also divert water to gatherings of plants in different areas of the garden.

There are some technical aspects to the laundry to landscape system that may warrant consultation with a trained greywater installer, including the kind of hardware to use, and the need for an air vent when the system is inside. Fortunately, people are being trained in this technology even as I write this sentence, and will soon be living in a city near you. For a more technical treatment of the laundry to landscape system, including a full list of all the hardware you will need, please see www.oasisdesigns.org, Laura Allen's DIY manual, "San Francisco Greywater Design Guidelines" at www.sfwater.org, or www.cleanwatercomponents.com. Though specific to certain locations, these designs represent the best practices in greywater reuse, and are available free online.

Diverting the Water from the Shower/Tub

Unlike the laundry to landscape system that does not require any change in basic plumbing, diverting the water from the shower involves a simple plumbing hack that directs the water from the drain into the landscape. This is a system more suited to owners than renters, as it involves the manipulation of the plumbing system to move the water from sewer to landscape by placing a diverter on your drainpipe. This should not be done without obtaining a permit from your local municipality, and in many places you will not be able to get one easily. If there is no permitting in place for bathtub diversions, this is a great opportunity for local pro-activism on the issue of greywater reuse.

This outdoor washing machine provides an easy way to divert water to the landscape.
Note red valve on left side of washer—allows for choice about whether water will divert to landscape or drain. Run a single or multi-trunk line, with or without valve or branches, to the landscape. This system can also be constructed on an indoor washing machine. The mechanism is the same.

Pipe from the laundry outflow is positioned throughout the garden. Valve boxes (black circular containers) are placed intermittently where greywater will be directed and help sink it more deeply into the ground. *Photo by Dustin Kahn/Greywater Action*

When laying line from the washing machine, dig a trench to bury the line. This protects the plastic from the sun, and brings the water to a deeper level of the soil, which is better for plants and people. *Photo by Dustin Kahn/Greywater Action*

Distribution plumbing of one inch should be used to get water to different parts of the landscape. The outlets are directed toward mulch basins filled with organic matter that further sinks the water into the ground.

Covered water outlets help sink water into the landscape.

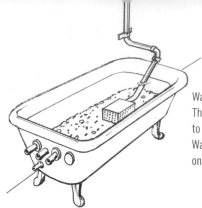

Water coming from above (laundry or bathtub) drops into bathtub used as a catchment tank. The end of the inlet is covered with a box (like a plastic milk crate) enclosed in tight mesh to prevent plant roots from blocking the inlet. The box is then covered with rocks and gravel. Water-loving plants are added as an additional bio-filter for the greywater. This system relies on gravity to disperse water through the garden.

As with the laundry to landscape system, once you divert the greywater from the drain, you have numerous choices about where it can go—into a branched drain throughout the garden, into a mulch basin, or into a catchment tank where it can be purified by a variety of water plants. This "biofilter" catchment tank, or constructed wetlands, can be a beautiful addition to a garden, adding a water habitat and another place for plants to grow. Once the water is filtered, it can be moved into the landscape, either through a branched drain, or more simply with a hose attached to the bottom of the bathtub or container. Appropriate plants for the constructed wetlands include papyrus, iris, cattails, euphorbia, equisetum (horse tail), and scirpus (rush). Use plants that will thrive in your location, and be careful—some of these plants spread. After a few years you will have to thin them out and pass them on to someone else.

All of these greywater systems are simple to install and easy to use. Some of them demand a little more interaction on the part of the user; some save time while saving water. When you create your greywater system, you'll want to factor in the water needs of your garden, the total volume of water use in your household, your ownership status, financial resources, and the level of direct action plumbing that meets your needs.

Systems for Cold Climates Including Wetlands

Year-round greywater reuse in cold climates can be a challenge. What to do with all that water when the garden's under snow? Will greywater freeze in the pipes? Many greywater users divert greywater to a sewer or septic system during the cold season, but it's possible to ecologically dispose of greywater even in the dead of winter. Successful greywater systems from Vermont to Montana and onward to North Dakota are evolving the form.

Suggestions for cold climate greywater systems:

1. Include a diversion valve and run the greywater to the sewer when it would otherwise freeze outside.
2. Let bathtub water sit around and exchange heat with the room, leaving you toasty long after your hot bath. Use the standing water to bucket-flush your toilet.
3. Cover all exterior pipes in black foam insulation, paint all possible surfaces black, and surround wetlands with big black rocks. This will insulate the pipe to some extent.
4. Insulate any raised surfaces of your greywater system with hay bales. Again, this will give moderate protection.

While greywater reuse may be less compelling or necessary in climates that have enough water throughout the year, reusing greywater yields benefits to taxed municipal water

Constructed wetlands in salvaged bathtub catches water from the shower or laundry, and plants purify the water before diverting it to the garden.

systems, household plumbing, and to the groundwater. Greywater reuse is more prevalent in dry regions than wet regions, but reusing all the water on site is an excellent conservation practice, whatever landscape you inhabit. There simply isn't enough fresh water to meet all of our needs; for this reason, it just makes good sense to conserve water wherever possible. In addition, allowing greywater to seep back into the watershed rather than sending it to the sewer compensates slightly for all the rainwater permanently lost to the watershed because it falls on paved streets and parking lots and roof tops and is diverted to storm drains.

Rainwater Harvesting

Rainwater is the other source of water on the homestead that's too frequently squandered. When you live in a climate where the rain comes only during one time of the year but you need to use it throughout the year, catching it when you can is an important way to conserve municipal resources. As with greywater, there's an astonishing amount of rainwater available for capture. There are a number of simple ways to catch and store the rainwater that falls on your house—swales and earthworks, simple rain barrels attached to gutter downspouts, cisterns, and a daisy-chain rain-barrel system are the systems that work best in the urban landscape.

on the ground: permaculture artisans

Erik Ohlsen is an inspired permaculture designer who runs a landscape and consulting company, Permaculture Artisans. Generally impersonating the Energizer bunny, Erik has enormous energy and passion for the work of renewing the Earth. He has worked in the field of permaculture since high school.

"Instead of going to college, my friends and I started a nonprofit called Planting Earth Activation. We learned about Monsanto's terminator seed technology and felt a passion to respond. Our thing was to give away heirloom seeds and gardens. Planting Earth Activation would go to a neighborhood and choose a street where we would get neighbors with front yards to receive a garden from us. In one weekend we'd organize volunteers, roving musicians, and potlucks on the block. And the neighbors would come and we'd put in gardens down the whole street in a day. We'd plant fruit trees, perennials, annuals.... We'd have piles of compost dumped on the street. This was ten years ago—just like what's now called victory gardens and Food Not Lawns."

He lives at a cohousing community of three houses with the back fences taken down, which he shares with four adults and five children. The site is being developed as a model of what's possible in urban and suburban landscapes. "The first thing we did was rip up the cement and throw down seeds. We didn't bring in a lot of dirt until we were ready to plant. We earth-worked the property and did a lot of drainage and water storage systems." A drain from the houses and streets above Erik's house comes out at the top of his property and pours water across the whole yard. Rather than have the water take up the topsoil and drain it out to the sea, Erik and his neighbors have designed an extensive water catchment system that ranges over the whole property. They estimate that over 400,000 gallons of water run across the property in a single winter, which is like having a spring in the middle of the garden, a rarity in water-scarce California. "We wanted to slow and catch water on our narrow sloping property. When it really rains, these swales are full, and the water drains into the soil over time. There are catch basins filled with mulch, but before the water spills out, this little basin has to fill up. We're slowing the water down, little by little, and sinking it into the ground."

As we walk around the garden, tasting and smelling, Erik shows me what he's growing. "Here's a great guild—we have a daikon radish, a dynamic accumulator biomass creating plant; the autumn olive, which is a nitrogen fixer; and lavender, an insectary, which is edible and culinary and medicinal. That's a lot of food, biomass, habitat, and herb for a little space. You could potentially grow something like this in a big barrel, or a small lawn. It doesn't need much space. The understory

This constructed streambed directs the water flowing over the landscape into swales and the pond below. The shape of the streambed slows down the water flow and helps direct and sink the water into the ground, rather than draining into the sewer system.

The pond receives the overflow of water from the diversion of the constructed stream bed.

of every tree in this yard is a potential vegetable garden. Eventually our perennials will grow so much that the area will get crowded out for annual vegetables. But for now, I go to the cherry espalier to maintain my lettuce patch."

Repair is the mantra of Erik's life. "We're not up to sustainability yet. We have a lot of repair work to do. . . . To me, caring for and building soil, slowing and infiltrating water, and growing plants that provide food for people and wildlife, create what I call a regenerative system. Sustainability is the ability to sustain cycles, but we really need to learn how to sustain the living cycles we have inherited. So what we're doing is more about earth *repair*. This was all asphalt and cement, compacted gravel—nothing could grow here. Look at it now. And in five years, it'll be amazing! We have to ask ourselves about the legacy we leave to humans, and the whole ecology in terms of soil, water, and habitat. Our actions in that direction are the ultimate expression of sustainability. We need to grow an economy that produces resources for human beings that are a by-product of healing the earth. That's the sustainability part, that kind of repair of the world."

Calculate the Rainfall

The first step in figuring out which catchment system will work best for you is to calculate the amount of rainfall that lands on your roof. You will need to know the square footage of your catchment area (roof), and the average amount of rainfall in a given year to make these calculations.

A = length x width of roof (square feet). The pitch of the roof does not matter in this calculation.

R = rainfall in inches per year.

Using the following formula: **(A x R)/12 = W** (in cubic feet) gives you the amount of cubic feet of rain your roof collects each year.

A cubic foot is about 7.5 gallons. To convert cubic feet to gallons, calculate **W x 7.5.** This equals the amount of rainfall in gallons per year that fall on your roof.

For example: A 35 x 30 foot roof covers 1,050 square feet. If the average rainfall is 25 inches, A x R = 1,050 x 25 = 26,250. Divide this by 12 = 2,188 cubic feet of water. Convert to gallons by multiplying 2,188 x 7.5, which equals 16,410 gallons of rainwater coming from the roof. Even in the Western dry lands, this is a good part of what anyone needs to irrigate a landscape throughout the summer.

Your obvious next problem, as you can see, will not be one of shortage, but of storage, the single biggest rainwater harvesting issue for urban homesteaders. Where are we going to stash all that water, and is it worthwhile to catch just a little bit of it? Despite our inability to store it all, it's always worthwhile to catch water whenever it falls. We can always use it, and we rarely have enough. This is pure California-ese where we can't afford to waste a drop. In other parts of the country, water shortages are not the same issue, but we still think it's better to catch the rain in mulch basins, swales, and other earthworks to return it to the soil and ultimately the watershed rather than having it run down the street, collect petroleum toxins, and end up in the storm drain. Undoubtedly, homesteaders in more water-rich communities will make different choices about catching and saving water than those made in the water-scarce West.

Catching rainwater for reuse isn't legal everywhere, though relative silence on the issue in some places confers consent for the practice. In many dry states, storing rainwater is not technically illegal, there's just no code yet written about it. Outlawing the storage of any pure rain water in a largely dry state is insane. Think with your heart on this one.

Swales

One great way to harvest rainwater is simply to sink it into the ground. You do this by digging swales on the contour of the land in your yard to direct the water flow where you want it to go. When you dig a deep swale between your garden beds, these trenches will catch the rainwater when it rains, and more water will sink into the ground over time. This recharges the groundwater in the garden, and will add to the garden's productivity. Less water running off the property and more sinking into the ground is the way to go.

Additional ways to catch water in our gardens include using plenty of mulch and lots of organic matter in the soil. Another way is to de-pave the driveway and replace it with a permeable surface. Water will then sink into the ground rather than running off the landscape.

Swale on contour. Water caught in the swale seeps into the ground. This recharges the groundwater and is a conservative way to manage water, especially in drought-prone climates.

Digging a swale as part of a Boys and Girls Club lawn transformation project, Petaluma, California. Swales replace pathways between the beds and allow for the water to sink into the ground during the rainy season.

Swale during the rainy season catches overflow water and sinks it into the ground.
Photo by Erik Ohlsen

Paving stones and grass patches replace a solid asphalt driveway. Water sinks into the ground rather than running across the landscape and down the drain.

Interestingly enough, these water-catching methods of mulching and organic matter work equally well in overly wet as in overly dry gardens. They function to maintain the balance of water in the ecosystem and do valuable work at both ends of the water spectrum.

Down Spouts and Diverters

To catch water off the rooftop, you need to divert it from the downspout coming from the gutters, and have a place to store it.

1.   Attach the gutter to a 1 ½ inch ABS pipe directed towards the barrel.
2.   Cut a hole in the top of the barrel and cover it with a screen. This will allow water in and keep dirt particles out.
3.   To prepare the barrel itself, install a spigot to the bottom of the barrel by drilling a hole with a ¾ inch bit. Seal the spigot with aquarium or other waterproof glue so the barrel will not leak. Attach a hose to the spigot so water can exit the barrel as needed.

Once the barrel is connected to the downspout, and tricked out so you can get the rainwater out of it when you need it, you're ready to go. When gathering the first rainwater of the year, create a simple "first flush" diverter by turning the drainpipe to the side before the first rain of the year. That way the dirtiest first rain is not collected in the barrel. This is especially important if you are going to be using the rainwater for drinking or cooking purposes. After the first rain, turn the pipe back over the screened barrel top and allow the water to drain into the barrel.

Daisy Chain Storage

The unfortunate part of the rain barrel's good design is that it just doesn't hold enough water. To increase the storage capacity of your system, add more barrels. The engineering and physics of the downspout and storage is the same, but the volume you catch is much greater. Stash the barrels in any narrow place alongside the house—an alleyway, or where your garbage cans usually sit.

Materials

Electrical male adapters (1 per barrel)
Rubber washers (not an O-ring)

Rainwater diverted from roof into rainwater barrels. To increase storage capacity, barrels are connected to form a daisy chain. Barrel on the far right has a spigot for water distribution when needed. Pipe coming from the top of the right-hand barrel allows the water to overflow when barrels have reached maximum capacity. Direct this pipe toward the landscape and away from the foundation of the building.
Photo by Rachel Kaplan

Female threaded barbed Tee fittings (1 per barrel)

1-inch vinyl tubing to connect the barrels (1 per pair of barrels)

1 brass shutoff valve

Separate from the cost of the barrels, the total cost of the system should not exceed $50.

1. The first barrel has a hose connection to the second barrel. Install a female threaded Tee to this first barrel.

2. On the side which points out and away from the barrel, attach a brass male hose thread to male pipe thread (MHT to MPT) connector, with a hose shut off valve connection. The side that points toward the next barrel uses a male pipe thread with barbed fitting and the vinyl tube connecting the two barrels is hose clamped over the barb.

3. When the rainwater is needed for irrigation, a garden hose is attached to the first part of the female Tee and the shut off valve is turned "on." The middle barrels have a female barbed Tee with a one-inch vinyl tube hose clamped on each end.

4. The last barrel in a string of barrels ends with the tube tied off with wire. This allows for easy connection of more barrels in the future.

5. If no further connections are desired, the last barrel could end with a female thread ninety-degree bend, with a MPT by barbed coupling. The one-inch vinyl tubing fits over the barbed fittings and is hose clamped. It is important to use electrical male adapters and not plumbing adapters because the plumbing fittings have rounded edges and it's harder to create a watertight seal with them. When trouble shooting and maintaining the system, check the screens and make sure they are intact.

Cisterns

If you have the space for it, use a cistern or larger storage container for more rainwater storage. They can be found at hardware stores or stores that service ranches and agricultural properties. They hold a lot of water, but take up a lot of space. They are too large for most city environments, though many

Barrels are connected with Tee fittings, one-inch distribution line, electrical male adapters and O-rings inserted into the barrel itself. Spigot at the end of the line attaches to garden hose and water can flow from there into the landscape. *Photo by Rachel Kaplan*

Raintree, an all-season garden sculpture by Christina Bertea, proves that anything worth doing should accomplish more than one purpose. It's a rain catcher in the autumn and winter, and a clothesline in the spring and summer. Made from mahogany plywood scraps, galvanized pipe and fittings, salvaged post base, removable repurposed vinyl, clothesline, and Velcro.

This 1,500-gallon backyard cistern routs water from the roof of the house and outbuilding and stores it for a sunny day.

buildings in cities have old cisterns on the roof, harkening back to an age when almost every building had its own rainwater tank.

Locate your storage basin upslope and in an inconspicuous place. A screen inside the top of the cistern keeps out leaves and other debris. A garden hose can be connected to the base of the cistern for use with outdoor irrigation.

Wherever you live—dry or wet land—conserving, reusing, and harvesting water are important home-steading practices. While most of us in the United States currently have access to affordable, clean water, all of us do not, and the limitations of the clean water supply are in evidence around the globe. Recognizing the natural limitations of the hydrolic system, global water scarcity, and the current threat of the privatization of water will keep us stepping toward remaking a just and diversified water system. This is a simple way to relate deeply to the water we need to live, and reminds us of our dependence on this precious, endangered river of life.

TENDING TO WATER

When thinking about refining water conservation and use, plan to make changes over a realistic period of time.[87]

Six-Month Transition

    Monitor and measure current water usage by month

    Implement basic conservation measures:

        Low flow showerheads

        "Mellow yellow"—only flush when you need to

        Add bulk to toilet tank to lower gallons flushed

        Check for all leaks

        Dishwashing basics—no dishwasher, limit water use

        Wash clothes less frequently

    Source food from soil-building farms

    Install downspout diversion to catch rainwater

Twelve-Month Transition

    Implement basic greywater systems—buckets, siphon, laundry diversion

    Harvest rainwater (if possible) using bladders, tanks, cisterns

    Create basic composting toilet (see Chapter 17, Zero Waste)

    Install drip irrigation under heavy mulch

Thirty-six to Sixty-Month Transition

    Install potable rainwater harvesting

    De-pave the city—sink the water into the ground

    Become a greywater policy advocate

    Install an outdoor shower

    Make or buy a solar hot water heater

    Get plumber's license and start greywater worker-owned co-op

    Plant perennial crops

    Establish food rotations with limited drip irrigation under mulch

Zero Waste

When you say you are throwing something away, where is that? There is no "away."

—*Julia Butterfly Hill*[88]

Natural cycles function without producing waste. The sun provides the energy for the plants, driving the process of photosynthesis that transforms matter into more complex forms of life, like food or forests. Dead matter from the forest (leaves, branches, trees, animals) is then recycled by microorganisms and small animals and transformed into soil to be used by the next cycle. Nature expresses this "closed system" over and over again. A popular expression of this concept is "Waste = Food."[89] In contrast to this natural law, our industrial cycle has broken these normally closed loops by removing both renewable and nonrenewable resources from the earth, transforming them further (usually with energy created by nonrenewable resources), distributing them through a global marketplace via more nonrenewable resources. Consumers purchase, use, and discard much of this valuable resource and add it to the growing landfill pile. The "openness" of this system does not adhere to the natural principle of Waste = Food. Rather, it tells an unsustainable story of stuff that threatens to overwhelm the systems of the earth.

The result of overstepping the boundaries of the natural system so significantly is that we are drowning in our own wastes. The Earth simply cannot absorb all the garbage we create. But we can reform these nasty habits and get real about our excessive consumption by radically reducing use and managing the resources available to us with maximum efficiency. Zero Waste is the Holy Grail of the sustainable city. There are many creative ways to reduce and reintegrate the resources we use at home, at our children's schools, and at work that set us on the path of radically reducing waste. It's actually easy and fun to challenge yourself to make do with less, and reuse more.

You probably have "Reduce, Reuse, and Recycle" tattooed on the inside of your head by now. If you're a compost devotee, Rot has been tagged onto the three R's as well. This is all good, but it doesn't really address the small world reality: there's simply nowhere for the waste to go. We need to move beyond Reduce, Reuse, and Recycle to Rethink ("waste" is a resource), Redesign (from open to closed system), Reduce (consumption and use), Repair (systems and objects themselves), and Recover (compost what's organic, and return it to the earth as nutrients).[90] Once we've internalized the ethics of earth care, people care, and fair share, zero waste practices become second nature, a central strategy for meeting our need to be of service to the Earth.

RUN THE EXPERIMENT: HOW MUCH GARBAGE DO YOU MAKE?

In our family of three, we experimented with how many weeks we could go before we really had to pull the garbage can out to the curb. We went for eight weeks (using a twenty-gallon can) before we put it out, and then only because the garbage was sprouting maggots. Rather than recoiling in too much disgust, we took it as a challenge to use less and reuse more. Apart from the maggots, we were glad to notice that even after eight weeks the can wasn't full. We've been holding onto the recycling bin for about ten weeks, and think we can eke out another two before we have to take it to the curb. That means we're taking our recycling to the curb only four times a year. Could we do better? Probably, but it's a start.

Setting clear goals and meeting them can inspire you to set the bar higher next time. This month, challenge yourself to reduce your use of recyclable glass by 10 percent. If you can reduce your use that much within the month, challenge yourself to reduce your use by 20 percent next month. You'll be amazed at how your spirit of creativity and positive competition gets engaged in the process and pushes you along the path to better resource use in your home place. See how much you can reduce your consumption, and keep raising the bar. Before you know it, you'll be taking your maggots out to the curb, too.

Reduce

Of the three R's, Reduce is absolutely the most important. Our country's over-consumptive buying habits are literally trashing the globe. You might be surprised how simple and enjoyable it can be to jump off the runaway train of consumption and start thinking about, and using, objects in a different way. How little can I use and still have what I need, and then some? If I identify what I need, and can only come up with an object to buy, am I really meeting the need, or cycling in an addictive system of fear and lack? Can I make something today rather than buying it at the store? Can I find a use for an item rather than tossing it into the trash?

Committing to the Reduce ethic is solid homesteading, a way of taking your energy (and money) away from the consumer economy, and making peace with what you've got (which we venture to bet is an awful lot anyway). About stuff we think we need to buy but probably really don't, we have a simple mantra: Just Say No. Every little trick you use stretches existing resources farther, reduces the need for more consumer products, and cuts down on landfill pressure. Every reused item, every bit of waste eliminated, everything you don't buy but make yourself, is a step toward creating a life-serving economy.[91]

This 100 percent glass house made from old windows is a great example of reusing materials to make something affordable that is also beautiful and useful. Christiania community, Copenhagen, Denmark. *Photo courtesy of Seier and Seier*

Crushed urbanite and other bits of "garbage" reformed into a wall on the side of a bicycle shed in the Haut House backyard. Rather than sending it to the landfill, creatively reused objects often thought of as garbage can be remade into something useful.

Recycle

The benefit of recycling is getting less and less clear, as reports about the mishandling of recyclables continues. What are alternative uses for your recyclables? Cardboard boxes can be broken down and used for sheet mulching. Beer and wine bottles can be reused when you take your excess honey from the bees and turn it into mead, or make beer from the hop vines climbing above your roof, or apple cider vinegar from the leftover apples you foraged from your neighbor's amazing tree. How much of the waste stream can you reuse? How much of it can you eliminate? Approach it in the same way: give yourself a use reduction goal and a time frame, and try to meet it. When you do, raise the bar, and meet the new low standard.

Other resources for recycling are the ever-popular Craigslist, which has online bulletin boards across the globe and is a dynamic source for used items. At www.freecycle.org people pass on, for free, an unwanted item to another person who needs it. From silverware to mobile homes, people are choosing to freecycle rather than discard. The practice frees up space in landfills and cuts down on the need to manufacture new goods. It also builds community. You can find a local freecycle group online, or start one yourself. In communities around the country, people are holding events like Really, Really, Free Markets and Freemeets. These events are like flea markets where people bring items to share with others. They give and take but not a dollar is exchanged. Any kind of community barter or exchange program enriches local folks, while lessening the load at the landfill.

Reuse

Gathering used things or giving away things you don't need are also solid reuse practices. I cannot tell you how many living rooms I furnished in my younger days from the Garbage Can of the Street. They were beautiful rooms, easy on the pocketbook, and simple to change when they wore thin. (Okay, I admit it. I found a great chair on the street just this week, and it's now in the living room. The cushions even match our rug. How cool is that?)

Thrift stores do a brisk business in almost everything, and accept donations as well. Many of them give their proceeds to a local nonprofit, like a homeless shelter, hospice, or school foundation. Your local homeless shelter will never say no to an old coat or other clothing. Reusing is simply another creative art game. How can I turn this thing into something else, give it new life, and save it from the ever-mounting pile at the dump?

Bits of stone, bricks, and leftover building materials can be reused to make a garden bed.

Urbanite (broken-up asphalt) treated with an iron sulphate fertilizer makes a patio area in this small backyard. *Photo by Ken Foster/Terra Nova Ecological Landscaping*

You'd be amazed at how you can convert something you think of as garbage to meet another need. An old cabinet with a glass window can be turned into a cold frame, a plastic bucket that once held cat litter can be used to gather rainwater, or can serve as a fine pot for growing plants. Plastic bottles can be turned into art. An old tire can be a great planter for potatoes. Reusing is also an opportunity to spread the wealth. If you don't use something for some time (outside limit, one year), send it to the thrift store or give it to someone who will use it.

Recover: A Kitchen Sink Recycling System

Dirty kitchen-sink water can be toxic for plants and humans, especially when it contains meat and dairy products whose fats and oils are not beneficial for plants or for the soil in your garden. But rather than sending the water down the drain to further contaminate the waterways, you can construct a biofilter to remediate this water and direct it to further uses on your homestead. This is a perfect example of the idea that one creature's waste is another creature's food.

This system will use the worm bin you built when you were making a vermicompost setup for your back porch (See Chapter 6, Urban Dirt Farmer).

"It is an interesting dichotomy to try and make something beautiful out of polluted materials that have attractive characteristics, like the light shining through a mass of clear plastic bottles. Plastics are one of the worst polluting dangers to the health of all living beings, as they leach many poisons including arsenic into the environment. By attracting an audience using the poetic and healing aspects of art, we hope to make them aware of its very physical presence and thereby introduce them to the dangers." Art collaboration by Persian New Art Group and Alida Line. *Photos courtesy of Raheleh Zomorodinia*

1. Add a third level to the two existing levels of your worm bin by getting another container that fits into the already constructed worm bin.
2. Drill holes in the bottom of the box for ventilation.
3. Place straw in this box and inoculate it with oyster mushroom spawn.
4. Pour the blackwater from your kitchen sink into this chamber.
5. Continue to place kitchen scraps in the second level of the worm bin as before.
6. Mushrooms will grow in the upper box. You can eat them, or plant them out in your garden to spread the mycelium to the trees and plants you are growing.
7. The straw in this level of the system will absorb some blackwater. The rest will drop into the next level of the box where the worms live and are fed by kitchen scraps. The kitchen sink water keeps this part of the system well watered, and brings more nutrients for worms and other microorganisms to eat.
8. The water will continue to filter down to the final, bottom layer of the worm bin.
9. The water has now gone through two levels of remediation. What drains from this final level of the worm bin is a very fine vermicompost tea, a golden elixir for the garden.
10. Drain this water and return it to the garden.

This triple-stacking-function system converts the blackwater from the kitchen sink into a great organic fertilizer for your plants in about two square feet on your back patio. The mushroom spawn and worms have transformed a waste product into food for themselves and the garden.[92]

on the ground: eco machine

Rick Taylor runs Elder Creek, a sustainable landscaping business that's been in operation for the last fifteen years. He lives in a duplex in Sebastopol, California, with his partner and child, and shares the building with a co-owner. His lot is small (about 5,200 square feet), but hosts a number of fruit trees, vegetable gardens, front garden landscaping, a small studio in the back with a living roof, an outdoor bar and shower system, and an ecological machine based on John Todd's work that transforms the blackwater from the kitchen sink into clean water for the garden.[93]

"This ecological machine has been my laboratory for learning about recycling water and waste," Rick says. "It started out as a quest to figure out how to do simple greywater systems that would work with an automated irrigation system, something I could sell to my clients. But it's turned into a way of saying, 'Why don't we clean our own wastes?' It's so do-able."

How does it work? "If we follow ten gallons of water from the kitchen, the water enters the anaerobic reactor, displaces the ten gallons of water that's in the reactors, and then goes into the first covered tank. The water keeps displacing through the open tanks, and when the fourth tank displaces the water it goes into the clarifier, which has a cone bottom that helps facilitate the solids settling out of the system. From the clarifier the water goes through a little rip-rap river system, which is a pipe cut in half, and then through a mycological filter, which is gravel and straw with mycelium woven in it. This further purifies the water as it goes through the system.

"Then the water goes into a constructed wetlands in the floor of the eco-machine. That original ten gallons of water then moves into an underground storage tank of about fifty gallons, which then pumps it through the irrigation system. There is a pump down below that takes the displaced clean water and pushes it through a drip irrigation system. It works on a gravity system, but there is also a pump in the fourth tank that feeds it back into the first, so the machine keeps the water spinning all the time. This movement keeps bringing oxygen into the water, which purifies it further. The water goes through the system multiple times, constantly moving through the rhizosphere of these plants and being cleaned. By the time it exits the system, it's clean enough to be used on the plants in the vegetable garden.

THE HOUSE

240

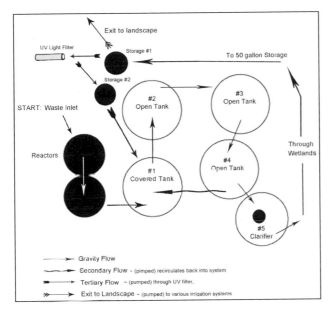

Water flow in the Ecological Machine.

Tanks three and four of the Eco-Machine.

"Is this sustainable? On its own, no. Every household doing it, yes. Having a certain level of your own food production, slowing down water, living a little bit more in community, reclaiming some of the water from the house, those are all things that go beyond gesturing toward sustainability. I think of it as *stepping* toward sustainability. Having this sense of transitional ethics, that we are all just doing the best we can, has really made a difference for me."

Shit Happens

One of the toughest problems for cities and for people over the course of history has been what to do with our own excrement. Different cultures have evolved different solutions, some more effectively than others. While we continue to contaminate the waters by flushing our poop out to sea, composting toilets are rooted in the history of cultures around the world. Traditional Yemeni apartment buildings collect feces, which are dried and burned as fuel. In nineteenth-century London, "earthen closets" safely processed feces to remove pathogens and create fertilizer. Gardeners the world over have proven that the safe composting of human excrement can add fertility and tilth to the soil. Rural China and Japan used this technique for centuries, applying "night soil" to their crops. This allowed them to grow food and maintain the fertility and viability of the land over many centuries. This is a perfect example of a closed loop, zero waste solution.

Our current health code is based on a strong desire to avoid the contamination caused by waste products—good—but is also rooted in a fear of illness, death, and the body—not so good. It turns out that flushing our poop into our waterways damages the ecosystem in two ways—the waters are contaminated, and we are wasting precious organic material that can be used safely to replenish our topsoil. We need to balance our social needs for hygiene and public health with the planetary needs for healthy topsoil, clean and abundant water resources, and a closed loop system overall.

It's time to reclaim our shit! Human excrement, or "humanure," is an unused resource, not a waste product, and recycling it is good for our gardens, the waters, and the earth. Our current sewage system is broken, poisoning the waters with bodily wastes that could be used to rebuild our diminishing topsoil. As there is no record of any civilization in the history of the world surviving beyond the life of its topsoil, it's high time we tended to this urgent need.

Americans use over 2 billion gallons of fresh water a year to flush away their waste products. Meanwhile, 3 million people die yearly of diphtheria, dysentery, and parasites for lack of clean water. This water, once flushed, is removed to a local municipal system that processes the water with chemicals, and usually flushes it out to sea or into other waterways. As our water infrastructure crumbles, and our soil fertility declines, there are simpler, more resource-based solutions to the problem: Recycle the pee. Recycle the poop. There are a few options for alternative waste removal in the city that are simple to do, good for your plumbing, good for the garden, and good for your soul. It's good work to take care of your shit rather than flushing it away, out of sight. Each of the options we discuss in this section, when done properly, is safe for humans, their pets, and the environments they occupy.

Composting toilets come in many makes and models—some simple and DIY, and some premade and fairly affordable options that are available on the open market. Dry-compost toilets not only conserve water, they also protect rivers and oceans. By circumventing modern sewers, dry-compost toilets avoid diverting nitrogen, potassium, and phosphate-rich wastes from the land (where they enrich the soil) to rivers and oceans, where they cause algae blooms, oxygen-robbing eutrophication, and oceanic "dead zones."

You can build yourself a composting toilet in the backyard in an afternoon. Think of all those quiet hours you could have, pooping at your outdoor toilet, listening to the sirens go by, or watching the stars. Most ecological toilets are built so that poop and pee go in different places (one bucket for each). There are also urine diverting inserts that can be integrated into the system that split the pee and poop into different places, making it possible to sit and do your business without having to separate liquid from solid waste on your own. Some homesteaders site their composting toilets outside in a little nook in the garden, protected from prying eyes by bamboo stands or other beautifully smelling plants. It's nice to make the privy private, even when it's outside, as cultural taboos about excrement are extremely powerful and inhibiting to the goal of zero waste.

How to Build a Simple Composting Toilet

Materials

Medical toilet seat

Two 5-gallon buckets

One plant pot that fits inside 5-gallon bucket

Subsoil

Compost

Red worms

The simplest way to make a composting toilet for your home is to purchase a medical toilet seat (the kind where you can easily sit down to pee), and keep two buckets nearby. One bucket is for pee, the other bucket is for poop. Both buckets are layered with sawdust in between deposits. You can also purchase the Separette Privy Kit, which separates urine and feces for easy collection. This lowers the work load, eliminates odors, and makes waste composting simple.

Place the toilet seat over the buckets, and do your business as usual. Sprinkle a handful of sawdust into the bucket after each use. The urine your household produces is safe to use without treatment. When the pee bucket is full, dump it on the compost pile, or layer it in the garden as mulch. The nutrient-rich sawdust will enhance

The simplest backyard humanure setup: a medical toilet seat, a five-gallon bucket, carbon matter, your deposits and…voilà—compost!

either garden or compost pile. Urine contains nitrogen, phosphorus, and potassium and makes an amazing plant fertilizer. Urine is typically sterile, and if separated from feces, can be easily and safely reused.

Urine can become contaminated if it comes into contact with feces. There are also a few diseases that can be transmitted through urine—leptospirosis and schistosomiasis (bilharzia), which are found almost exclusively in tropical aquatic environments, and typhoid, which is inactivated shortly after excretion. If urine has been contaminated with feces, or if it came from strangers who may carry these diseases, it should be purified before it is used. To purify urine, all you have to do is wait. Urine leaves the body fairly acidic and the pH increases rapidly until pathogens are unable to survive. This process takes from fifteen days in warmer Mexico to over three months in the chilly Scandinavian winter. Make sure you know the specifics for your location.

You can use urine directly by diluting it, one part urine to four parts water, and pouring it into the soil around your plants. Don't use it on young seedlings, and water alternately with rainwater or city water to flush salts from the soil. Better yet, apply it before a rain. You can compost it and let it rot, or add it to your greywater system.

In the simple two-bucket system, the poop bucket is constructed slightly differently than the pee bucket. For the poop bucket, take another five-gallon bucket (which has a cover) and find a plant pot that fits within it. Usually the black plastic pots that come with small trees or large plants will fit perfectly. Line the bottom of that pot with about one inch of subsoil, packing it in so that it keeps the fecal matter from leaking out. This creates a block from any disease vectors escaping, but allows for a little bit of drainage. Add a layer of compost with live worms in it, about two inches deep. You can then start making deposits in the bucket, covering each deposit with a bulk agent, like wood chips or sawdust. Keep the bucket covered when not in use.

When the pot is nearly full (about three inches from the top) remove the plant pot from the five-gallon bucket, and add another layer of worm-rich compost. Plant a tree in the center of the pot, and let the pot sit in the shade under a tree for a few months. Keep it watered, but not drenched. This will keep any contaminants from leaking into the soil. Rodents don't get in because they won't dig through the subsoil. The trees love this soil, which has no pathogens because the worms have been processing it for the month or so it takes for one person to fill up the pot.

After you let it sit for a few months, you can transplant the plant into the ground, or give it to someone you love.[94]

Indoor Composting Toilet

This is a simple, indoor option that everyone can use. The basic mechanism is still the same; the only difference is that this composting toilet is inside, rather than outside, which makes it more convenient to use throughout the year. Urine is separated into one bucket and is put into the compost bin or diluted with water and used in the garden. Feces are separated into another bucket and allowed to decompose for a year before being returned to the garden.

An aestheticaly pleasing indoor composting toilet. Deposits are separated into two buckets and gathered from below the house.

Add bulk to the composting toilet to speed up the decomposition process and mitigate any odors.

You need three things to make this system work: a toilet receptacle, a compost bin, and carboniferous materials such as straw, leaves, or wood shavings to mix with the poop. Once the bucket of humanure is full, bring it to a dedicated compost bin, where it will be covered with straw or other carboniferous materials to aid the decomposition process. In this method, the heat of the compost pile is crucial. In order to instigate thermophilic (heat-based) composting, the compost pile needs to reach temperatures of at least 113 degrees F, or hotter. When it reaches this heat for a consistent amount of time, humanure will break down within months. Purchase a thermometer to track the temperature to make sure that thermophilic composting is taking place. This is the best way to ensure that any pathogens in the human waste are destroyed.

If you do not have the space to create a dedicated compost pile for humanure, you can seal the bucket of poop once it is full, and allow it to sit in a sunny spot in the garden for at least one year. To aid the process, this bucket can also be turned periodically the way you turn a compost tumbler. Make sure the bucket is tightly closed to inhibit insects or vermin from visiting. Once composted, this soil can be safely used in the garden.

We understand that not everyone lives in a place where this is feasible. Unless your apartment building installs a giant composting toilet in the basement (not impossible, and a great idea by the way), apartment dwellers with no yards will be hard pressed to comply with this shitty regime. Although North America doesn't have a sewerless city (yet), there are a few sewerless buildings. One example is the public restroom at the Bronx Zoo in New York City. Faced with an exorbitantly expensive sewer connection, they decided to go sewerless, installing large composting toilets underneath the bathroom and using all greywater on the landscape. China, long familiar with the collection of night soil for fertilizer, has begun to standardize

The deposits in this composting toilet are accessed from a hatch alongside the house. Underneath the toilet, two separate buckets gather urine and feces for composting, and eventual integration into the garden.

urban ecological sanitation systems and apply ecological sanitation principles on a large scale. In 2007, in the drought-ridden north, an "eco city" was built with modern urine-diverting toilets and systems for treating household wastes. The urine and feces—as well as kitchen compost—are collected from each home and processed in an eco-station, much the same way garbage and recycling are collected in parts of the United States. Large-scale ecological sanitation based on this model could be integrated into any urban environment.

But if you live in a place where you have a yard and a bit of extra space, you can safely compost and reuse your excrement to great effect. Humanure toilets are limited to situations where adequate and appropriate coverage materials are available. They are a knowledge-based sanitation system, sometimes referred to as "the thinking person's toilet." When properly built and managed, however, they provide a low-cost, hygienically safe, environmentally friendly, and pleasant sanitation option that produces a wealth of soil fertility.[95]

Our biggest obstacles to recycling and reusing human wastes are neither technical nor health related, but simply in our minds. We live in a culture that hates and fears the body and the shadows of the body. Composting this story—the disconnection from our bodies, our fear of death, our separation from the world around us—is as important as literally composting our bodily wastes and reusing them as precious fertility for our soil.

The photographs show some great examples of aesthetically designed urine and feces composting setups. Partly experimental and partly art, these backyard installations illustrate the maxim Waste = Food with wit and beauty.

Made from reclaimed cable spool, wood chips, humus-rich topsoil, half a plastic barrel, coco coir, veggie seedlings, and a waterless urinal, "Pee Pee Ponics" by Nic Bertulis grows vegetables from the nitrogen rich fertilizer of urine deposits. Urine is full of inorganic nitrogen, good for plants but hard to keep around. By adding nitrogen to woodchips, the inorganic nitrogen is converted to organic nitrogen, making it available for slow release throughout a plant's life cycle. By isolating this reaction under a layer of soil, not only are offensive odors minimized, but the heat, moisture, and nutrients can wick upwards via osmosis and capillary action to feed the roots of veggies and flowers. Any pathogens in urine (which is almost always sterile) would have a hard time living in this reactor.

"Poo Poo Ponics" by Nic Bertulis is a small scale bio-gas digester. The poo is flush-pumped into the black digester, and holes in the top give flies access to the poo. They lay eggs and larvae hatch and help digest it. Any excess liquid drains into the plant below. Then the larvae seek water and drown themselves in the pond (behind). This feeds the fish, which then poop and give further nutrients to the plants in the water. Ideally, the fish feeding off the larvae would be edible by humans. This is an experiment in process and not fully operational, but a great expression of "Waste = Food" and the power of stacking functions.

HOME SCALE ZERO WASTE SOLUTIONS

When instituting zero waste policies at home, think about the changes you want to make within a reasonable timeframe.[96]

Six-Month Transition

Monitor and measure trash and recycling

Research basic composting methods

Remove toxins from household—cleaners, paints, etc.

Bring plants inside to clean the air

Create space at home by giving stuff away

Consume less

Accumulate less

Thrift more

Retrofit with reused materials

Begin to research natural building techniques and materials options

Begin to research crafts/hobbies for things you need and would enjoy making yourself

Twelve-Month Transition

Focus purchases of material from "waste" stream:

    Reused or recycled or "waste" materials

    Locally offered goods and services

    Support worker owned co-ops as places to get food and other resources

    Try to eliminate packaging

Use libraries or create your own library for sharing media with your community

Set up vermicomposting and/or other composting as appropriate

Research cooperative housing and become member of community land trust

Build and begin using composting toilet (indoor or outdoor)

Thirty-six to Sixty-Month Transition

Communal sharing of materials and tools

Communal living at appropriate scale

Work on a natural building project

Advocate for land trust set-asides

Research alternative land use policies

Start craft exchange with community

In one year, Permaculture Artisans turned a parking lot into this productive food garden and lush (sub)urban wildlife habitat. Pond provides flood irrigation, recreation, and a generative microclimate for a wide variety of creatures. With good design and elbow grease, we can transform pavement to paradise.   *Photo by Jill Lawless Photography*

Self-Care,
City-Care

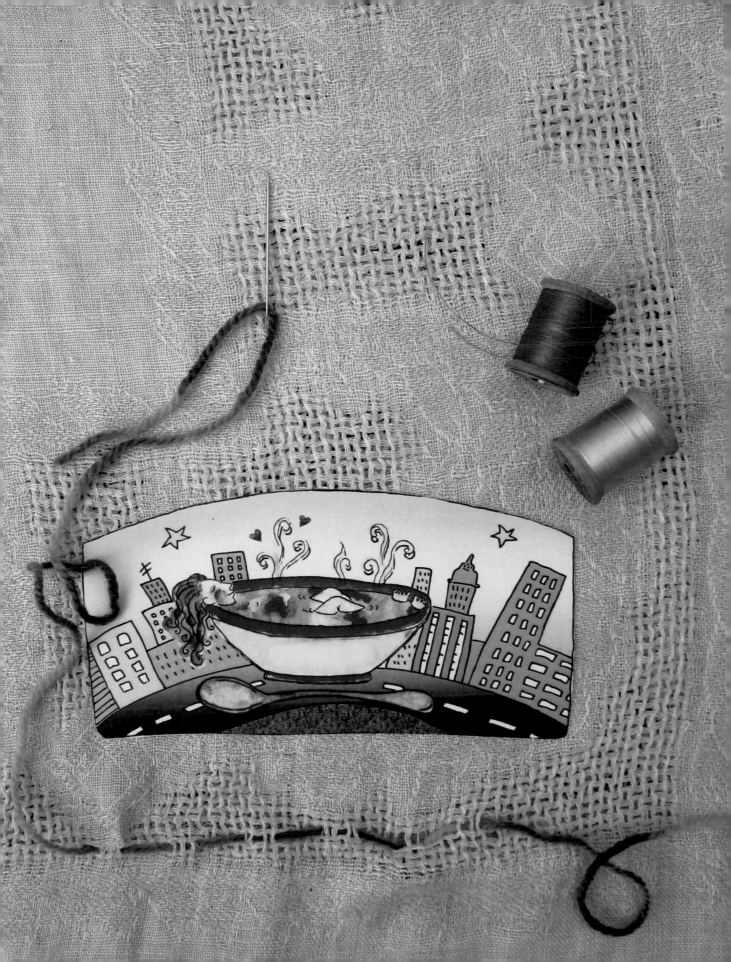

Personal Ecology

The spiritual crisis of the West is the cause for the many sufferings we encounter. Because of our dualistic thinking that god and the kingdom of god is outside of us and in the future, we don't know that god's true nature is in every one of us. So we need to put god back into the right place, within ourselves. It is like when the wave knows that water is not outside of her... Everything we touch in our daily lives, including our body, is a miracle. By putting the kingdom of god in the right place, it shows us it is possible to live happily right here, right now.

—Thich Naht Hahn[97]

Our bodies, our minds, our emotions, and our spirits together are our first homesteads—unique, demanding, mountainous, oceanic, rocky, fertile, windblown, and ultimately compostable. It's smart to sustain the most precious resource of the self. Rather than thinking of self-care as an indulgence, we see it as the beginning of cultivating the ground from which we take positive actions in the world, and a reflection of how we care for the world around us. Continuing to separate our personal bodies from the body of the earth just doesn't work. Instead of getting to ourselves last, or sacrificing ourselves to "save the world," cultivating self-love and care as an outgrowth of our concern for our world and our future is a priority.

We talk about energy a lot when we talk about sustainable living, but focus mostly on efficiency or solar energy or biofuels or bicycles. Managing human energy should be part of the conversation if homesteaders are to be successful over time. Our physical energy—the ability to do, to build, to dig, to eat—is supported by exercise, rest, and play. Our emotional energy—happiness, or at least contentment and satisfaction—needs support and appreciation to flourish. Our mental energy or clarity—the ability to plan, make decisions, and solve problems—is often balanced through non-linear processes like art making, ritual, and ceremony. Our spiritual energy—our awareness of the life force flowing within and around us—is fed by relationships with nature, meditation, and other spiritual practices, and our engagement in community. Tending to these aspects of our personal lives is like tending to the garden in different seasons, or the chickens in the morning and the bees on a sunny afternoon.

Creating sacred space out of a garage wall. Cob art by Miguel Micah Elliott.   *Photo by Miguel Micah Elliott*

Mind, body, emotions, and spirit are useful but artificial distinctions for aspects within each of us that we largely experience as the Self, or just "how we are." These distinctions further break down when we practice self-care, as exercising our bodies can change our minds or our mood, and clarifying our thinking can help us feel better emotionally or more connected spiritually. This is another way of noticing that the work we do on ourselves will have an impact on many levels, and affect all of our practices in the art of living.

on the ground: the personal is the political

Trathen Heckman is the founder of Daily Acts, a nonprofit committed to inspiring and empowering people to rise to the cultural moment with their best offering toward regenerating the world. Daily Acts offers tours and workshops and produces **Ripples**, an inspirational journal that has long been a public forum for Trathen's writing. Daily Acts also works with diverse green alliances and local city and county agencies to change and apply policy on issues such as water, food production, and community resilience.

There are many inspiring things about Trathen, from his lush garden to his enormous commitment to social justice and a reverent relationship to the world. Another inspiring part of Trathen's practice is something he does quietly, every morning, at home. Trathen's background in diverse spiritual practices, movement studies, and personal ecology informs and enlivens his powerful work in the community, and gives him the energy to do his work, amidst many challenges but fed by the joy of service. We met in his backyard garden for our interview, across the street from a Boys and Girls Club whose lawn has been transformed into a food forest by Daily Acts volunteers.

Trathen, on personal practice: "If you look at the first permaculture principle, 'observe and interact,' getting connected inside yourself, tuning in and listening, and connecting with nature's rhythms is really critical. My movement practice came first—I used to study and teach T'ai Chi, which is like permaculture for your body, just like permaculture is T'ai Chi for living in rhythm with life. Both come out of mimicking and understanding natural patterns and systems. As I started growing more aware of our daunting planetary problems, starting a personal practice was really critical for maintaining my own sanity. I needed to be able to live it in my body before I could go out and teach it in the world since my internal state of being greatly impacts everything I do and how people respond to me.

"I do the same thing every day—I tune in with my senses and accept what I'm vexed with. I stretch and check in with my body, go through my personal mission statements for myself and for each important part in my life. I re-center on the essence of what's important to me. What are my strengths, visions, and passions? What are the key issues I am struggling with? When we have a practice of pausing and remaining mindful and choosing our responses, then we have the chance of breaking some habits and creating new, more enriching ones. Our imagination and emotions are the most powerful things we have as physical creatures. We need to focus them around a clear vision of what is important to us and how to get there.

"It's like walking a path in the garden. People generally aren't going to walk through the bushes if there's a path, because there's a clearly defined route that guides you and is easy to follow. It's the same thing in your body, with your mind and emotions. We can follow the dysfunctional, disempowering paths our world has given us, or we can establish new patterns and wear a more inspiring, beautiful path through our *chosen* responses to any given day. Once you establish a path, it is easier to follow and also to establish new ones. It's just simple daily actions and the choice to make a habit out of the richest, tastiest, most empowering feelings, perceptions, and responses to whatever life throws our way.

"It boils down to what the Peace Pilgrim said: 'Find your highest light. Schedule it. Live it.'[98] Reconnect to nature and community, and find and live your part. It's of course easier said than done, but I just keep showing up and living up to my inspiration

Photo by Rachel Kaplan

the best I can. My personal practice has become an inseparable part of living a life of sacred service while somehow continually stepping up to what this time asks of me and doing it from a place of peace and reverence.

"Having a really good personal compass is essential for navigating life's complexity while staying true to your purpose and priorities and nurturing your vital relations. Are you willing to do the work to find out what inspires you and keep it in front of you day after day? Why not consistently choose to make something amazing of the materials and the moment? It's not about resources or having enough time, money, or skill. It's about resourcefulness, relationships, and getting clear about what matters. Leadership is about leading with how you live, inspiring others to rise to the potential of this amazing time."

Refilling the Well

City living is compelling and dynamic, with its cultural diversity, commerce and entertainment, services, and opportunities. The resources of cities are staggering, if you can put up with the grit. But cities can also be challenging places. They move fast, they're loud, they're dirty, with lots of people with lots of different needs stuffed into small spaces. Our incredibly fast-paced world demands more attention than most of us have, and more and more people report levels of debilitating anxiety, depression, and fear as they just try to get by. For those of us who live with an awareness of the threat that humanity and other life forms are facing, a deep feeling of exhaustion can arise from shouldering a disproportionate amount of earth care. We swim upstream in a culture that appears to be largely oblivious to the urgency of the situation. To walk this path we need to cultivate deep reserves of personal resilience.

Everyone benefits from crafting a set of personal practices that serve the goal of refilling the well. Some of them are as simple as slowing down and doing nothing, or learning to play a musical instrument, or cooking a meal for friends. The homegrown life is nourishing because it works with our human energy, and by its nature, counters the prevailing cultural forces of speed, need, and greed. A strong connection with the natural world is a reliable source for replenishing our energy when the well runs dry, and many homesteading practices serve the dual functions of being good for the planet and good for us.

The rest of this chapter outlines some practices we use and love to keep ourselves in balance; there are of course many more such practices that will work just as well, and you will undoubtedly find, or already have, ones that suit you best.

Body

We all have some basic strategies for maintaining good physical energy, like exercising, sleeping, and eating good food. Taking a hot bath. Playing basketball with a friend. Jumping up and down. Running around the block. Joining the gym. Moving yards of compost from one end of the yard to the other. Whatever your preference, keep the energy flowing in your body—exercise is a proven anti-depressant and will make you feel better. Contemplative, body-centered practices help us make more subtle contact with our physical and emotional bodies, helping us slow down and re-tune to the world around us. Taking time to get still and listen inside can be a powerful ally for our nervous systems, allowing us to become aware of the infinite power of earth that supports us even in the midst of city living. Stillness and silence feel good.

Finding a Sit Spot in Nature

It's good to have a place to go where you can listen to yourself, even when the world around you is filled with cars and sirens, or when you feel noisy and altered and are rushing around inside. Finding a sit spot in nature can serve this need. It's a simple process of finding a place in the city that calls to you—a big tree, a bench in a tucked away spot, a corner of your backyard where you can relax—and going there every day to sit and listen.

Sit down in your spot. Bring your awareness to your breath. Bring your awareness to your skin. Bring your awareness to your bones. Use your imagination to picture your bones inside your skin. Bring your awareness to your spine, that column of electrical energy in your body. Imagine your head resting on the top of your spine, and the bottom of your spine descending down into the bowl of your pelvis. Feel your spine and the back of your body. Does the energy in your spine flow freely? Get stuck in places? Bring your attention to the stuck places, and breathe through them. Repeat until the energy flows freely.

Listen to the sounds around you. If you are lucky enough to have a sit spot where there is some quiet or the sound of birds or running water, let that sound wash over you. If you sit in a place that is dense with urban noises, let those sounds sink into you as a well. Feel yourself a part of the place where you sit. Become aware of where you are, how you are sitting, the wind on your skin, the sun on your face, and the heat of the day. Bring attention to all of your skin—behind your ears, at the bottom of your feet, your inner thighs, the tips of your fingers. What draws your attention?

Find your sit spot in nature.   *Photo by Erik Ohlsen*

You know you are on a planet hurtling through space, but can you feel it? Imagine the motion underneath you. Notice what arises when you sit.

Connect to the natural pace of your breathing. Feel the air going inside your body with an inhale, and out into the world with your exhale. Inhaling brings the world inside you, and exhaling releases your energy out into the world. See if you can track your breath for ten inhales and ten exhales. Don't be bothered if you cannot—it takes time to build up your attention for this practice. When your attention wanders, just bring it back to your breath. Try to do this every day for ten minutes. It will help you ground your energy, and connect you to your place.

Sensitizing ourselves to the life force streaming around us all the time but which we are often too busy or distracted to experience deepens our practice in the art of living. This helps recalibrate our systems and brings us closer to what is meaningful to us, to our own truth. This supports us acting in alignment with what matters, and how we want to live in the world. These practices of sensation are akin to the work of preparing a field for seeds, clearing off your desk before starting a big project, or sweeping the house at the end of a busy week. Clearing inner space in this way makes room for new growth.

OTHER WAYS TO TEND TO OUR BODIES

Get a good massage.

Sleep eight hours a night, for a month.

Take one day of rest a week—no car, no phones, no e-mail.

Take a long walk in the early morning.

Learn a martial art like T'ai Chi or Aikido.

Practice yoga. Salute the sun.

Emotion

Once you have established a practice for caring for your body, you will see how your emotions are not really separate from your body, and how challenging feelings can make you feel physically constricted, tired, or incapacitated. Sometimes you can change your feelings by moving your body, but often you need to address the source of your discomfort directly in its own language. We recommend body-centered therapy, co-counseling, circles of like-minded comrades, or the therapeutic use of psychedelic drugs as ways and means to release challenging emotions.

Making an altar out of natural objects sets an intention and creates a beautiful environment where practical magic can happen. Allow the elements of the natural world to influence you. Altar by K. Ruby Blume.

HOMEGROWN RITUALS TO CATALYZE CHANGE

We can make rituals from the elements around us—water, fire, earth, and air—as well as the power of our intention, and our hopes and dreams. Homegrown rituals are unique and often happen only once. Sometimes they happen with groups of people; other times they are moments of alignment we have with the natural forces around us. A homegrown ritual can acknowledge transformation, grief, joy, perseverance, or the change of seasons. Don't underestimate the power of your mind and the dreams of your heart to remake the world. The magical power of intention can shape your life.

Ritual to Shed, Heal, and Cast Intention

Go to the beach or a river or a pond,
a place with living water.

Declare what you need to shed.
Write it down.
Burn the paper.
Taste the ashes.
Scatter the ashes.

Declare what you need to heal.
Write it down.
Burn the paper.
Taste the ashes.
Scatter the ashes.

Declare your intention for the future.
Give it to the water.

Take off your clothes.
Walk into the water.
Submerge yourself over your head.
Float.

Emerge.

Anger, sorrow, fear, and grief are never in short supply, and deeply affect our capacity to feel good in our lives. Venting, pounding, and screaming in a safe environment help release anger. Smashing a set of used dishes against a wall provides relief. Rituals involving the burning of objects or words that express our rage or sorrow can also help. This kind of ritual can symbolize the process of transformation we seek in our homesteading lives. Finding ways of dealing with our painful emotions without causing any one else pain can prevent a violent and outward expression of anger, rage, and grief, which are too often turned against ourselves, or one another.

Eco-Anxiety

If you have really conceptualized how dire our situation is environmentally, what it means for the web of life and your place in it, your most natural reaction will be fear. This feeling has a lot of faces—immobilization, manic re/activity, exhaustion, and insomnia—all completely natural responses to a terrifying possibility. This eco-anxiety needs to be managed in order for us to continue (or begin) to take effective personal and cultural action. Managing our feelings and our minds at this time is key to continuing on a pro-active path.

Anxiety is relieved by a few easily applied practices: purposeful action, positive thinking, and good community relations.

1. **Action alleviates anxiety.** Purposeful action feels good. If you are aware of the fragility of our food system, the depletion of our water supplies, the inequities between rich and poor, and the diminishment of fossil fuel resources, then growing your own food, catching your own water, and sharing your skills and abundance

Even when it's simple or 100 percent amateur, making art is good practice in tending to yourself and living well.

are all made-to-fit solutions. What you do to address your fears for tomorrow will make you feel better today.

2. **Positive thinking.** Even when things seem dire, holding a vision of hope is important. Worry is praying for the worst. What is the story you are telling yourself? When we walk around thinking, "The end is near," we can't have much energy to motivate action toward a generative future. When you walk around thinking, "It doesn't matter what I do," or "People will never change," you've delivered a prescription for cynicism and inaction. If you walk around thinking, "How interesting to be alive at such a challenging time," or "How thrilling that there should be so many opportunities to do good work," you'll find yourself curious about your visions for the future and how to bring them into reality.

    The other night at bedtime my eight-year-old daughter said, "Sometimes I just can't believe I am alive. It seems so amazing. I get chills just thinking about it!" she crowed. "I can do things!" Life wears down the passion of the eight-year-old, who cannot know all the challenges she will face. As adults we can re-pattern that hope within ourselves to remember the work that is ours to do, if not for ourselves, then for the children who follow behind us.

3. **Community relations.** Talk to other people. Find your community, your affinity group, your tribe. Most of us work and think too much in isolation. Contact with others should not be underestimated. A small circle of like-minded friends who meet weekly or monthly or spontaneously as needed can be a great support. A group of other homesteaders can be helpful when you are taking on this path. We've initiated projects like a Homesteading Grange, where urban homesteaders come to share tricks of the trade; or the Homegrown Guild, where we share everything from dinners to tomato starts to mulch piles to labor. Find friends who feel as you do. Talk to them. Listen to how they are working out their solutions and what inspires and gives them strength.

Unplug/Plug In: Unmediated Living

We are affected by the impossible amount of information we have to process, which in any given day is far more than our systems can handle. The simplest solution is to unplug—turn off the computer, turn off the television, turn off the radio, and stop reading the newspaper. Take a full-on media fast. Empty your colonized mind of apocalyptic stories, and fill it with music you like, or silence.

After unplugging from the dominant datastream diatribe, find a way to plug in to community. Share a meal with friends. Find an event that interests you. Gather a group of friends together for a meal, or for no reason at all, just because it feels good to be with people. Don't underestimate the value of celebrations for all sorts of life passages, and relaxation. Plugging in to people and place will make you feel better. Get creative—art is solid practice for surviving the city. It helps us transform one thing into another, bringing beauty into our lives and stretching our creative muscles. It doesn't matter what it looks like when it's done—it's the process that's

A reuse and recycle project: old bits of glass and the aesthetic expression of what lives inside in this ceramic and mosaic piece. Art by K. Ruby Blume.

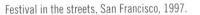

important. Try painting, collage, sculpture out of used and found objects, clay—all wonderful, tactile mediums that express our needs, our longings, and our passions in non-verbal, non-linear ways. Make your own entertainment. Learn a musical instrument and please yourself and your friends.

Why Art Matters

Art does not change laws, stop wars, or feed and clothe the hungry. If we expect art to make change in the same way that revolution does, we will be sadly disappointed. On the other hand, art has the potential to deeply touch individual art-makers and viewers. It has the potential to change consciousness and to feed the soul. The process of making art puts us in direct connection with the physical world. It is a spiritual act, which expands our capacity to manifest our dreams and visions. Making art collaboratively with others gives us the opportunity to be in a conversation about how we think and to communicate what is collectively important to us. Viewing art is a subjective process that leads us into self-knowing and an understanding of how others perceive the world. Art makes change through the evolutionary processes of reception and reflection. It is all around us, pervasive. It seeps under the skin and into the spaces between synapses, nagging at the unconscious to reveal injustice, paradox, emotion, and desire.

Ceramic art by Anne Aronov.

Festival in the streets, San Francisco, 1997.

Citizens and clowns interacting on the city streets, San Francisco.

Giant puppets by Wise Fool Puppet Intervention used in theatre, demonstrations, processionals, and celebrations in the streets.

Art-making gets us to look and see and feel outside the lines, both physically and metaphysically, preparing us to meet the status quo and challenge injustice wherever it lives. Art expands our brains into nonlinear intuitive patterns of thinking and seeing, in a process of finding visual and physical metaphors and solutions for the pain, joy, anguish, or grief we feel. Making art helps us access our emotions and express them, which is itself a healing act.

Art brings voice to the unseen. The process of crafting with the hand is ancient; art expresses what is in our hearts and in our bodies. The process of art-making grows the mind's capacity for creative problem solving and intuition and the hands' capacity for precision and skill.

When we make art in parks, gardens, schools, or the street, it enlivens our common places and gives us experiences of connectivity, inspiration, challenge, and joy. If people are part of creating the art in the space, it increases their sense of ownership. Parades, parties, fairs, and festivals celebrate the people and the land, marking the seasonal shifts with artful imagery. The performing and community-based arts bring voice to the human experience and help us understand and define who we are as a species and where we are going. Organizing events that bring people together in an artful expression of life can revitalize a neighborhood, resist gentrification, inspire new relationships, and bring the excitement of collective creativity to diverse groups of people. Group creativity, especially around issues that affect the participants, holds a promise of personal and political healing.

on the ground: for the next generation

The connection between environmental sustainability, art, and education flourishes in the homesteading way. Two of the artists we interviewed work towards similar goals of sustainability through their ongoing experiments in art and education.

Lauren Elder has been making gorgeous place-based art for decades—in the theater, on the cliffs overlooking the Pacific Ocean, and on the streets. Recent projects merge art and ecology and bring students of different ages along. "Art and gardening are so completely synergistic that once the children experience what's around them, their natural creativity comes out. Art and gardening also become tools for them to observe and learn and draw and read and write. They are natural additions to any school curriculum. Children understand that nature is experimental and creative and has all these thriving forms that are inspiring to them. A leaf is not just a heart-shaped thing—there is a whole vocabulary within the shape of leaves. This helps them recognize that there are many viable responses to the challenges of life.

"I like making school garden projects because it creates an intergenerational environment and asks for an ongoing commitment. That's really what community is—an ongoing commitment to people and a place. It isn't about somebody giving it to you; it's about making it for yourself. Creating a garden is easy. It's maintaining the long-term vision that's hard. I particularly like to bring college students from the art school to work as mentors with teenagers at the high school. This brings together different cohorts of students, an affecting combination."

Lauren's work highlights things people can do for themselves. A recent backyard installation of DIY projects focusing on the issue of water was designed to inspire creativity around the issue of water, and help people understand it doesn't take much time or money to make their places beautiful and efficient. Any time you make a garden you are confronted with the issue of water. "School districts have tried to stop garden projects by not giving us any water. No water, no garden. This forced me to focus a lot more on water as a global and regional issue and as an art medium and as a coming crisis for a lot of people." Lauren's impetus for doing the installations was to highlight the different things we can all do right now to make a difference.

"I think those of us living in the United States need to look outside our own borders and take responsibility for our behaviors and choices. Our country is the worst model of overconsumption and waste on the planet. We could set a standard for how to be environmentally minded, but there is a pernicious lobby of folks who want to keep plundering and producing. It's up to each of us to change our perceptions and our habits. The issue of consumer addiction and unnecessary waste that's wrecking the planet is the main thing we have to address right now, each one of us.

"Education can interrupt this cycle—showing people, engaging people in the process of making things for themselves out of simple materials, using what they have available. Using their intelligence and creativity. We're up against a lot in terms of advertising and industry, but we can start where we are."

Working in the back of the school garden brings on the great lessons of life: cooperation, experiential learning, and love.
Photo by Lauren Elder

Andree Thompson's art has centered on the theme of healthy survival on the individual, community, and planetary levels for many decades. Many of her mixed media installations include community participation and are often collaborations with other artists. She teaches a class at a local community college called EcoArt Matters, which stems from an earlier program taught at a self-esteem camp for inner city girls, and decades working as an artist in the Richmond elementary schools.

"The camp offered a literacy art program for girls eight to fourteen, many of whom had never been in the country or out of the city before; they characterized what we now call nature deficit disorder. We read and wrote stories about the environment, science, and Native and African American stories since most of the girls were African American. Then we made ceramics and art related to the subject. For example, after learning about Indian symbols and about the source of oxygen, they made clay planters decorated with symbols representing their own lives. These were raku fired and then filled with sweet dark dirt from last year's lunch (worm poop compost) and planted with plants that they took home so their city apartments would have more oxygen from living plants."

The EcoArt Matters class brings together creative thinking and skills with urgent information about environmental and social justice issues. Thompson brings science, activist, and EcoArtist lecturers into the first part of the class to create a context for the learning. During the second half of the class, students create pieces and install a public exhibition. Field trips are always included to expose students to local art and eco-related sites. Working in a community garden each spring is always an inspirational experience for the class.

Thompson said, "My own work was motivated by a need to heal from a traumatic past. Art was a safe place to confront and do battle with demons and the hard work of transitioning from being a victim to a survivor. My past included a highly dysfunctional family, partly due to the Holocaust that destroyed most of my father's family in Hungary. When I returned from a trip to Hungary to find survivors, I started a twenty-year project that began as a monument to those who were lost, but became a tribute to the miracle of surviving severe trauma with love and compassion still intact. My inspirations, in addition to those in my family who made it, are all survivors who understand the process and meaning of survival. I seek out the rare individual who has made a difference. It is no accident that I have always chosen to work with the underbelly of our culture where there is so much healing to be done.

"It is important to inspire young people to become politically active. I think it is impossible *not* to become political once they understand and feel the urgency of these times. One of the most important things we can teach students is to return to the earth, learn to love dirt and growing things, understand the importance of where food comes from, and all that is connected to growing and eating in a healthy way. I think it may save lives and will be a survival skill eventually. I can only hope that these are little drops of *tikkun olam* (care for the world) in a big pond. I continue to have hope that we humans have the necessary intelligence, creativity and compassion to create solutions for healthy survival."

Mind

The quality of our thinking is affected when we are run-down, stressed out, and just don't take the time to check in with ourselves. Making a little time during your week to inquire deeply into your own personal process yields many positive benefits, and can give rise to important insights that will guide you along your path.

Writing Practice

Writing practice, coupled with sitting in your sit spot, can help clarify thoughts or feelings that arise when you sit. Each time you start writing, clear your mind with a "free write," a short time of writing whatever comes to you. No themes, no punctuation, no capital letters, just let your hand race across the page and the words fall out. The free-write structure is so stripped of the "rules" of writing that what comes from it can inform and surprise and delight you.

Once you have done a free write, open up to what's on your mind—if you find yourself preoccupied or upset, use your writing to deepen your understanding of what's going on inside. If you find yourself inspired, write it down as a way commit to your vision. It's also just a really nice way to express yourself creatively, which is always nourishing.

Writing practice helps us notice patterns and trends and cycles in our thoughts; it also helps us clarify values. As we clarify our values, we edge closer to what we think of as our calling. Some people are obsessed with water. Other people deeply love plants. Some people are inspired to make ways for other people to rise to their highest light, or to heal from their past hurts. Clarifying values can help you design strategies to take purposeful action.

In our culture, the default strategy for getting what we want is to buy it, even though we know on a deep level that this doesn't work. When we take the time to really inquire into our own process, the movement of energy, thought and feeling in our bodies, we come into

The early November processional Dia de los Muertos (Day of the Dead) reminds the living of the dead. By walking together in honor of our ancestors at the time of year when the veil between the realms is thin, we link ourselves to the evolutionary cycle of life.

contact with actions that are more authentic and fulfilling. Once our values are clarified, actions that reflect these values become second nature, simple, and pleasurable.

Spirit

Most difficult to talk about is the spiritual side of our lives, our connection to the source of life, our ability to sense ourselves as part of the whole. It is the process of staying connected to what we are working for—the renewed, enlivened world we love—and drawing support and nourishment from it. Sensation practices help with this awareness, as does connecting with others in community. Prayer is an excellent vehicle—find your best way to give thanks every day for what you have, and for the work in front of you. We can't really say enough about gratitude, how it fills us with the energy to move forward and continue. Alongside gratitude, forgiveness practices and practices of reconciliation also connect us to this higher path of connection to the larger whole.

Don't forget ritual, ceremony, and celebration. These are essential for personal health and the health of our communities. They bring us into a place of gratitude for life and connection with one another. They are also ways to connect with life as it is lived over generations—with the dead, with the living, with the beings yet to come. Anything that reminds us of the cycle of life—how finite our own lives are and the endlessness of life itself—brings us into relationship with spirit.

Practice this kind of reverence in your life by building an altar and changing it every season. Express gratitude for the people you love, every day. Think of your garden as an altar. Or your body. Or your kitchen. Or your bed. Repeat after Rumi: "Life is sacred. I am. I am. I am."

Let the World Sink into Your Body

Friends and mothers and daughters honor the dead.

When I worked as a performance artist and teacher, I would do an exercise with my students called "Let the World Sink into Your Body." I practiced this as I traveled around the world, and found myself, small, foreign, alone, trying to understand where I was. It was beyond my mind's capacity to understand all that I saw—the language was different and the smell was different and the people were different and their assumptions were different. I was left with my body to try to sense where I was. I would walk along the rice paddy paths in Indonesia and let the world sink into my body. I would feel the wet humid air, the work of the place worn into the hills over time, the people in the fields, and imagine it sinking into my body. It was a life-changing practice. When I encouraged my students to try it, they began to access a deeper sense of awareness from inside their bodies. They were able to identify with what they were seeing because they felt it as well. Letting the world sink into our bodies became a practice for growing our understanding and our empathy, and for breaking down some of the boundaries we habitually put between the world and ourselves.

Here's how to do it

- In a place where you feel safe, imagine the cells in your body opening up to receive what you see, hear, and feel.
- Imagine your body as a giant receiver, a large instrument of sensation with a vast capacity to take in information.
- Imagine yourself as part of a tree—or perhaps a tree as part of you. Imagine yourself, in all your cells, being part of everything around you. No boundaries. Notice where that's simple to feel, and where it's challenging.
- Notice when you've stopped being able to practice. Remind yourself to practice again.
- Over time, you'll be able to do this for longer and longer periods of time.

The risk of letting the world sink into your body is that you feel the broken fragmented gorgeousness of our place. The benefit is that you feel the broken fragmented gorgeousness of our place. A friend of mine in Germany, where I taught this exercise a lot, accused me of being crazy for making my students do this exercise. "Why would you want to let the world sink into your body? What would be the benefit of that?" he snarled, sucking down one filterless cigarette after another. And so it goes—we get the life we choose.

Metta Practice

The metta practice from Vipassana meditation is a practice of loving kindness that confers a sense of peace on the practitioner, as well as the extended family of humanity. We offer a stripped-down version here; this deep

Every day opportunities for reverence are anywhere and everywhere. Use them to nourish yourself.

practice from the Buddhist lineage is ancient, and practiced around the world in many different ways. These spiritual practices of forgiveness and loving kindness are good medicine for the intensity, distortion, and complexity in our world.[99]

Sitting in mediation, begin to track your breathing. Do not change your breath; just notice what it is doing.

When you feel settled in sitting, you will begin the mantras of the metta meditation:

May I be happy. May I have physical well-being. May I have mental well-being. May I have ease of well-being. Allow yourself to sink into the solace of these words. Repeat the sequence to yourself for the time you have allotted for meditating.

The metta practice evolves to include people in your life, people you think of as "enemies," and people you do not know. Once you feel you have settled into the first level of the practice where you direct loving kindness to yourself, you can direct loving kindness to someone you know. In traditional Buddhist practice, it can take a practitioner many months to go from one level to the next. Do not rush the practice. You have your whole life to offer loving kindness to yourself and others.

When extending the practice beyond yourself, it is best to start with someone you feel neutral about, an acquaintance or work associate, rather than your most intimate people. The mantra would then be: *May so-and-so be happy. May so-and-so have physical well-being. May so-and-so have mental well-being. May so-and-so have ease of well-being.*

The next level of practice is to direct metta to your loved ones. And when you feel that you have settled into a pattern with your metta practice, you can direct this loving kindness to people for whom you feel antipathy, anger, or rage—your "enemies." This is more challenging than directing metta at yourself or people you love.

Metta practice can be extended to people you don't know, or to the whole of humanity: *May all beings be happy. May all beings have physical well-being. May all being have mental well-being. May all beings live in ease.*

This chapter in no way shares a conclusive list of ways to tend to your most intimate zone, the zone of the self. The practice of personal ecology is just that: personal—each one of us can evolve practices that support our work in the world and tend to the different parts of ourselves. In some ways it doesn't matter what you practice, it simply matters that you do. As Trathen says: "You're only as good as the soil you're growing on." Make sure you remember to tend your own soil or the seeds you plant just won't grow.

Don't forget to celebrate your life and the lives of others around you through song and dance and play. Life is rich and abundant with choices and chances, ceremonies and celebrations. Life is a delight. Savor it.

The Art of Our Common Place

The most important thing about localism is getting along with the locals.

—Erik Gage, from the band White Fang[100]

In some ways, this whole book is about creating community structures to support us all as the dominant systems of our culture continue to spiral out of control. It's about remaking a generative local economy to counter the destructive forces of globalism. The stronger our relationships are to the places where we live and the people we live with, the more we can reclaim the means of production and reduce our dependence on the global economy, and the more secure and stable our lives and communities will become.

We've focused on the many different things we can do at home to direct our actions toward the tasks of repairing the urban earth, and in every instance we've highlighted projects you can do yourself and with the people around you. We've put forward ideas for community-based strategies like food co-ops, energy collectives, seed banks, tool sharing collectives, educational forums, natural building projects, community gardens, and homegrown guilds. Existing groups that can be retooled to suit a regenerative agenda include neighborhood associations, condo associations, or school boards. Working together toward a common goal builds greater community integrity while flexing our atrophied "learning-to-share" muscles.

Building true community and making relationships with neighbors around common needs even when we don't have common backgrounds, ways of doing things, or agreement can be a huge challenge. We live in a diverse, multicultural, and divided world, but we all need the same things to live healthy lives: a renewed ecosystem; right livelihood; respect for our beliefs, needs, and feelings; nutritious food and medicine; and safe homes where we can live with dignity and joy. We've been raised in a culture that over-idealizes independence and this self-reliant streak shows up as much in people who identify with politics on the

Students mixing together straw and clay to insulate a cob oven. Many hands make light work.
Photo by Sasha Rabin/Vertical Clay

left as on the right. On a cultural level, we have little experience with sharing, asking for help, or offering our hands to one another. Embedded in our understanding that our single biggest mistake lies in separating ourselves from nature is the idea that our single most important action at this time in human history is to turn around, forgive and reconcile the past, and learn to work together.

Community building, like all the practices outlined in this book, is best started small so that one success can build upon another. In the localized future, working with our neighbors is a manageable and achievable goal, and one that can bring us the human and material resources we need. It's good to learn to grow your food, raise the animals you want to eat, identify plants in your area, and source and eat what's local. But in the city, we are our own greatest resource. The most important things we can do are to learn to get along; to communicate; to listen; and to respond wisely to our collective needs. The more we craft networks of support, camaraderie, friendship, and family, the more successful our urban redesigns will be and the more resilient our cities will become. Urban homesteading, while a positive response to the need for food security, healthy food, and water and waste management, is a partial solution, at best. The irony is that as we seek self-sufficiency, we come up again and again against how interdependent we are, and how self-sufficiency is not our true goal.

Here's another secret: Growing food is easy. Dealing with people is hard. It might take you two or three years to understand important things about gardening and start producing much of your

Community adobe brick making party in the parking lot of the local Chevron gas station adjacent to where the bricks were eventually laid.
Photo by Sasha Rabin/Vertical Clay

own food. But learning to communicate effectively, and to shift the fear and hatred that exist between people, is a practice that will occupy you for the rest of your life.

In the midst of writing this book, I had a run-in with my neighbor's chickens. I noticed that something had been digging and pecking in my front garden. I couldn't tell what it was from the markings, but it was making a serious mess. A few days later, my partner told me he'd seen the neighbor's chickens strutting down the driveway. I went to ask them to pen the chickens in and offered to help if they needed. They promised they'd do it, but a few days later, a chicken was back in the yard, pulling up the tender vegetable shoots. Did I go to my neighbor and ask her to get the chicken? I did not. I cornered the chicken in an alleyway alongside the house and flung her in an empty cage on the back patio. She sat squawking for many hours before I removed her and placed her with *my own flock* around the corner. Did I go tell my neighbor I had stolen her chicken? I did not. Here I was, Little Miss Community Homesteader, *stealing chickens* from my perfectly nice neighbors who were just too busy to pen in their birds.

I knew I'd done something wrong when I wouldn't tell my daughter about the chicken, and I left the marauder with my flock for almost a full week before my chicken coop partner called and told me my neighbor had come to collect the chicken. How did she know where it was? What would she say now? I had to suck it up and apologize. I told myself I had been intending to return the chicken anyway, but I wasn't sure if that was really true. Did I want to apologize to her? I did not. Was it the right thing to do? You bet. Did she give me hell? She sure did. Has it affected our relations? It certainly has. Are they repairable? Maybe, over time, but I truly wish I had taken a breath before I acted, rather than stealing that chicken.

I tell you this embarrassing tale to underscore the simple fact that it's easier to get mad than it is to be good. It's simpler to seek vengeance than justice. Protecting ourselves when we feel threatened is an automatic response, but this kind of reaction is the enemy of change. I can laugh about it now, but looking at this minor skirmish as just one small bit of the conflict between humans gives insight into how wars start and never end. Magnified one thousand times over hundreds of years, the weight of human conflict is almost too heavy to bear. We have to do better than this. Community change ultimately begins inside, with each one of us. Every day is an opportunity to confront our prejudices, our desire to control, and our fear of the other. A big challenge in front of us is the inside work we need to do, so that we can start looking at other people as assets, rather than liabilities.[101]

So we end on much the same note as we began: change begins at home, within each one of us. Learning to rig up a greywater system is important, but not if you're running on the fumes of fear. Resolving conflict is as important as growing your food, and making relationships with the people and creatures around you is the ultimate practice of truly living in place.

on the ground: mariposa grove

The story of Mariposa Grove, the first urban retrofit intentional community on the West Coast, is not so much about how much food it's possible to grow on a city lot, or how much water can be saved, or how many chicken eggs can be gathered. It's more about learning how much we can share with one another. How much of our preconditioning around relationships, private property, and privacy can we shed? How much are we willing to show up, to know and communicate about ourselves, in the hard work of learning to live with others? Intentional communities impart a model of human relationship. They show us a way to live more lightly on the land by living with others, and point us toward the invisible communication structures, celebrations, and re-education that are part of the reimagination of culture.

Mariposa Grove when it was first purchased—derelict and depressing. *Photos by Hank Obermayer*

Mariposa Grove after ten years of retrofitting, de-paving, sheet mulching, and urban gardening.

Mariposa Grove's founder, Hank Obermayer, had a vision of living in community when he bought three former crack houses in Oakland. The houses were in sorry shape, the neighborhood was rough, and the workload he purchased stretched before him for longer than he knew. What was once an abandoned and trashed landscape—buildings in great disrepair; plumbing, roofs, and electrical systems in need of replacement; cement courtyard and falling down fences—is now verdant with vines, architecturally sound buildings housing multiple families, a backyard theater space, and a yard filled with growing soil, for plants, children, and their parents. The community has expanded to include another house (whose back expanded the shared yard and which fronts on another street), and is home to sixteen adults, one teenager, and five children.

Most intentional communities start with a core group of people who invest in the property and commit to doing the work of growing the community from the ground up. Mariposa Grove didn't start this way. Hank purchased the property first, and found people second. Hank was able to shape a core part of the mission because he planted the first seeds, but being the community's first farmer was an emotional, financial, and personal burden for many years and affected him, as well as the development of the community. It took the community nearly six years to change from sole to joint ownership, which was a process of new members stepping up to belong, and Hank letting go of control for the development of the project.

There is always a tension in group projects between the visions of founders and the needs of people who enter the process later on. Members have to answer the question of how to own the project, together. After much negotiation and legal processing, a condo association structure was finally chosen for joint ownership. This gave each owner the right to their unit and to the common spaces, as well as shared responsibility for repairs of structures that everyone holds in common (walls of buildings, plumbing, garden, outdoor space, common space). It also means that individuals hold their own mortgages, and can sell their units at will.

Further support for keeping the homes affordable was an agreement with the Northern California Land Trust, which owns the land under the houses (though not the houses themselves). This legal maneuver keeps the units more affordable, and expresses a commitment to sustainable urban land use. There are other creative co-ownership models—tenants in common, limited liability partnerships, ownership through a nonprofit, or condo association. The best one for a community is the one that's legal, familiar to local real estate agents and bankers, and takes into account the needs of individual shareholders, affordability, and access.

These legal structures sometimes supercede the human structures, but to live in community you need both. The human social structures are the most evolved on this homestead. Dialogues about how to share resources, deal with money,

THE ART OF OUR COMMON PLACE

267

communicate about feelings and experiences, settle disagreements, share food and meals, and develop garden and buildings occupied many meetings in the first ten years of the community. Members use e-mail lists to coordinate many community events, and Google calendars for guest room reservations and the scheduling of group meals or sharing childcare. There is no "typical" day at Mariposa Grove, but residents have many informal interactions about living in the community.

Breaking ground on a lawn transformation project.
Photo by Trathen Heckman/Daily Acts

There is a sense of mutual support, especially among the parents in the community. There are usually two meetings a month—"more than enough," according to Hank—and a quarterly meeting of the whole community's condo association. The group shares two or three meals a week in the common house kitchen cooked by different people each night. The community designated one level of one house as common—the unit has a kitchen, living room, dining room, art studio, two guest rooms, office space that houses many community functions, and a back porch romper room enjoyed by children and adults alike.

Now that the buildings are all legally retrofitted and owned jointly through the condominium association, the community turns its collective intention to the work of growing food, caring for chickens and bees, developing rainwater and greywater systems, and the ongoing, joyous work of educating children into the life of the earth. Mariposa Grove's ability to provide for its own food or water resources is still minimal. Because of the number of people and buildings on site, and the marginal amount of land with sun, growing a large percentage of food will never be possible.

"Our gardens produce fava beans, red scarlet runner beans, peas, bush beans, nearly every brassica ever cultivated, raspberries, gooseberries, strawberries, tomatoes, spinach, chard, lettuce of any type, potatoes, and several varieties of squash. Trees provide plums, lemons, and oranges. We cure olives from the olive tree. We've started many other plants to see how varied a diet we might be able to produce. It is amazing what we have been able to accomplish. Although we have improved our soil fertility and are using some great intercropping techniques to increase yields, we are a long way from being able to feed ourselves from the yard."[102] Producing food on-site has been as much a community-based process—sorting out dreams from needs, and possibilities from realities—as it has been mastering the art of gardening.

Organizing at the Grassroots

In order to re-organize our culture, we're going to have to work with one another in a different way than we've mostly been taught. Organizing at the grassroots and working for social change demands inspiration, outreach, connection, and communication. It also calls for a democratic power structure that gives voice to different perspectives and peoples, and makes it possible for people to take independent action on behalf of the group. Community works best when we listen, invent freely, empower one another to act, respect leadership, decentralize power, and resolve conflict with respect and non-violence.

When we begin projects aimed at generating community resilience, the first step is to outreach to find our collaborators. It is important to bring people together with a clear intention. Community projects all begin by meeting with interested members; sharing and listening to needs, concerns and intentions; and starting to make step-by-step plans toward accomplishing the goal. Integrating a practice where people can share their feelings and concerns in a non-judgmental environment helps pave the way for deeper relationships, good working alliances, and a personal feeling of satisfaction and inclusion. Implementing a conflict resolution process that people can learn and practice is absolutely essential.

Outreach

There are many ways to do outreach to maximize participation in a project. Business cards, brochures, flyers, banners, bumper stickers, newsletters, or public presentations broadcast intentions loud and clear to the community. The connectivity of the Internet is extremely valuable for sharing information and bringing people together. You can start a blog or a listserv, or use other social networking sources to communicate with interested people about events, resources, and gatherings. Articles and 'zines are also DIY forums for sharing the good word, as are street theater events, mass bicycle rides that shut down the streets, and canvassing door to door to enlist people's aid. Depending on the nature of your project and how

The Institute of Urban Homesteading
OAKLAND, CA

art gardening canning herbs bread beekeeping brewcraft chickens

Affordable classes in the art of living
Preservation of a slower, more sustainable & more pleasurable way of life
Reclamation of the lost arts of the garden, the kitchen & things done by hand
Imbuing everyday tasks with wonder & beauty

www.sparkybeegirl.com/iuh.html

big you want your group to become, you can use any and all of these available options to bring people to the table.

Making Connections

Beyond printed or technological outreach, we make connections through our own inspired presence. Talk about your vision when you are out in the community. Take every public speaking opportunity that comes your way, even if it makes you nervous. Generate a mail list, so you can keep people in touch with what you are doing. Throw a potluck or a party to bring people together in a pleasurable way. Motivating others is a gift, but it isn't something that some people have and some people don't. It's a power you can access by living in the moment and stepping up to your own inspiration. This inevitably inspires others. If you want to build community, moving outside your circle of comfort to talk to new people, share your ideas, speak up at a community gathering or a school presentation or a workshop, or put out a brochure or article announcing your intention, all have the power to begin something new.

Group Process

While bringing people together has its challenges, working with them over time is ultimately the greater work. Functional group process has at its center a quality of respectful attention and listening, a commitment to taking responsibility for ourselves when we are upset with other people, a commitment to truth-telling in a posi-

350 GARDEN CHALLENGE

The number 350 has been suggested as the parts-per-million (ppm) of CO_2 in the atmosphere that is a safe upper limit to avoid a climate tipping point. The current record level is 391 ppm of CO_2, an almost 40 percent increase from the pre–Industrial Revolution level of 278 ppm.[103] Bill McKibben, a environmental writer and activist, founded an organization called 350.org to encourage people to meet the challenge to bring the CO_2 parts per million in the atmosphere under 350. In 2009, numerous groups around the country took up the challenge to "green" their communities. In Petaluma, California, the lawns of City Hall were taken over in a huge sheet-mulching project, which ultimately saved the city hundreds of thousands of gallons of water and tens of thousands of dollars in maintenance costs. Over 250 citizens came out to support the initiative and spent the day transforming the City Hall lawn into a model water-wise and edible landscape.

In 2010, in response to the challenge, three nonprofits in Sonoma County joined in an intention to plant 350 gardens in one weekend. Through canvassing, Internet outreach, public talks, and a coordinated approach to enlisting participants, volunteers and organizers managed to enlist not just 350 gardens in that one weekend, but a total of 632 gardens. Friends were made, plants were planted, orchards were installed, sheet mulch was laid, and good energy toward the goal of sustainable living was spread. On the next 350.org challenge day (10/10/2010), global activists coordinated events in every state in our country and throughout the world to meet the challenge. Undoubtedly this good work will continue at an ever-escalating scale, as global citizens take on the task of doing the work that must be done while reminding our governments of the role they must play in mitigating climate catastrophe.

Moving a mountain of mulch, City Hall Lawn Transformation, 350 Challenge Day, 2009.
Photo by Trathen Heckman/Daily Acts

A caravan of drought-tolerant plants brought in by volunteers, City Hall Lawn Transformation, 350 Challenge Day, 2009.
Photo by Trathen Heckman/Daily Acts

Planting perennials in a sheet mulched lawn as part of the 350 Garden Challenge, 2010.
Photo by Trathen Heckman/Daily Acts

tive and nonviolent way, and a willingness to forgive, reconcile, and get back to work when the going gets tough.

In trying to build a democratic process, it is important to have people who can facilitate meetings so that everyone's voice can be heard. Sitting in a circle is usually a better form than having a facilitator face the group, as in the classic teacher/student relationship. Circles spread the power around, and give participants a sense of inclusion. Check-ins at the beginning of meetings often set the stage for everyone having a say. Some people will always talk more in the group than others. As long as the quality of talking is respectful and doesn't dominate the group in a detrimental way, that can be all right. A good facilitator keeps a meeting process moving quickly along by establishing a simple agenda people agree to, managing time in such a way that each item is discussed fully and acted on to the extent possible, and then moving the group forward to the next agenda item.

High school students working together in a circle, passing energy, laughing, and learning.   *Photo by Rachel Kaplan*

POTLUCKS WITH PURPOSE

The Transition Town movement, originally conceived by Rob Hopkins in the UK, is based on a recognition of climate change, peak oil, and economic instability as three driving forces toward localism in the coming decades. The Transition Town movement includes many community-based initiatives for change that include many of the kinds of projects this book has been discussing. Sand Point, Idaho is the second Transition Town in the United States. With 8,000 people, the town is the right size for a true transition initiative. With a diverse and politically polarized population, Sand Point is also an example of the particular challenges facing the American application of the Transition Town initiatives.[104] Learning to listen to different values, ways of thinking, and ideas for action across the political spectrum, and finding common ground for moving forward, is one of the difficulties we face at this time in the U.S.

In Sand Point, the Transition Town group started a potluck series called Potluck with Purpose. These are opportunities for people to share concerns and inspirations with one another over time about how to grow their community with sustainability in mind. Potlucks with Purpose further breaks down into workgroups of about ten people on the themes of building and design, education, energy, health, folk school, waste, food production, and natural resources. A core group meets monthly with a representative from each group who come together to share forms of communication and support, coordinate schedules, apply for grants, and craft action plans. The working groups support the overall mission of creating a resilient, vibrant community in Sand Point.

BUILDING THE HOMEGROWN ECONOMY

When taking on the work of community and local-economy building, imagine changes taking place over a period of time.[105]

Six-Month Transition

    Self-inquiry into right livelihood

    Identify people who are doing what you want to be doing (i.e. "reference points")

    Host a potluck at appropriate scale

    Begin planning street closure event for celebration or community action

    Identify local neighborhood organizations and attend meeting

    Begin to explore Non-Violent Communication (NVC), or similar conflict resolution training

    Consume less

    Reduce all expenditures if possible

Twelve-Month Transition

    Meet with reference points, mentors, and others for livelihood guidance

    Host a regular, periodic potluck

    File block party application and go door to door on block

    Create a record for block with asset and skills inventory (under guise of disaster planning)

    Join board of local neighborhood group

    Support localization initiatives

    Practice NVC

Thirty-six to Sixty-Month Transition

    Organize urban eco-village resource share

    Form worker owned cooperative or independent cottage craft "career"

    Work in and from your "village"

    Evolve neighborhood group to seek autonomy from large-scale municipal systems

    Advocate for municipal decentralization in all essential services—energy, water, food, education, waste

    Initiate service projects in your neighborhood with your community to provide support for those requesting help

    Practice NVC and celebrate!

When we're working in a group, we want to do things together, but we also want to empower individuals to access their creativity and personal joy in any given group context. Group meetings are essential for getting things started and for people getting to know one another, but it is also useful to empower individuals or small groups to take actions that move a project along, without getting bogged down in a larger group process. Once a group determines a course of action, a smaller group can move it forward, reporting back to the main group on their successes, challenges, and further needs. In this way, a larger group can be split into a diversity of functions, spreading the wealth of group energy throughout the project.

Conflict Resolution

When problems arise in groups (and they will), it's important to have simple, effective ways to address them. The first step in trying to reconcile conflict is for people to take responsibility for their part. Blame is off the table. Check your own self at the door and see how or if you've gotten upset, what your expectations might have been that were not met, or how you may be telling a personal story that has nothing to do with what's going on. Once all people have checked in with themselves about what's happening on a personal level, coming together and listening to one another is a next step. Often, one person speaking and the other person listening and feeding back what she has heard works best. It rarely works to have two people try to tell their own part of the story at the same time, and really listening to someone else's side of things always helps to clear the air.

If apologies are in order, make them. If apologies are offered, accept them. If two people cannot come to terms on his or her own and just feel stuck, asking for help from a facilitator or mediator is a good idea. People trained in nonviolent conflict resolution are extremely valuable community members. Their capacity to listen and to set up situations where people can be heard is vital to the work of growing community.

This is another life work, like gardening and beekeeping and seed saving and self-care. None of these skills is learned overnight, but all of them grow with use and practice. If you're motivated to be a peacekeeper, you are needed! Get yourself some training and experience and head out into our world that calls out hungrily for true understanding at all levels. To improve our system, we have to improve our relationships. This is a basic tenet of ecological design, but it is also a central means for enhancing community living. And it's not something that seems to come natively to most of us—we've learned the habits of fear and aversion, and in this challenging time, we have to learn to reach out our hands to one another and work together.

Start Your Own Homegrown Guild

A group of committed citizens coming together to share labor, resources, time, and bounty can begin the process of community resilience in the place where you live. These homegrown guilds can be neighborhood groups, or affinity groups—parents in the same preschool or elementary school, people from work, any people with like-minded concerns. However we bring these guilds together, they are positive forums for social change in the place where you live.

Here's a simple structure for getting started.

1. Set up a potluck where people can talk about the sustainability issues that interest them, actions they want to take, or projects they want to initiate.
2. Discuss what actions you can take to move these dreams toward reality. Try

K. Ruby Blume worked with low-income seniors to build this MLK puppet, which made its public debut in a march against poverty on Martin Luther King Day at the Oakland Federal Building.

breaking into smaller groups according to affinity and desire. (For example, community resilience may be the larger goal, but some people may want to focus on starting a folk school, or managing water resources, or working on a legislative angle to create change. Different groups with different themes can be part of the overall project of building community resilience.)

3. Create an online resource that interested people can join. This is a simple way to share information about when events are happening, and provide answers to ongoing guild questions.

4. Plan events such as a Seed Swap, a Cider Pressing Party, a Summer Solstice event, or a Winter Bounty dinner. Use the listserv and other outreach tools to organize these events, and get interested members to activate a phone tree for those people who don't have access to the Internet.

The Art of Our Common Place

Some of the most interesting urban homesteading projects take a community-based approach to the problem. They recognize how systemic failures affect us individually and collectively and that change can take place on both levels. Food security projects, intentional communities, burgeoning transition town initiatives, community supported agriculture, food co-ops, garden wheels, and educational projects all have the intention to share information, resources, and relationship in the quest to rebuild the city.

on the ground: eco-workforce

The North Bay Institute of Green Technology (NBIGT) is the kind of agency that embodies the values of the world that is coming into being: holism, cultural respect, personal resilience, and inclusivity. Evelina Molina and Cris Oseguera, the all-volunteer leaders of this organization, met during a community conversation about how to continue a small pilot program for young men's empowerment in the mostly Latino neighborhoods of Santa Rosa, California. Funding was scarce, but need was high. The youth and their parents requested more groups of this sort, and Evelina and Cris moved forward with their mission.

NBIGT serves young men (ages eighteen to twenty-four) who have huge barriers to employment by bringing them education and experience in the green-collar sector jobs, connecting "the workers of today with the careers of tomorrow." Most of these young men have a history of violent gang-related felonies that occurred when they were minors or young adults. They've spent many years behind bars, moving from juvenile hall to the California Youth Authority (CYA) to the maximum-security prisons of Pelican Bay and San Quentin. Before ending up in the "justice" system, many of these young men fell through the widening cracks in the "education" system. Many have learning disabilities, mild mood disorders, and undiag-nosed and untreated PTSD from street violence and incarceration. And many of these young men cannot function without a support system. NBIGT provides the needed support through educational programs, funding for further education, and the initiation that genuine mentorship provides.

Academia Quinto-Sol (Fifth Sun Academy for Young Visionaries), NBIGT's initial flagship green pre-vocational/ technical program, incorporated a strong focus on cultural values and personal growth. In the nine-week summer program, fifteen young men voluntarily gathered to gain skills in energy and resource conservation along with employment-readiness skills needed for the green-collar jobs sector. This program's competency-based approach to learning helped students attain the skills necessary to successfully accomplish their goals of becoming viable working partners in the green economy.

As funding got more scarce, Cris and Evelina evolved the Eco-Workforce Collective (a social enterprise partner to NBIGT) as a way to further connect this population of young men with the skills, education, and mentors (contractors, retired tradesmen, and other professionals) they need. "We offer many options in the growing field of alternative energy, water conservation, and drought-tolerant landscape redesign and installation, with special focus on modification of existing systems to make them more efficient overall. We offer a cafeteria-style selection of opportunities so these young men can find something they love."

Evelina, a human rights and climate change activist, shares the *why* of the program with the young men. "We show them the reality and consequences of global climate change and its connection to their own lives and to poverty. These men, who lost their freedom at a young age, are very receptive to issues of environmentalism, social justice, and care for mother earth. We encourage these young men to hang on to the *why* (they want to do work in environment and energy conservation) for their lives. It makes the *what* (they will do as a career) so much more meaningful."

Cris, a single dad who works a full-time job at a youth advocacy organization alongside his work with NBIGT, brings the *what* and the *how* to the table. He connects the young men with mentors in the classroom environment—contractors, teachers, solar installers, greywater activists, environmentalists, and farmers—so that they can get the experience and education they need to work productively in the growing green economy. Cris says, "Our instructors meet the young men and build relationships. Then it's no longer about what you look like on paper, or the challenges you've had to overcome, it's about who you are as a human being."

Evelina said, "We've both been where they are, we're just down the road a little bit. When we share that with the young men, they are amazed. They wonder 'Why would anybody do something without pay for people like us?' And when they see that this is the case, they start to believe that maybe they are really special. And they are!" Cris and Evelina call that commitment and vision; some would just call it love.

Recently NBIGT has been moving in the direction of training in organic farming and gardening. When they received a small grant from a foundation, the young men got to choose where to donate the money. They chose a local organic farm where they could work with others and learn many aspects of organic farming. "We've successfully brought together young men who were once rivals at the street level to begin working side by side at an organic farm, preparing and tilling the soil for a habitat garden for the children in a low-income neighborhood," Cris says. "They understand how a simple butterfly garden matters in the overall picture of the world. And when they get it, when they really get it, their eyes sparkle, they stand up a little taller, they look you in the eye and shake your hand. They say, 'I'm with the Eco-Workforce, and I'm helping save the world.'"[106]

Education—Planting the Seeds for the Future

One of the surest ways for the green revolution embodied by the urban homesteading way to ripple outward into the world is through the power of education. This is happening in children's classrooms and adult folk schools, nonprofit organizations with a mission to "train the trainer" and bring information to different communities, as well as nonprofits that exist to teach sustainability skills in a wide variety of disciples. Many skilled homesteaders spread the wealth by teaching small groups of people how to preserve food, or care for a chicken, or butcher a rabbit. There are organic gardening classes and beekeeping classes, and opportunities to learn about the relationships between art, activism, and social justice.

Some of the most resilient projects in the urban homesteading way are educational in nature, and many of the people we interviewed feel that teaching is an important part of their work. Daily Acts has been educating people about sustainability for the last ten years through tours and workshops. The Institute for Urban Home-steading is devoted solely to teaching people heirloom skills and a DIY attitude toward life. The Spiral Garden offers Summer Sundays in the Garden, as well as ongoing support for the creation of home and community gardens. Chris Shein works as a teacher at a local college; Ben Macri teaches ecological living at City College of

Children learning to identify local plants in environmental awareness classes.   *Photo by Dori Midnight*

San Francisco. Laura Allen of Greywater Action spends most of her time teaching; the women of Villa Sobrante teach children and adults the joys of natural building.

The list goes on. The more research we did, the more examples we found, to the point of wondering if we'd have to write another book to cover them all. (We would! So we chose instead to feature a few that made us hopeful and happy.) It goes without saying that giving back to the community as a teacher, especially about something inspiring and life-giving, is a profoundly *social* thing to do, an action that reweaves the web and has the potential to bring everyone to the table with common information and a similar set of skills.

on the ground: urban resilience in oakland

Ashel Eldridge, an educator, organizer, and artist who wears several hats, leads resilience efforts in the Bay Area with Alliance for Climate Education; Communitree; Art and Action; Oakland Resilience Alliance; as well as serving on the Board of Transition US. He is also participating in a program called Green Life in San Quentin State Prison, bringing garden and sustainability projects to for-life inmates. His work is aimed at getting "people strong, physically, energetically, and environmentally, so we can start looking at the systems we have and get back to a way of living which is indigenous to all of us."

The Alliance for Climate Education offers hip-hop multimedia presentations in high schools that inspire students to do projects that green up their schools—such as gardens or solar panels—and raise ecological awareness. Art and Action, founded in 2001, works with youth from disenfranchised communities through arts activism summer camps and other events. Art in Action is also the convening organization that catalyzed Oakland's Green Youth Arts and Media Center. The newly formed Communitree is a network of individuals making change through the combination of art, music, culture, and ecology. Transition US is addressing adaptation and resilience for the coming economic and earth changes. Ashel is working with them to address much-needed issues around inclusion and diversity.

"A few years ago, I connected with the Transition movement and found it powerful, and wanted to acclimate it to the culture in Oakland," said Ashel. But parts of the "green" movement are alien to the African American community of

Oakland. "Some of our folks aren't even sure if they belong to the *old* economy, much less the new one," he said. "And while 'energy descent' is a really powerful concept, it doesn't always play in Oakland, where people are trying to get more power for themselves and working to get out of poverty... What's the right model for a place like Oakland, where people don't always have what they need? Perhaps the co-op model, where many people work together toward a common goal, and share the challenges and the profits, will work. Ultimately we need to ask: what do we need to do to break down the walls?

"You don't bring transition to Oakland. Transition's already happening here," says Ashel. "A lot of well-meaning white people came into Oakland with transition ideas and didn't even take the time to learn the history of the place, or meet the people who were already doing the work. You have to learn about the history to be effective. So we started the Oakland Resilience Alliance, and began to organize around the issues the community really takes to—paying jobs, food, and healthy homes. That's what's important in Oakland. We set up different projects in the community to help people bring out their skills and begin an intergenerational dialogue.

"Part of being in an urban resilience movement in Oakland happens in meetings and community events and gatherings and through traditional activism. Some of it happens by planting carrots and serving beans or giving our extra zucchini to our neighbors next door," said Alli Chagi-Starr, co-founder of Art in Action and a founding staff member of Green for All. "People are renewing their relationship to the land, and making a return to a more sustainable culture in lots of different ways."

"There's a lot of deep personal work that needs to be done to handle the spiritual destitution people feel. People who can heal themselves at the root of this injury can really bring power to themselves, and make a difference," said Ashel. He works to facilitate forgiveness and reparations and to help people work together. "Only by really looking at our history will we be able to heal so we can create a viable future together. The 'oneness' mentality is great, but we have to get comfortable with actual diversity.

"What do people need to do? Know your neighbor. Create community spaces where people can share. Throw a potluck. Help your neighbor dig in her garden. It's the lack of communication and creativity that continues to create obstacles. We haven't created an energy future because we haven't been able to harness our communication and creativity in the halls of power.

"We have to work together. There is no other way. But to really work together is hard because there's all this baggage from the past that's in the way. Ultimately, people who have skills in working with this, and an awareness of our history are going to be much needed as communities become more dependent on each other...

"Healing is about owning the wounds of the past and reconciling today. What are reparations without reconciliation? The blood of the slave and the slave owner is in the soil. A project started by Belvie Rooks and Dedan Gills, Growing a Global Heart, proposes that we plant one million trees along the Transatlantic Slave Route to commemorate the million souls who perished in the African holocaust. Now that's a solution that fits our times."

Still Knitting

The world is smaller than it once was, the abundance of resources a dream of the past. The need to stream-line and coordinate our actions and resources and energy is essential and becoming more so every day. Our timeline for turning things around for our civilization is very short—some scientists and ecologists believe that the next ten years are the make-it-or-break-it moment for planet Earth and her current inhabitants. This means that all of us have a role to play in waking up and stepping up to the tasks at hand. We need to break down the walls of silence within and between us and unify against the enemies of life: greed, complacency, cynicism, and fear.

Yet, even as we scale towards sustainability in our cities, we have to face the fact that the likelihood of reaching it in our lifetimes in slight. This should not deter us from doing all we can to participate in the great work of our time—renewing the earth and reclaiming our culture from the ashes of hundreds of years of grossly misperceiving our place in the garden of life.

Growing our skills not only in food production, but also in making decisions with a diversity of people who may have different ethics and values will be one of the challenges of the coming decades. The final frontier for our country cannot be the conquest of land as it was for the first wave of homesteaders. It is the work of learning to collaborate fully with one another, tending the earth within a living web of relationships, and building resilient communities where we do not leave anyone behind. This is the knitting together that must be done if we are to survive and thrive as a species. Without a healthy environment or reparations in our relationships or a reclama-tion of the skills that link us to the land, what is left for us? We need to engage in the profound work of rewriting our story on earth—from dominator to collaborator, from destroyer to sustainer.

Urban homesteading is part of the global project of knitting together, unraveling, and knitting some more. We have before us an opportunity to take what works, and leave the rest behind. We have left the garden through our own design. We can re-enter it with our creative imagining of another way to live in relationship to the earth and one another. The goal is radical interdependence, a revitalized economy that serves our families and our communities, and a way of living with nature that honors and respects the intelligent patterning of which we are a part. Even as the global situation grows more urgent, engaging in the work that is being done toward this goal, and all the work we can each do at home, cultivates possibility.

The fabric is being rewoven with the threads of care, camaraderie, and community. In every city in this country, people are opting out by digging in, turning their life energy toward composting the dominant paradigm into something life-giving and fine. There is a place for you at the table, in the garden, in your body, on the piece of urban earth you inhabit. There is so much work for us to do, together. Join us.

Endnotes

Knit It Up

1. Rich, Adrienne. "Natural Resources," *The Dream of a Common Language*. New York: WW Norton, 1993, 23.

2. Hawken, Paul. *Blessed Unrest: How the Largest Social Movement in History Is Restoring Grace, Justice and Beauty to the World*. New York: Penguin Books, 2007, 7.

The Empire Has No Clothes

3. Paley, Grace. *Just as I Thought*. New York: Farrar, Straus and Giroux, 1998.

4. Blasing, T. J. "Recent Greenhouse Gas Concentrations," CDIAC: Carbon Dioxide Information Analysis Center. July 2009.

5. Hawken, *Blessed Unrest, 100*.

6. Pinchbeck, Daniel. *2012: The Return of Quetzalcoatl*. New York: Jeremy Tarcher/Penguin, 2006, 7.

7. Tomlin, Kathyleen and Helen Richardson. *Motivational Interviewing and Stages of Change: Integrating Best Practices for Substance Abuse Professionals*. MN: Hazelden, 2004. *See also,* Hopkins, Rob. *The Transition Handbook: From Oil Dependency to Local Resilience*. White River Junction, VT: Chelsea Green Publishers, 2008, 84–88.

8. Lieberman, Alicia and Patricia van Horn. *Psychotherapy with Infants and Young Children: Repairing the Effects of Stress and Trauma on Early Attachment*. San Francisco: Guildford Pub., 2008.

9. Hermann, Judith Lewis. *Trauma and Recovery*. New York: Basic Books, 1992.

10. Howard, Beth. "The Secrets of Resilient People." www.aarpmagazine.org/lifestyle, 2009.

11. Hayes, Shannon. *Radical Homemakers: Reclaiming Domesticity from a Consumer Culture*. New York: Left to Write Press, 2010, 13.

12. Hawken, *Blessed Unrest*, 165.

Permaculture—Peace with Creation

13. Macy, Joanna. http://www.gratefulness.org/readings/MacyGratitude.htm, 2009.

14. King, Franklin Hiram. *Farmers of Forty Centuries: Or Permanent Agriculture in China, Korea and Japan*. New York: Courier Dover Publication, 1911.

15. Smith, Joseph Russell and John Smith. *Tree Crops*. New York: Island Press, 1987.

16. Holmgren, David. *Energy and Permaculture*. Reprinted from *The Permaculture Activist #31*.

17. Fukuoka, Masanobu. *The One Straw Revolution: An Introduction to Natural Farming*. Emmaus, PA: Rodale Press, 1978.

18. Anderson, Kat. *Tending the Wild: Native American Knowledge and the Management of California's Natural Resources*. Berkeley: University of California Press, 2005.

19. Holmgren, David. *Permaculture: Principles and Pathways Beyond Sustainability*. Hepburn, Australia: Holmgren Design Services, 2002.

20. Mollison, Bill. *Permaculture: A Designer's Manual*. Tazmania: Tagari Press, 1998.

21. Anderson, Bart. "Zones and Sectors in the City", *Permaculture Activist,* www.permacultureactivist.net/articles/urbanzonesectr.htm. Much of the following discussion of urban zones and sectors, as well as the diagrams, comes from Bart's smart thinking.

Creating a Personal Sustainability Plan

22. Hawken, *Blessed Unrest*, 100.

23. Lanzerdorfer, Joy. "Plot Grows Thicker." *North Bay Bohemian*, June 30, 2010.

24. US Energy Information Administration. http://www.eia.doe.gov/oiaf/aeo/electricity.html, Energy Demand, 2010.

25. Leonard, Annie. *The Story of Stuff*. www.storyofstuff.com.

26. Annual Water Quality Report. City of Petaluma Department of Water Resources and Conservation. Summer, 2010.

27. City of Petaluma, *Water Quality Report*.

28. Food and Agriculture Consumer Protection Department of the United Nations. www.fao.org/ag/magazine/0612spl.htm, 2006.

29. Welsh, Bryan. *Mother Earth News*, June 21, 2010. www.motherearthnews.com.

True Growth Economy

30. Berry, Wendell. *Sex, Economy, Freedom and Community*. New York: Random House, 1992, 24.

31. Oliver, Mary. *Journey: Dreamwork*. New York: North Atlantic Monthly Press, 1986.

32. www.instructables.com/id/The-Dearthbox-A-low-cost-self-watering-planter/

Urban Dirt Farmer

33. Midler, Bette, *Los Angeles Times*, May 8, 1996.

34. Diamond, Jared. *Collapse: How Societies Choose to Fail or Succeed*. New York: Viking, 2005.

35. Berry, Wendell. *The Art of the Commonplace: The Agrarian Essays of Wendell Berry*. Washington: Counterpoint, 2002.

36. Seymour, John. *The New Self-Sufficient Gardener: The Complete Illustrated Guide to Planning, Growing, Storing and Preserving Your Own Garden Produce*. New York: Dorling Kindersley, 2008, 14.

37. Jeavons, John. *How to Grow More Vegetables Than You Ever Thought Possible on Less Land Than You Can Imagine*. Berkeley: Ten Speed Press, 1974.

38. Jeavons, *Grow More Vegetables, 3*.

39. Seymour, *The New Self-Sufficient Gardener*, 106–107.

40. Kevin Bayuk in private conversation, 3/10/10.

41. http://whatcom.wsu.edu/ag/compost/easywormbin.htm.

42. Stamets, Paul. *Mycelium Running: How Mushrooms Can Help Save the World*. Berkeley: Ten Speed Press, 2005, 2.

43. Stamets, *Mycelium Running*, 7.

Seeds to Stem

44. Stevenson, Robert Louis, found on http://www.iwise.com/lx6de.

45. For a calendar of what to plant when in your region: www.motherearthnews.com/Organic-Gardening/What-To-PlantNow.aspx.

46. Food and Agriculture Organization. "Agricultural Biodiversity in FAO Factsheets," www.fao.oprg/docrep.010/i0112e/i0112e00.htm.

From Patterns to Details

47. Powell, Richard. *Wabi Sabi Simple: Create Beauty, Value Imperfection, Live Deeply*. Avon, CT: Adams Media, 2005.

48. Riotte, Louise. *Carrots Love Tomatoes: Secrets of Companion Planting for Successful Gardening*. Pownal, VT: Storey Communications, 1975.

49. Hemenway, Toby. *Gaia's Garden: A Guide to Home-Scale Permaculture*. White River Junction, VT: Chelsea Green Publishing, 2001 (from Ianto Evans, 144).

50. Seymour, *The New Self-Sufficient Gardener*, 99.

Life Is with Creatures

51. Gandhi, Mahatma, found on www.thinkexist.com.

52. Einstein, Albert, found on www.stumbleupon.com/ url/www.spaceandmotion.com/albert-einstein-god-religion-theology.htm.

53. *Henderson's Handy Dandy Chicken Breed Chart*, www.ithaca.edu/staff/jhenderson/chooks/chooks.html#9.

Food Is a Verb

54. Berry, *The Art of the Commonplace*, 83.

55. www.mobetterfood.com.

56. www.peoplesgrocery.org.

57. www.cityslickerfarms.org.

58. www.freegan.info.

59. Carlson, Tucker. "Freegans choose to eat garbage." MSNBC. http://www.msnbc.msn.com/id/11154276/, February 3, 2006.

60. www.fallenfruit.org.

Urban Farm Kitchen

61. Fukuoka, *The One Straw Revolution*.

62. Bayuk, Kevin. Transitioning Cities presentation, 2008.

Living Cultures

63. Fukuoka, *The One Straw Revolution.*

64. Fallon, Sally, with Mary Enig. *Nourishing Traditions: The Cookbook that Challenges Politically Correct Nutrition and the Diet Dictocrats.* Washington: New Trends Publishing, 1999.

65. Fallon, *Nourishing Traditions,* 87.

66. Fallon, *Nourishing Traditions,* 596.

People's Medicine Chest

67. Berry, Wendell. *The Unsettling of America: Culture and Agriculture.* Washington: Counterpoint, 1977.

68. www.motherearthnews.com/Medicinal-Herbs-Common-Uses.aspx,http://www.motherearthnews.com/Modern-Homesteading/2008-06-01/Homegrown-Medicine-Grow-Medicinal-Herbs.aspx.

69. Potts, Billie. *Witches Heal: Lesbian Herbal Self-Sufficiency.* New York: Hecuba's Daughter's Inc., 1981.

70. From Dori Midnight, kitchen witch, teacher, and herbalist. Dori teaches a course through the Institute for Urban Homesteading called The People's Medicine Cabinet. For this book, she graciously gifted me with the great title for this chapter, and the beautiful homemade remedies that follow.

From the Ground Up

71. Schumacher, E. F. *Small is Beautiful: Economics as if People Mattered.* New York: Harper and Row Publishers, 1973.

72. Theriot, Tracy, in private conversation, 3/10/10, www.tactileinc.com.

73. Eliott, Miguel Micah www.livingearthstructures.com.

74. Bee, Becky. *The Cob Builder's Handbook.* http://weblife.org/cob/cob_049.html.

75. Eliott, Miguel Micah, www.livingearthstructures.com.

76. Crews, Carole. *Clay Culture: Plasters, Paints and Preservation.* Self-published and available at http://carolecrews.com.

Powering Down

77. Obama, Barack. www.Newsweek.com, November 5, 2008.

78. www.epa.gov/seahome/energy.html.

79. www.ehow.com/how_2083_make-solar-oven.html.

80. www.cityofdavis.org/cdd/sustainability/lowcarbondiet/.

81. Bayuk, Kevin. Transitioning Cities presentation, 2008.

Sourcing the River

82. Fuller, Thomas, *Gnomologia,* 1732. Full citation not available.

83. Bangs, Richard and Christian Kallen. *Rivergods: Exploring the World's Great Rivers.* San Francisco: Sierra Club Books, 1985.

84. Pflanz, Mike. "World Water Day, Dirty Water Kills More People than Violence," *Christian Science Monitor,* March 23, 2010.

85. Greywater Action, www.greywateraction.org.

86. Leopold, Luna. *Waters, Rivers and Creeks*. Sausalito: University Science Books, 1997.

87. Bayuk, Kevin. Transitioning Cities presentation, 2008.

Zero Waste

88. Hill, Julia Butterfly. "We Live in An Age of Disposability Consciousness." http://www.huffingtonpost.com/2010/07/09.

89. McDonough, William. "Buildings Like Trees, Cities Like Food." www.mcdonough.com/writings/buildings_like_trees.htm, 2009.

90. Hill, Julia Butterfly. *One Makes a Difference*: *Inspiring Actions that Change Our World*. Harper SF, 2002.

91. Hayes, Shannon. *Radical Homemakers: Reclaiming Domesticity from a Consumer Culture*. New York: Left to Write Press, 2010, 207.

92. Kevin Bayuk, personal conversation, 3/16/10.

93. John Todd has been working for over thirty years in the field of natural wastewater treatment design, general aquatic management, and ecological design. His Ecological Machine, like Rick's, can be a tank-based system traditionally housed within a greenhouse or a combination of exterior constructed wetlands with Aquatic Cells inside of a greenhouse. The system often includes an anaerobic pre-treatment component, flow equalization, aerobic tanks as the primary treatment approach followed by a final polishing step, either utilizing Ecological Fluidized Beds or a small constructed wetland.

94. Kevin Bayuk in personal conversation, 3/6/10, as taught to him by Ben Jordan.

95. Jenkins, Joseph. "Humanure Sanitation." www.humanurehandbook.com.

96. Bayuk, Kevin. Transitioning Cities presentation, 2008.

Personal Ecology

97. Hahn, Thich Naht. "Let Go of the Need to Save the Planet," from article by Matt McDermott, planetgreen.com, August 31, 2010.

98. From 1953 to 1981 a silver-haired woman calling herself only "Peace Pilgrim" walked more than 25,000 miles on a personal pilgrimage for peace. She vowed to "remain a wanderer until mankind has learned the way of peace, walking until given shelter and fasting until given food." In the course of her twenty-eight-year pilgrimage she touched the hearts, minds, and lives of thousands of individuals all across North America. Her message was both simple and profound and continues to inspire people all over the world today.

99. Sharon Salzberg was the teacher for this particular meditation.

The Art of Our Common Place

100. Yardley, Michael. "The Pride and Prejudice of 'Local'" c, July 8, 2010.

101. Karen Lamphear, personal conversation, May, 2010.

102. Van de Walle, Robert. Permaculture Activist, *Local Food Issue*, Spring 2010.

103. Serna, Laura. "Drawing the Future, Stomatal Response to CO_2 Levels," Plant Signaling and Behavior 3034, 213–217; April 2008; Landes Bioscience.

104. Karen Lamphear, personal conversation, May, 2010.

105. Bayuk, Kevin. Transitioning Cities presentation, 2008.

106. For more information about these superheroes and to support their work, please go to www.nbigt.org.

Resources

Homesteading

Carpenter, Novella. *Farm City: The Education of an Urban Farmer*. New York: Penguin Group, 2009.

Coyne, Kelly and Erik Knutzen. *The Urban Homestead: Your Guide to Self-Sufficient Living in the Heart of the City*. Port Townsend, WA: Process Self-Reliance Series, 2008.

Gussow, Joan Dye. *This Organic Life: Confessions of a Suburban Homesteader*. White River Junction, VT: Chelsea Green Publishing, 2001.

Hayes, Shannon. *Radical Homemakers: Reclaiming Domesticity from a Consumer Culture*. Richmondville: Left to Write Press, 2010.

Kellogg, Scott and Stacey Pettigrew. *Toolbox for Sustainable City Living: A Do-It-Ourselves Guide*. Boston: South End Press, 2008.

Kingsolver, Barbara. *Animal, Vegetable, Miracle: A Year of Food Life*. New York: HarperCollins Publishers, 2007.

Nearing, Helen and Scott. *Living the Good Life: How to Live Sanely and Simply in a Troubled World*. New York: Schocken Books, 1954.

Nyerges, Christopher and Dolores. *Extreme Simplicity: Homesteading in the City*. White River Junction, VT: Chelsea Green Publishing, 2002.

Robinson, Ed and Carolyn. *The "Have-More" Plan: A Little Land—A Lot of Living*. Norotom: Country Bookstore, 1943.

Permaculture

Bell, Graham. *The Permaculture Garden*. Permanent Publications, 2004.

Fukuoka, Masanobu. *The One Straw Revolution*. New York: New York Review of Books, reprint 2010. Original copyright, 1978, published by Rodale Press.

Hemenway, Toby. *Gaia's Garden: A Guide to Home Scale Permaculture*. White River Junction, VT: Chelsea Green Publishing, 2000.

Holmgren, David. *Permaculture: Principles and Pathways Beyond Sustainability*. Australia: Holmgren Design Services, 2002.

Mollison, Bill. *Introduction to Permaculture.* Tazmania: Tagari Press, 1991.

Mollison, Bill. *Permaculture: A Designer's Manual.* Tazmania: Tagari Press, 1998.

Organic Gardening

Coleman, Eliot. *The New Organic Gardener.* White River Junction, VT: Chelsea Green Publishing, 1989.

Coleman, Eliot. *Four Season Harvest: How to Harvest Fresh Organic Vegetables from Your Home Gardens All Year Long.* White River Junction, VT: Chelsea Green Publishing 2007.

Flores, H. C. *Food Not Lawns: How to Turn Your Yard into a Garden and Your Neighborhood into a Community.* White River Junction, VT: Chelsea Green Publishing, 2006.

Jeavons, John. *How to Grow More Vegetables Than You Ever Thought Possible on Less Land Than You Can Imagine.* Berkeley: Ten Speed Press, 1974.

Jeavons, John. *Lazy Bed Gardening.* Berkeley: Ten Speed Press, 1992.

Riotte, Louise. *Carrots Love Tomatoes: Secrets of Companion Planting for Successful Gardening.* Pownal, VT: Storey Communications, 1975.

Ruppenthal, R. J. *Fresh Food from Small Spaces: The Square Inch Gardeners Guide to Year-Round Growing, Fermenting and Sprouting.* White River Junction, VT: Chelsea Green Publishing, 2008.

Seymour, John. *The New Self-Sufficient Gardener.* London: DK Press, 2008.

Solomon, Steve. *Gardening When It Counts: Growing Food in Hard Times.* Canada: New Society Publishers, 2005.

Indigenous Wisdom/Tending Wild Lands

Anderson, M. Kat. *Tending the Wild: Native American Knowledge and the Management of California's Natural Resources.* Berkeley: University of California Press, 2005.

Imhoff, Daniel, *Farming with the Wild: Enhancing Biodiversity on Farms and Ranches.* Healdsburg, CA: Watershed Media Book, 2003.

Margolin, Malcolm. *The Earth Manual: How to Work on Wild Land without Taming It.* Berkeley: Heyday Books, 1975.

Wilson, Gilbert (as told to). *Buffalo Bird Woman's Garden.* St. Paul: Minnesota Historical Society Press, 1987 (originally published in 1917).

Dirt First

Applehof, Mary. *Worms Eat My Garbage.* Kalamazoo, MI: Flower Press, 1997.

Campbell, Stu. *Let It Rot: The Gardener's Guide to Composting.* North Adams, MA: Storey Publishing, 1975.

Rodale, Robert. *The Complete Book of Composting.* Emmaus, PA: Rodale Books, 1975.

Seed Starting and Seed Saving

Ashworth, Suzanne and Kent Whealy. *Seed to Seed: Seed Saving and Growing Techniques for Vegetable Gardeners.* Denver: Seeds of Change, 2000.

Ausubel, Kenny. *Seeds of Change; The Living Treasure.* San Francisco: Harper San Francisco, 1994.

Bubel, Nancy. *The New Seed-Starters Handbook.* New York: Rodale Inc., 1988. Download PDF version at www.homegrownevolution.org.

Small Animal Husbandry

Damerow, Gail. *A Guide to Raising Chickens: Care, Feeding, Facilities*. Pownal, VT: Storey Communications, Inc., 1995.

Damerow, Gail. *Barnyard in the Backyard*. North Adams, MA: Storey Publications, 2002.

Kessler, Brad. *Goat Song: A Seasonal Life, A Short History of Herding, and the Art of Making Cheese*. New York: Scribner, 2002.

Storey Animal Handbooks at www.Storey.com, for whatever animal you wish to raise.

Beekeeping

Bonney, Richard. *Beekeeping: A Practical Guide*. North Adams, MA: Storey Publishing, 1993.

Buchmann, S. L. and G. P. Nabhan. *The Forgotten Pollinators*. Washington, DC: Island Press/Shearwater Books, 1996.

Conrad, Ross. *Natural Beekeeping: Organic Approaches to Modern Apiculture*. White River Junction, VT: Chelsea Green Publishing, 2007.

Flottum, Kim. *The Backyard Beekeeper: An Absolute Beginner's Guide to Keeping Bees in Your Yard and Garden*. Gloucester, MA: Quarry Books, 2005.

Hubbell, Sue. *A Book of Bees*. New York: Houghton Mifflin Company, 1988.

Langstroth, L. L. *The Hive and the Honey-Bee*. Mineola: Dover Publications, 2004 (from an 1878 work).

Lovell, John. *Honey Plants of North America*. Medina, IL: A.I. Root Company, 1926.

Root, Amos Ives. *The ABC & XYZ of Bee Culture*. Medina, IL: A.I. Root Company, 1990.

Water

Ludwig, Art. *Water Storage: Tanks, Cisterns, Aquifers and Ponds*. 2005, www.oasisdesigns.org.

Ludwig, Art. *Create an Oasis with Greywater: Choosing, Building and Using Greywater Systems*. www.oasisdesign.org, 1991.

Lancaster, Brad. *Rainwater Harvesting for Drylands, Volumes I and II*. Tucson: Rainsource Press: 2006.

Waste

Hill, Julia Butterfly. *One Makes A Difference: Inspiring Actions that Change Our World*. San Francisco: Harper SF, 2002.

Jenkins, Joseph. *The Humanure Handbook: A Guide to Composting Human Manure*. Joseph Jenkins Publishing, 2005.

Van der Ryn, Sim. *The Toilet Papers: Recycling Waste and Conserving Water*. Berkeley: Ecological Design Press, 1995.

Natural Building

Bee, Becky. *The Cob Builders Handbook: You Can Hand-Sculpt Your Own Home*. Cottage Grove, OR: Groundworks Publishing, 1998.

Chiras, Daniel D. *The Natural House: A Complete Guide to Healthy, Energy-Efficient, Environmental Homes*. White River Junction, VT: Chelsea Green Publishing, 2000.

Evans, Ianto. *The Hand-Sculpted House: A Practical and Philosophical Guide to Building a Cob Cottage*. White River Junction, VT: Chelsea Green Publishing, 2002.

Kennedy, Joseph F., Michael Smith, and Catherine Wanek, eds. *The Art of Natural Building: Design, Construction, Resources*. British Columbia: New Society Publishers, 2001.

Natural Plasters

Crews, Carol. *Clay Culture: Plasters, Paints and Preservation*. Self-published. Available at http://carolecrews.com.

Edwards, Lynn and Julia Lawsell. *The Natural Paint Book*. Emmaus, PA: Rodale Press, 2003.

Guelberth, Cedar Rose and Dan Chiras. *The Natural Plaster Book: Earth, Lime and Gypsum Plasters for Natural Homes*. British Columbia: New Society Publishers, 2003.

Weismann, Adam and Katy Bryce. *Using Natural Finishes*. New York: Green Books, 2008.

Energy

Chamberlain, Sean. *The Transition Timeline (for a local, resilient future)*. White River Junction, VT: Chelsea Green Publishing, 2009.

Denser, Kiko. *Build Your Own Earth Oven: A Low-Cost Wood-Fired Mud Oven; Simple Sourdough Bread; Perfect Loaves*. White River Junction, VT: Chelsea Green Publishing, 2001.

Gershon, David. *The Low Carbon Diet: A 30 Day Program to Lose 5,000 Pounds*. Berkeley: Empowerment Institute, 2006.

Pahl, Greg. *Natural Home Heating: The Complete Guide to Renewable Energy Options*. White River Junction, VT: Chelsea Green Publishing, 2001.

Potts, Michael. *The New Independent Home: People and Houses That Harvest the Sun*. White River Junction, VT: Chelsea Green Publishing, 2001.

Food and Nutrition

Fallon, Sally with Mary Enig. *Nourishing Traditions Cookbook: The Cookbook that Challenges Politically Correct Nutrition and the Diet Dictocrats*. Washington, DC: New Trends Publishing, 1991.

Haas, Elson. *Staying Healthy with the Seasons*. Millbrae, CA: Celestial Arts, 1981.

Henderson, Robert. *The Neighborhood Forager: A Guide for the Wild Food Gourmet*. White River Junction, VT: Chelsea Green Publishing, 2000.

Madison Area Community Supported Agriculture Coalition. *From Asparagus to Zucchini: A Guide to Farm Fresh Seasonal Produce*. Madison: Wisconsin Rural Development Center, 1996.

Nabhan, Gary. *Coming Home to Eat: The Pleasures & Politics of Local Food*. New York: Norton, 2002.

Peterson, John and Angelic Organics. *Farmer John's Cookbook: The Real Dirt on Vegetables, Seasonal Recipes, and Stories from a Community Supported Farm*. Salt Lake City: Gibbs Smith, 2006

Pinkerton, Tamzin and Rob Hopkins. *Local Food: How to Make It Happen in Your Community*. London: Transition Books, 2009.

Pitchford, Paul. *Healing with Whole Foods, Oriental Traditions and Modern Nutrition*. Berkeley: North Atlantic Books, 1998.

Pollan, Michael. *The Omnivore's Dilemma: A Natural History of Four Meals*. New York: Penguin Press, 2006.

Prentiss, Jessica. *Full Moon Feast: Food and The Hunger for Connection*. White River Junction, VT: Chelsea Green Publishing, 2006.

Mushrooms

Stamets, Paul. *Mycelium Running: How Mushrooms Can Help Save the World*. Berkeley: Ten Speed Press, 2005.

Stamets, Paul. *The Mushroom Cultivator: A Practical Guide to Growing Mushrooms at Home*. New York: Agarikon Press, 1984

Stamets, Paul. *Growing Gourmet and Medicinal Mushrooms*. Berkeley: Ten Speed Press, 2000.

Preserving the Harvest

Anderson, Lorraine and Rick Palkovic. *Cooking with Sunshine: The Complete Guide to Solar Cuisine with 150 Easy Sun-Cooked Recipes*. New York: Marlowe and Company, 1994.

Bubel, Mike and Nancy. *Root Cellaring: Natural Cold Storage of Fruits and Vegetables*. North Adams, MA: Storey Publishing, 1991.

Carroll, Rikki. *Home Cheese-making Recipes for 75 Delicious Cheeses*. North Adams, MA: Storey Publishing, 2003.

Carroll, Rikki. *Cheese-making Made Easy*. North Adams, MA: Storey Publishing, 1983.

Emery, Carla and Lorene Edwards Forkner. *Canning & Preserving Your Own Harvest*. Seattle: Sasquatch Books, 2009.

Gardeners and Farmers of Terre Vivante. *Preserving Food Without Freezing or Canning*. White River Junction, VT: Chelsea Green Publishing, 1999

Greene, Janet. *Putting Food By*. New York: Plume, 1973.

Katz, Sandor. *Wild Fermentation*. White River Junction, VT: Chelsea Green Publishing Company, 2003.

Preserve Food (an online canning resource): www.preservefood.com

National Center for Home Food Preparation: http://www.uga.edu/nchfp

Cheese-Making Sites and Supplies

Fankhouser's Cheese Page: biology.clc.uc.edu/Fankhauser/Cheese/Cheese.html

Books, recipes, supplies, and more: www.fiascofarm.com/dairy/index.htm

New England Cheese Supply: www.cheesemaking.com

The Cheesemaker: www.thecheesemaker.com/supplies.htm

Leener's (kits and supplies): www.leeners.com

Herbalism

De Bairacli Levy, Juliette. *Nature's Children*. Woodstock, NY: Ash Tree Publishing, 1996.

De Bairacli Levy, Juliette. *Common Herbs for Natural Health*. Woodstock, NY: Ashtree Publishing, 1979.

Buchanan, Rita. *Taylor's Guide to Herbs*. New York: Houghton Mifflin Company, 1995.

Gladstar, Rosemary. *Family Herbal: Guide to Living Life with Energy, Health and Vitality*. North Adams, MA: Storey Publishing, 1996.

Kowalchik, Claire and William Hylton. *Rodale's Illustrated Encyclopedia of Herbs*. Emmaus, PA: Rodale Press, 1998.

Tierra, Leslie. *Healing with the Herbs of Life*. Berkeley: Ten Speed Press, 2003.

Wood, Matthew. *The Book of Herbal Wisdom*. Berkeley: North Atlantic Books, 1997.

Self-Care

Chodron, Pema. *The Wisdom of No Escape*. Boston: Shambala Classics, 2001.

Heckler, Richard Strozzi. *The Anatomy of Change*. Berkeley: North Atlantic Books, 1985.

Heckler, Richard Strozzi. *Holding the Center: Sanctuary in a Time of Confusion*. Berkeley: North Atlantic Books, 1997.

Kabat-Zinn, Jon. *Wherever You Go, There You Are*. New York: Hyperion, New York: 1994.

Palmer, Wendy. *The Intuitive Body*. Berkeley: North Atlantic Books, 1994.

Starhawk. *The Earth Path: Grounding Your Spirit in the Rhythms of Nature*. New York: Harper & Row Publishers, 2003.

Communities

Christian, Diana Leaf. *Creating a Life Together: Practical Tools to Grow Ecovillages and Intentional Communities*. British Columbia: New Society Publisher, 2003.

Hanson, Chris. *The Co-Housing Handbook: Building a Place for Community*. Vancouver, WA: Hartley & Marks Publishers, 1996.

Hopkins, Rob. *The Transition Handbook: From Oil Dependency to Local Resilience*. White River Junction, VT: Chelsea Green Publishers, 2008.

Mindell, Arnold. *Sitting in the Fire: Large Group Transformation Using Conflict and Diversity*. Portland, OR: Lao Tse Press, 1995.

Rosenberg, Marshall. *Nonviolent Communication: A Language of Life*. Encinitas: Puddledancer Press, 2003.

Great Thinking

Berry, Wendell. *The Art of the Commonplace*. Washington, DC: Counterpoint, 2002.

Carlsson, Chris. *Nowtopia*: *How Pirate Programmers, Outlaw Bicyclists, and Vacant-Lot Gardeners Are Inventing the Future Today!* Oakland, CA: AK Press, 2008.

Elgin, Duane. *Voluntary Simplicity*. New York: William Morrow & Company, 1993.

Hawken, Paul. *Blessed Unrest: How the Largest Social Movement in History Is Restoring Grace, Justice and Beauty to the World*. New York: Penguin Books, 2007.

Schumacher, E. F. *Small is Beautiful: Economics As If People Mattered*. New York: Harper & Row, 1973.

Seed and Organic Starts Catalogs

Baker Creek Heirloom Seeds, www.rareseeds.com, Baker Creek Heirloom Seeds, 2278 Baker Creek Road, Mansfield, MO 65704

Filaree Farm (best for garlic), www.filareefarm.com, info@filareefarm.com, 182 Conconully Highway, Okanogan, WA 98840

Peaceful Valley Farm and Garden Supply, www.GrowOrganics.com, P.O. Box 2209, 125 Clydesdale Court, Grass Valley, CA 95945, (888) 784-1722

Seeds of Change, www.seedsofchange.com, c/o Marketing Concepts, PO Box 152, Spicer, NM 56288, Gardening Hotline: (505) 699-1462

Territorial Seed Company, www.territorialseed.com, P.O. Box 158, Cottage Grove, OR 97424-0061

Seed Savers Exchange, www.seedsavers.org, 3094 North Winn Road, Decorah, Iowa, 52101

Crimson Sage Medicinal Plants Nursery, www.CrimsonSage.com, PO Box 83, Orleans, CA 95556, 530-627-3457

Journals

Countryside and Small Stock Journal, 145 Industrial Drive, Medford: WI, csyeditorial@tds.net

Storey Country Wisdom Bulletin. A series of small pamphlets on issues specific to homesteading, including "Building Your Own Underground Root Cellar," "Building and Using Cold Frames," "Easy Composters You Can Build," etc. Storey Publishing, www.storey.com, 210 Mass MoCa Way, North Adams, MA 01247

Permaculture Activist, www.permacultureactivist.net, P.O. Box 5516, Bloomington, IN 47404

Communities Magazine, Journal of Cooperative Living, www.ic.org, Fellowship for Intentional Community, Route 1, Box 156, Rutledge, MO 63563

Websites

Daily Acts. Inspirational sustainability non-profit in Northern California.
www.dailyacts.org

Institute for Urban Homesteading. Classes, resources, interviews, photographs.
www.iuhoakland.com

Fungi Perfecti. Information about mushrooms and source for buying mushroom spawn to grow your own.
www.fungi.com

Greywater Action. Resources on urban greywater and rainwater options, water security and water justice.
www.greywateraction.org

Mother Earth News. Smart tips on sustainable living, rural and urban.
www.motherearthnews.com

MykoWeb. Online guide to mushrooms.
www.mykoweb.com

Oasis Design. Art Ludwig's website with information about greywater and rainwater catchment.
www.oasisdesign.net

The Pollination Home Page. For information about pollination ecology.
www.pollinator.com

Regenerative Design Institute. Northern California permaculture farm offering classes, resources, and inspiration.
www.regenerativedesign.org

Transition Towns. Learn to move your community from fossil fuel dependence to local self-reliance.
www.transitionus.org and www.transitionnetwork.org

Urban Permaculture Guild. Inspiring permaculture blog. How-to, why-to, where-to.
www.urbanpermacultureguild.org

Index

About the Authors

Rachel Kaplan has been gardening in and around urban environments for over fifteen years and belongs to a bicoastal family of farmers and gardeners. She works alternately as a somatic psychotherapist, writer, teacher, permaculture designer, activist, and mother. She holds two masters' degrees, one in Multidisciplinary Art and the other in Counseling Psychology. She has written and edited numerous books, including two collections of her performance work, *The Probable Site of the Garden of Eden*, and *Diaspora: Stories from the Cities*. She lives in Northern California with her partner and daughter on their little homestead, Tiny Town Farm.

Photo by Adam Kinsey

Photo by Dafna Kory

K. Ruby Blume is an educator, gardener, beekeeper, artist, and activist, with 20+ years experience gardening in an urban setting. A life-long learner, she has studied everything from permaculture to pollination ecology and has taught herself cooking, canning and fermentation techniques, as well as how to set tile, install a sink, do electrical wiring, tend a beehive and repair a motorcycle. Ruby has worked extensively in the arts and is the co-founder and artistic director of Wise Fool Puppet Intervention, a community theatre project for environmental social change. In 2008 she founded the Institute of Urban Homesteading, a project dedicated to promoting localism, self-reliance and urban sustainability through low-cost adult education. Ruby lives and works in Oakland, California.